# Covid Wars

## A True Story from Both Sides

ii

*Other books by James R. Olsen*

I Ching of a Thousand Doors

Hazel's Great Adventure

Make Things in America

*Alternate Doorway: Neuroinvasion by SARS-CoV-2* by Dr. Disha Chauhan, CellSpace 2020 Artists, The Scripps Research Institute
810550 North Torrey Pines Road 8, La Jolla, California 92037 • License CC-BY-4.0.
https://ccsb.scripps.edu/new-ways-of-living/wp-content/uploads/sites/54/2020/11/DishaChauhan_AlternateDoorway-scaled.jpg

*Alternate Doorway (SARS-COV-2) by Disha Chauhan*

# Covid Wars
## A True Story from Both Sides

### How America got Public Health wrong when others were getting it right

---

## James R. Olsen

---

Illustrated by James R. Olsen
www.JamesROlsen.com

### Book 1 — November 2019 to January 6, 2021
COVID-19 Pandemic in Ravalli County, Montana, the United States, and the World

*COVID-19 Pandemic in Ravalli County Montana, the United States, and the World Series*
Book 1 — COVID WARS
Book 2 — VAX WARS
Book 3 — HOW TO DO PANDEMICS BETTER

### Science Editor

Alex Klattenhoff PhD Microbiology and Genetics — Assistant Professor Microbiology

### Guest Authors

Pamela Small PhD Microbiology — Professor Emeritus (Book I)
Hillery Daily ND — Naturopathic Doctor, Acupuncture (Book II)
Doctor Timothy Binder — Doctor of Chiropractic Medicine, Naturopathic Doctor (Book II)

Since this book contains a variety of viewpoints, guest author contributions do not imply agreement or disagreement with any other part of this book. Further, its presence does not imply agreement or disagreement by any other contributor.

*Breaking Wave Publishing*

Hamilton, Montana 59840

## Published by Breaking Wave Publishing
Hamilton, Montana
https://www.breakingwavepublishing.com

**www.JamesROlsen.com**

Library of Congress Control Number: 2025903478
Olsen, James R.
Illustrated by James R. Olsen
   Covid Wars — A True Story from Both Sides

   Includes Bibliography and Index
   ISBN: 978-1-7342332-7-8

Book 1 of the Pandemic Series,
Book 1: COVID WARS; Book 2: VAX WARS; Book 3: How to do Pandemics Better

   Paper Back Edition

Dedicated to
Those who suffered,
Those who cared for them,
          and
Those left behind.

*Abandoned by Fritz von Uhde*

In Memoriam
## Ruth Thorning

*Journalist, Writer*
Volunteered to edit:
100 pages: Comments to *Environmental Impact Statement (EIS)*,
*Rocky Mountain Laboratories (RML), Integrated Research Facility*, 2004

1947-2009

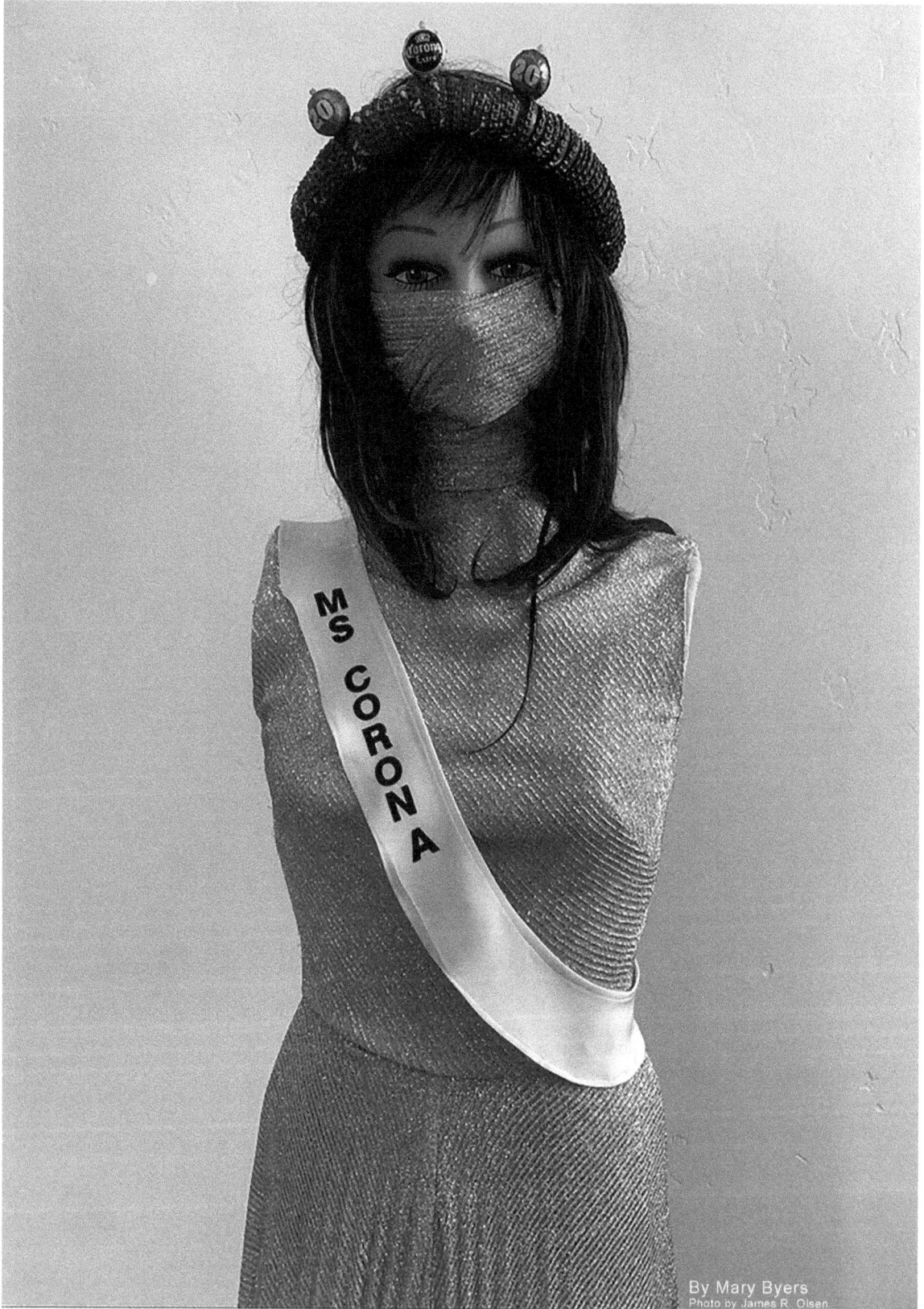

By Mary Byers
Photo by James R. Olsen

Ms Corona

# Contents

If this is a digital version  it is best to read in a two-page spread as the book is laid out to so that relevant artwork generally printed on the page or facing page. In addition book includes two-page graphics that span the gutter — spanning the two pages of an open book.

# List of Illustrations and Figures

## The People

All quoted text was said or written, quoted as accurately as possible. Some are from videos and soundtracks that are sometimes hard to make out, having been recorded in public meetings, so there may be some unintentional mistakes. The exception is that I removed ah's, repeats, and other non-informative fillers common in speech.

This book quotes real people who have put their thoughts on the record because they are public officials or spoke at a public meeting. Some spoke to a journalist or typed something on social media. Some I interviewed. Many spoke in public meetings in Montana — not always an easy thing to do. Public input becomes part of the official record (Montana Code Annotated (MCA) § 2-3-212 (1)). This book uses real names of National public figures, elected officials, heads of Federal and State institutions, local government officials, some people speaking on the record in public meetings, some journalists, some public commentators, and people who have provided written permission.

I firmly believe that the people of Ravalli County quoted in this book were speaking sincerely and advocating what they believed to be true — they have my respect, and I ask that they have yours.

## The Data

This book is about Public Health, making personal and societal decisions in the face of uncertain and changing information — studies, reports, and articles in medical and other journals, online postings, news articles, government decisions and guidance, and social media feeds.

This book will have more than the usual endnotes and references because there have been tens of thousands of documents and papers written on COVID-19, with very different views of the events. The reason is that I may well have missed something. The endnotes allow readers to see the information I used so that they can judge for themselves whether I used faulty logic, applied too much bias, or missed the point because of information I missed. All Arial Narrow Font citations are references listed in the References Cited section of this book. Endnote citations without a publisher and date are journal articles, which can be found in the References Cited section.

Information regarding the issues discussed in this book is rapidly changing and often contradictory. It is so voluminous that the author and publisher cannot do a complete review. This book is likely to contain errors and mistakes. Therefore, the authors, contributors, and publisher accept no responsibility for any inaccuracies or omissions and expressly disclaim any liability or risk, personal or otherwise, which is incurred as a consequence, directly or indirectly, of the use or application of the contents of this book.

## The Medicine

This book will present information about medical treatments and decisions. Readers are strongly cautioned to consult with a physician or other healthcare professional before using any information in this book. No book can substitute for professional care and advice. This book is likely to contain errors and mistakes. Accordingly, the author, contributors, and publisher expressly disclaim any liability for loss, damage, or injury caused by the contents of this book.

Ravalli County, Bitterroot Valley

To Missoula

Florence

Stevensville

Victor

Pinesdale

Corvallis

Hamilton

MAP US Geological Survey

Darby

Conner

To the Chief Joseph Pass

Sula

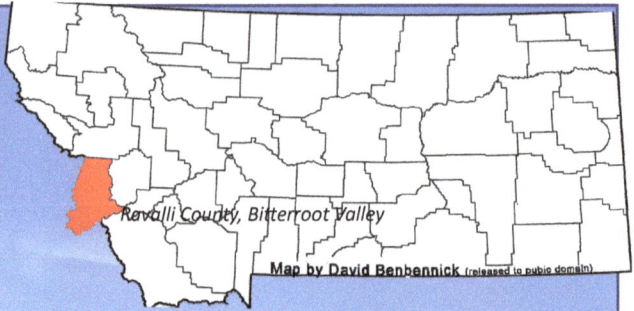

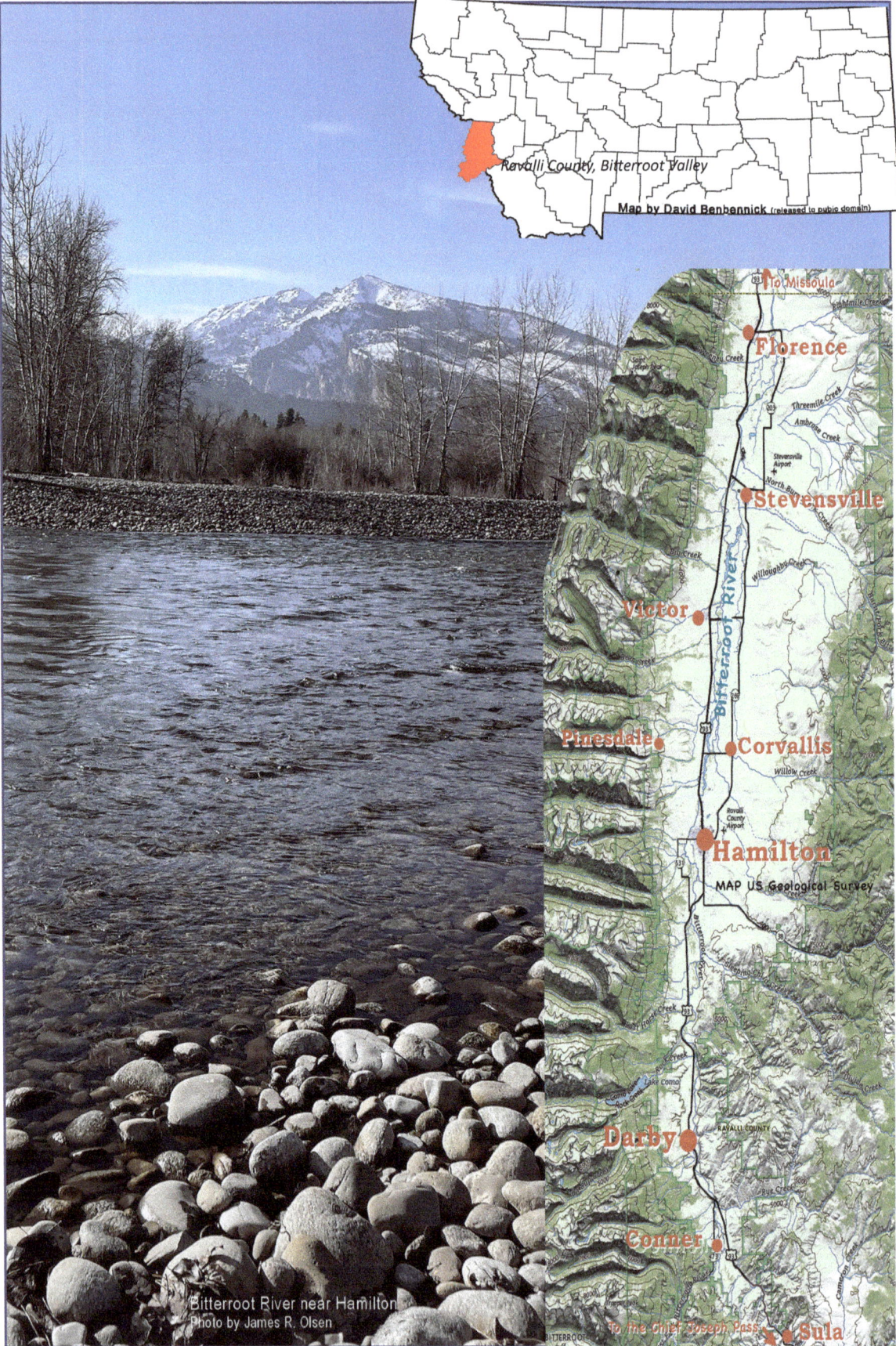

Bitterroot River near Hamilton
Photo by James R. Olsen

# PREFACE
## TRYING TO MAKE SENSE OF AMERICA'S RESPONSE TO COVID-19

*1860: Men put special meaning on words and phrases, so that
what sounded good to one sounded evil to another.
Certain slogans took on their own significance and became portentous...
Even the voices that called for moderation became immoderate.[1]*
— Bruce Catton

New words entered the English Language. The disease is known as **CO**rona**VI**rus **D**isease of 20**19** — COVID-19.[2] Coronavisuses include the second leading cause of the common cold,[3] MERS,[4] and SARS,[5] but not the flu. They are characterized by things that look like "coronas" that provide the entry mechanism to invade a cell. The agent that causes COVID-19 is called **S**evere **A**cute **R**espiratory **S**yndrome **C**orona**V**irus **2** — SARS-CoV-2. This distinction is often ignored in everyday language, using these and other terms synonymously: COVID-19, COVID, Coronavirus, 2019-nCoV, Novel Coronavirus, and SARS-CoV-2.

SARS-CoV-2 is a Ribonucleic acid (RNA) virus, single-strand, positive sense (+ssRNA), which means that it can get to work in a cell right away using the cell's resources without any intermediary steps. Too small to "see" with a regular microscope, a virus can be imaged with an electron microscope. More useful is the genome sequence, which is exposed using a high-tech process that subjects samples to chemistry and computer analysis.

The virus was initially called the "Wuhan Virus" until it was renamed. Statements by some politicians call the virus the "Chinese Virus." The *Epoch Times* newspaper calls it the " CCP Virus (Chinese Communist Party Virus)."[6] In both cases, the choice of terms is political — to remind people of the origin (in the case of President Trump) or blame the government of the People's Republic of China (PRC)[7] for its spread world-wide, based on claims of delays and cover-ups. But, no matter what the origin, the important thing is what the United States did in response to this new disease. That is the subject of this book.

This book takes up the challenge of Public Health with a capital "P." The public health response to a new disease includes developing new treatments and using old remedies. But there is much more. The effectiveness of public health is driven by how the public reacts to government pronouncements, how the media regurgitates the information, how the public consumes the media, how politics affects public health, and how public health affects politics.

Public health includes epidemiology. Epidemiology includes finding the origin of a disease. It is a first step, but epidemiology is much more than that. Congress stuck on the first step, the origin of COVID-19, to the detriment of examining the Full Monty of public health. Instead, this book answers Doctor Fauci:

> Something clearly went wrong.
> We weren't prepared.[8]

This book will illustrate why we weren't prepared and suggest how to prepare.

Docker Fauci became a pandemic icon, the hero who would liberate humanity from the disease for many, the jailer who locked them in their homes for others, and the villain who caused it in the first place — the adherents to these stories will find that it is a lot messier than they might hope.

The book does show both sides. It provides insight into Doctor Fauci's other parting comment,

> People pushing back. Those people could not all be wrong. We've got to understand what the underlying reason they were push-

ing back about what looked like completely valid public health principles. We've got to figure that out and reach out.[9]

The country is in a tough spot. Everybody has made up their mind, and that is not a good place to be. We are a split country, "The Great Divide," arguing over mandates, medicine, elections, guns, policing, and just about any issue you can name. We find ourselves in the same place as we were a century and a half ago, a 20th Century Civil War historian, Bruce Catton, illuminates in the quote above. His quote is about politics one year before the beginning of the American Civil War, a war in which the people of the United States killed 1 out of every 25 males on the battlefield.[10] We can't even agree on what to call it: Is it the American Civil War, the War Between the States, a war to preserve the union, a war to free the slaves, or the War of Northern Aggression?

A growing mistrust of the government by many in both the conservative and liberal camps plays out in the Covid Wars. It has had a long time to heat up, the molten metal of distrust, going by different names: "occupy," "sovereign citizen," "woke," "black lives matter," "freedom."

Freedom is the name it goes by in Ravalli County, an idea held like a precious jewel in the American psyche. Many things are seen to tread upon this idea. The fear of losing it simmers beneath the surface of civil discord, its partner, anger, occasionally coming to a boil, breaking the crust of a civil façade, as it did two score and two years ago in Ravalli County when some self-appointed, angry militia members converged on the county courthouse, guns on their hips, intimidating judges, one so fearful she spirited her children away to an undisclosed location.[11] Someone I knew.

I live in Ravalli County, Montana, a county famous in many circles, infamous in others, for arguing out nearly every issue. It is a microcosm of what many communities in America have gone

through. This is a true account, every quote spoken or written. This book crosses The Great Divide to present "both" sides of the story, the "narrative" of the CDC, CNN, and *New York Times* and the "counter-narrative" of the *Epoch Times* and other purveyors in many versions, the medicine of the AMA and traditional medicine, the world of epidemiology and mistrust of public health.[12]

COVID-19 has affected everybody and everything. In Ravalli County, both sides of the story have been put on the public record at local Health Board meetings and other forums, often articulate, sometimes surprisingly knowledgeable, always passionate. This, along with interviews and local news sources, gives voice to our community, telling both sides of the story. People, including me, speak for themselves. Science, health, and making a living get intertwined with politics.

This story does have heroes. In Ravalli County, there are two examples of the many people behind the scenes and across the county navigating the divisive politics to treat us as best they can — Public Health Director/Health Nurse Tiffany Webber, RN, and Health Officer Doctor Carol Calderwood, MD.

I am an engineer by trade. I have run large, complex engineering development defense programs. I once told a retired general in the acquisition business that I had run two programs with large software content that came in on time, on budget, as originally specified, as originally bid. To my surprise, he said I was the only one he had met who could say that.

It is not magic once you realize it is more about how people work together, often under stress, than technology. It is more than making correct decisions; it is about dealing with decisions that turn out to be wrong. There are always some mistakes when faced with risks and unknowns. One cannot always wait to have all the data. When events reach a cusp, they force your hand, when doing nothing affects the outcome as much as any action you may take.

This is the nature of the pandemic. We don't know everything, yet we have to make decisions. Sometimes, those decisions turn out to be wrong. The real question is how does our society deal with mistakes made.

I am a volunteer activist by avocation. In the process, I have acted on technical literature during campaigns involving public land use, wildfire, transportation, domestic abuse intervention, community-based mental health crisis intervention, local food systems, and in response to the Biosafety Level - 4 (BSL-4) upgrade to the Rocky Mountain Laboratories (RML). RML is located a few blocks from my house. The National Institute of Allergy and Infectious Diseases (NIAID), headed by Doctor Anthony Fauci, conceived the project.

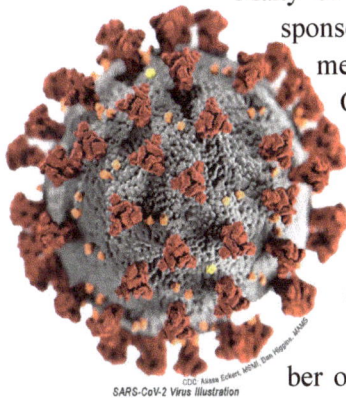

Many of these campaigns involve responses and comments to Environmental Impact Statements (EIS). Over the years, I developed a bad habit. If someone references a study, I actually read it. It is amazing how many agencies and blog sites assume I won't.

SARS-CoV-2 Virus Illustration
CDC, Alissa Eckert, MSMI, Dan Higgins, MAMS

We look at a good number of peer-reviewed papers as we go along. Because tens of thousands of documents and articles have been written on COVID-19, this book will have more than the usual endnotes and references so the reader can see what I used and what I missed. Writing about COVID-19 without some acronyms is almost impossible. For those that are more common, such as "HEPA," the Index also serves as a glossary.

I include some of my correspondence to the Health Board as this narrative unfolds, mistakes and all. No one knew everything in January 2020, so miscalculations are inevitable, often helped along by ego, always tainted by bias. If there is a subtext in this book, it is that mistakes are made, mistakes are sometimes corrected, and mistakes are sometimes left to play out. The question is how we navigate through a pandemic in the face of uncertainty.

This book posits that "everyone is biased, including me"; we would not be here if our ancestors weren't instantly biased when they saw orange fur with black stripes moving in the jungle.

This book's narrative is driven by people, including me, in Ravalli County navigating COVID-19 as well as events in Washington and elsewhere. On occasion, we'll peek behind the curtain to see what was going on at the time, even though we had to wait months to find out, as surreptitious camera footage, retrospective interviews, and Freedom of Information Act requests come to light. We'll see the contrast between what we are being told and what is being hidden.

As the narrative unfolds, we hear from a few people I know who have a variety of viewpoints, to add, in their own words, their expertise in epidemiology, naturopathic health, and vaccinosis.

While one could write about public health from a global perspective, I decided to write this book from the perspective of the United States so that I could present the effect of laws, politics, and customs more precisely. I hope the international reader can translate any worthwhile ideas to their situation.

Everyone wants to put COVID-19 in the rearview mirror — but the latest variant is upon us — riding on the hood with vax mandates crawling up the rear bumper. Even if it is over:

*We are going to get to do this again.*
*It does not need to be this hard.*

This book is written in the hope that we talk *to* each other rather than *at* each other.

*James R. Olsen,*
*Hamilton, Montana, February 9, 2022*

# SARS-CoV-2 Genome Sequence

```
   1 cttcccaggt aacaaaccaa ccaactttcg atctcttgta gatctgttct ctaaacgaac
  61 tttaaaatct gtgtggctgt cactcggctg catgcttagt gcactcacgc agtatataatta
 121 ataactaatt actgtcgttg acaggacacg agtaactcgt ctatcttctg caggctgctt
 181 acggtttcgt ccgtgttgca gccgatcatc agcacatcta ggtttcgtcc gggtgtgacc
 241 gaaaggtaag atggagagcc ttgtccctgg tttcaacgag aaaacacacg tccaactcag
 301 tttgcctgtt ttacaggttc gcgacgtgct cgtacgtggc tttggagact ccgtggagga
 361 ggtcttatca gaggcacgtc aacatcttaa agatggcact tgtggcttag tagaagttga
 421 aaaaggcgtt ttgcctcaac ctatgtgttc ataaaacgtt cggatgctcg
 481 aactgcacct catggtcatg ttatggtgga gctggtagca gaactcgaag gcattcagta
 541 cggtcgtagt ggtgagacac ttggtgtcct tgtcctccat gtgggcgaaa taccagtggc
 601 ttaccgcaag gttcttcttc gtaagaacgg taataaagga gctggtggcc atagttacgg
 661 cgccgatcta aagtcatttg actaggccga cgagcttggc actgatcctt atgaagattt
 721 tcaagaaaac tggaacacta aacatagcag tggtgttacc cgtgaactca tgcgtgagct
 781 taacggaggg gcatacactc gctatgtcga taacaacttc tgtgccctg atggctaccc
 841 tcttgagtgc attaaagacc ttctagcacg tgctggtaaa gcttcatgca ctttgtccga
 901 acaactggac tttattgaca ctaagagggg tgtatactgc tgccgtgaac atgagcatga
 961 aattgcttgg tacacggaac gttctgaaaa gagctatgaa ttgcagacac cttttgaaat
1021 taaattggca aagaaatttg acacttccaa tgggaatgt ccaaattttg tatttcccgt
1081 aaattccata atcaagacta ttcaaccaag ggttgaaaag aaaaagcttg atggctttat
1141 gggtagaatt cgatctgtct atccagttgc gtcaccaaat gaatgcaacc aaatgtgcct
1201 ttcaactctc atgaaatgtg ataacttggc tgaaacttca tggcagacac gccgatttgt
1261 taaagccacc tgcgaattt gggccactga gaattgacct aaagaagtg ccactacttg
1321 tggttactta ccccaaaatg ctgttgttaa aattattgt ccagcatgtc acaattcaga
1381 agtagagacct gagcatagtc ttgccgaata ccataatgaa tctggcttga aaccattctt
1441 tcgtaaaggt ggtcgcacta ttggctttaag gctgtgtg ttctcttatg ttggttgcca
1501 taacaagtgt gccattgggg ttccacgtgc tagcgctaac ataggttgta accatacgga
1561 tgttgttgga gaaggtcacg aaggtcttaa tgacaacctt cttgaaatac tccaaaaaga
1621 gaaagtcaac atcaatattg ttggtgactt taaacttaat gaagagatcg ccattatttt
1681 ggcatctttt tctgcttcca caagtgcttt tgtagaaact tggaaaggtt tggattataa
1741 agcattcaaa caaattgttg aatcctgtgg taattttaaa gttacaaaag gaaaagctaa
1801 aaaaggtgcc tggaatattg gtgaacagaa atcaatactg atgtctcttt atgcatttgc
1861 atcagaggct gctcgtgttg tacgatcaat tttctcccgc actcttgaaa ctgctcaaaa
1921 ttctgtgcgt gttttacaga aggccgctat aacaatacta gatgaattt cacagtattc
1981 actggagactc attgatgtca tgtatcac atctgattg gctactaaca atctagttgt
2041 aatggcctac atttacaggtg gtgttgttca gtgctgctaa ccagtggctaa ctaacatctt
2101 tggcactgtt tatgaaaac tcaaacccgt cctgattg cttgaagaga agttaaggaa
2161 aggtgtagag tttcttagag acggttggga aattgtaaa tttatctcaa cctgtgcttg
2221 tgaaattgtc ggtgaacaaa ttgtcacctg tcaaaggaa attaaggaga gtgtcagaac
2281 attctttaag cttgtaaata aatttttggc tttgtgtgct gactctatca tattggtgg
2341 agctaaactt aaagcctga atttaggtga aacatttgtc acgctcactaa aagccccaa
2401 cagaaagtgt gttaaatcca gagaagaaac tggctctactc atgctctaa aagcccaa
2461 agaaattatc ttcttagagg gagaaacct tcccacagaa ggtgttaacg aagacttaca
2521 cttgaaaact ggtgatttac aaccattaga acaacctact agtgaagctg ttgaagctcc
2581 attggttggt acaccagttt gtattaacgg gcttgtgtt ctcgaaactca accacatacc
2641 aaagtactgt gccctgcac ctaatatgat ggtaacaaac aataccttca cactcaaagg
2701 cggtgcacca acaaggtta ctttttggtga tgacactgtc atagaagtgc aagttacaa
2761 gagtgtgaat atcactttg aacttgatga aggattaca aaagtactta ttagaagaag
2821 ctctgcctat aacagtgaac tcggtcagaa agtaactgat aataagaag ttcgcctgtg
2881 tgctgtcata aaactttgc aaccatgatc tgaattactt ccaccactgg gcattgattt
2941 agatgagtgg agatatggtc catactactt atttggtggt tgtaacttgc ttaaattggc
3001 ttcacatatg tattgttctt tctaccctcc agaaggagat gaagaagaag gtgattgtga
3061 agaagaaagg tttgagccat caactcaata tgagtatgct actgaagatg attaccaagg
3121 taaaccttg gaatttggtg ccacttctgc tgctcctgc cctcgaagag agcaagaag
3181 agattggtta gatgatgata gtcaacaaac tgttggtcaa caagcgca gtgaggacaa
3241 tcagacaact actattcaaa caattgttga ggtttcaact caattggga agaacttac
3301 accagttgtt cagactattg aagtgaatag ttttagtggt tatttaaaac ttactgacaa
3361 tgtatacatt aaagctgca acattgtgga agaagctaaa aaggtaaaac caacagtggt
3421 tgttaatgca gccaatgttt accttaaaca tggaggaggt gttgcaggag ccttaaataa
3481 ggctactaac aatgccatgc aagttgaatc tgatgattac atagctacta atggaaccaat
3541 taaagtgggt ggtagttgtg ttttaagcag acacaatct gctaaacact gtcttcatgt
3601 tgtcggccca aatgttaaca aaggtgaaga tattcaacct gtttaagtg cttatgaaaa
3661 ttttaatcag ctcgacaagtt cacttgcacc attattatca gctagtgatt ttggtgctga
3721 ccctacatct tctttaagag ttgctgtgaa aggtgatgct acaactgtct acttaagtga
3781 ctttgataaa aatctctatg acaaacttgt tcaagctttt tggaaatga agattggtaa
3841 gcaagttgaa caaaagatcg ctgagatttc taaagaggaa agtttaggcct tttataactga
3901 aagtaaacct tcagttgaac ataaaacaca agatagtaaa aaatcaaag cttgtttga
3961 agaagttaca acaactctgg aagaaactga gttcctcaca gaaaacttgt tactttatat
4021 tgacattaat ggcaatcttc atccatgattc gttagtgaca ttgacatcac
4081 tttcttaaag aaagatgctc catatatgtg gggtgatgtt gttcaagagg gtgttttaac
4141 tgctgtggtt atacctacta aaaaggctgg tgaacgtagc gaaacgtag cgaaagctttt
4201 gagaaaagtg ccaacagaca attatataac cacttacccg ggtcagggtt taaatggtta
4261 cactgtaggg gaggcaaaga cagtgcttaa aaagtgtaaa agtgcctttt acattctcga
4321 atcttattatc tctaatgaga agcaagaaat tcttggaact gtttcttgga atttgcgaga
4381 aatgcttgca catgcagaaa aacacgaaca attaatgtct gctgtgtgga aaactaaagc
4441 cataagttttca actatacagc gtaaattaaa gggtattaaa atacagagg gtgtgggttga
4501 ttatggtgct agatttttact tttacaccag taaaacaact gtagcgtcaa ttataacaac
4561 acttaacgat ctaaatgaaa ctctcttgtt ac agtgatgtaa cacatggctt
4621 aaatttggaa gaagctgctc gtgttatgga atctctcaa gtgcttatta cagtcttgtt
4681 ttcttcacct gatgctgta cagcgtataa tggtttatcc acttctttctt ctaaaacacc
4741 tgaagaacat tttattgaaa ccatctcact tgctggttcc tataaagatt ggtcctattc
4801 tggacaatct acacaactag gtatagaatt tctaagaaga cagatagaat gtgtatatta
4861 cactagtaat cctaccacat tccacctaga tggtgaagtt atcacctttg acaatcttaa
4921 gacacttctt tctttgagag aagtgaggac tattaaggtg tttacagaat taattgacaat
4981 taacctccac acgcaagttg tggacatgtc aatgacatat ggacaacagt ttggtccaac
5041 ttatttggat ggagctgatg ttactaaaat aaaacctcat aattcacatg aaggtaaaac
5101 atttttatgtt ttacctaatg atgacactct acgtgttgag gcttttgagt actaccacac
5161 aactgatccct agtttttctg gtagtttct gtcagcatta aatcacacta aaaagtggaa
5221 atacccacaa gttaatgtct taacttctat taaatgggca gataacaact gttatctttg
5281 cactggattg ttaacactcc aacaaatgga aaattggctt aatccacctg ctctagcaaga
5341 tgcttattac agagcaaggg ctggtgaagc tgctaacttt tgtgccatca tcttagccta
5401 ctgtaataag acttaggtg agttcaggtga acatgagact gtagagaaa acatgagtct
5461 acatgccaat tagaattctt gcaaaagagt cttgaacgtg gtgtgtaaaa cttctttttca
5521 acagcagaca acctttaagg gtgtaaagac tgttatgtac atggtgacac tttcttatga
5581 acaatttaag aaagttgttc agatactcg tacgtggtgt aaacaacgta caaaatatca
5641 agtacaacag gagtcactt ttgttatgat gtcagcacca cctgctcagt atgaactaa
5701 gcatggtaca ttactttgtg ctagtgagta caccggttaat taccagttga gtgcactaa
5761 acatataact tctaaagaaa ctttgtattg catagacgct gctttaggt gcttacttaa
5821 agaatacaaa ggtctcatta cggtgtttt ctacaaagaa aacagttaca caacaacttt
5881 aaaaccagtt aacttataa tggatggtgt tgtttgtaca gaaattgacc ctaagttgga
5941 caatttttat aagaaagcat attctttatt cacagacgaa aacaattgtc ttgatgatac
6001 ccaaccatat ccaaacgcaa gcttcgataa ttttaagttt gtatgtgata atatcaaatt
6061 tgctgatgat ttaaccaagt taactggtta taagaaacct gcttcaagac agcttaaagt
6121 tacatttttc cctgacttaa atggtgatgt ggtggactat cttcataagc agaaggggaa
6181 ctcttttaag aaaggagca attgttttaa aaatgtacat cttagcatg tttgggataatt
6241 aactaataaa gccactgttg aacaaatcag cctcgagac cctgtatata gatgttgatt
6301 accagttgaa acatcaaatc cgttcatga ctggaagtgc agaggaagaa cggtgaatgca
6361 taatctgcc tgcgaagatc taaaaccaga tctctgaagaa gtaggtaaca gaccagtctc
6421 acagaaagac gttctgattc gtaattgaa aactacctgac gttgtaggac attctcaccat
6481 taaaccacca gataaatctt tattaacaat tttaatttat aaagttggag tctattaatt
6541 tgcttatgta gacaattcta gtctactaca taagaaaact aatgaattat ctagagattt
6601 aggtttgaaa accctgatt agctgctgca agctgctgca ttgtatagtt tatatttaac
6661 tatagctaat tatgtctaagc ctttcttaa caagttgca tctttacaact caacatatatg
6721 tacacggtgt ttaaaccgtg ttgtactaca ttatatagaa aattttttgc ctttacttga
6781 acaattgtgt acttttacgtta gaagtacaaa ttctagaatt aaagcatcta tgccggactac
6841 tatagcaaag aatactgtta agagtgttgg agaagtgtcg gttgactttt tactttttttg
6901 tttgaagtca cctaattttt ctaaactgta aatattttgt attttttgt tactattaag
6961 tgtttgccta ggttctttaa tctactcaac cgctgctttta ggtgttttaa tgtctaantt
```

NUCLEOTIDES

a — Adenine
g — Guanine

**Purines**

c — Cytosine
t — Thymine (DNA Only)
u — Uracil (RNA Only)

**Pyrimidines**

Nucleotides by BruceBlaus CC BY 3.0 unported
https://creativecommons.org/licenses/by/3.0/deed.en

Each letter represents a nucleotide. The sequence is like a book, waiting to be read.

https://www.ncbi.nlm.**nih**.gov/nuccore/MT072688

*SARS-CoV-2 Genome Sequence*

# 1

## DARWIN'S DREAM

*It is not the strongest of the species that survives, nor the most intelligent,
but rather the one most adaptable to change.*[1]
— Leon Megginson paraphrasing Charles Darwin

**NOVEMBER 2019**

We call it COVID-19. No one will ever know when it emerged in our world, 30,000 nucleotides in a new arrangement, any picture relegated to images of electrons bouncing through the helix string of molecules.

Six miners in China, exposed to bat droppings, got sick with a SARS-like virus in 2012. Three died. A bat virus was found with a 94% genome match with the virus we call COVID-19. That means about 2,000 nucleotides were different.[2]

If it did *Run Silent, Run Deep*,[3] the nucleotides mixing and swapping in every cell it invades, it might have clicked on the right combination in seven years to become the novel coronavirus, COVID-19. Or COVID-19 could have originated somewhere else in some other way. Of the 1,497 blood samples from 2015 to 2017 from rural areas in southern China near bat caves, 9 were positive for a bat coronavirus.[4]

It seems no one will ever know when it first infected a person. The evidence from Wuhan is that the virus infected people in November, if not earlier.[5] Footprints have been found, which *may be* the new coronavirus as far afield as Brazil in samples collected in November 2019.[6] Some version of it could have been with us for years, as many diseases cross over from animals. Some, like the measles, crossed over so long ago we don't remember.[7]

We will soon learn that the novel coronavirus has found a deadly sweet spot. Very contagious, but many people don't even know they have it. Deadly, but not too deadly. Adaptable.

It's Darwin's dream, COVID-19.

12 December 2019 31

Cluster of patients in Wuhan begin to have shortness of breath

Covid Timelines - CDC Museaum , AJMC COVID Timeline, Washington Post Timeline Wuhan leak theories, MT Gov. Bullock Emergency Directives, and other sources

WHO told infections unkonwn cause. Cases associated with seafood mart

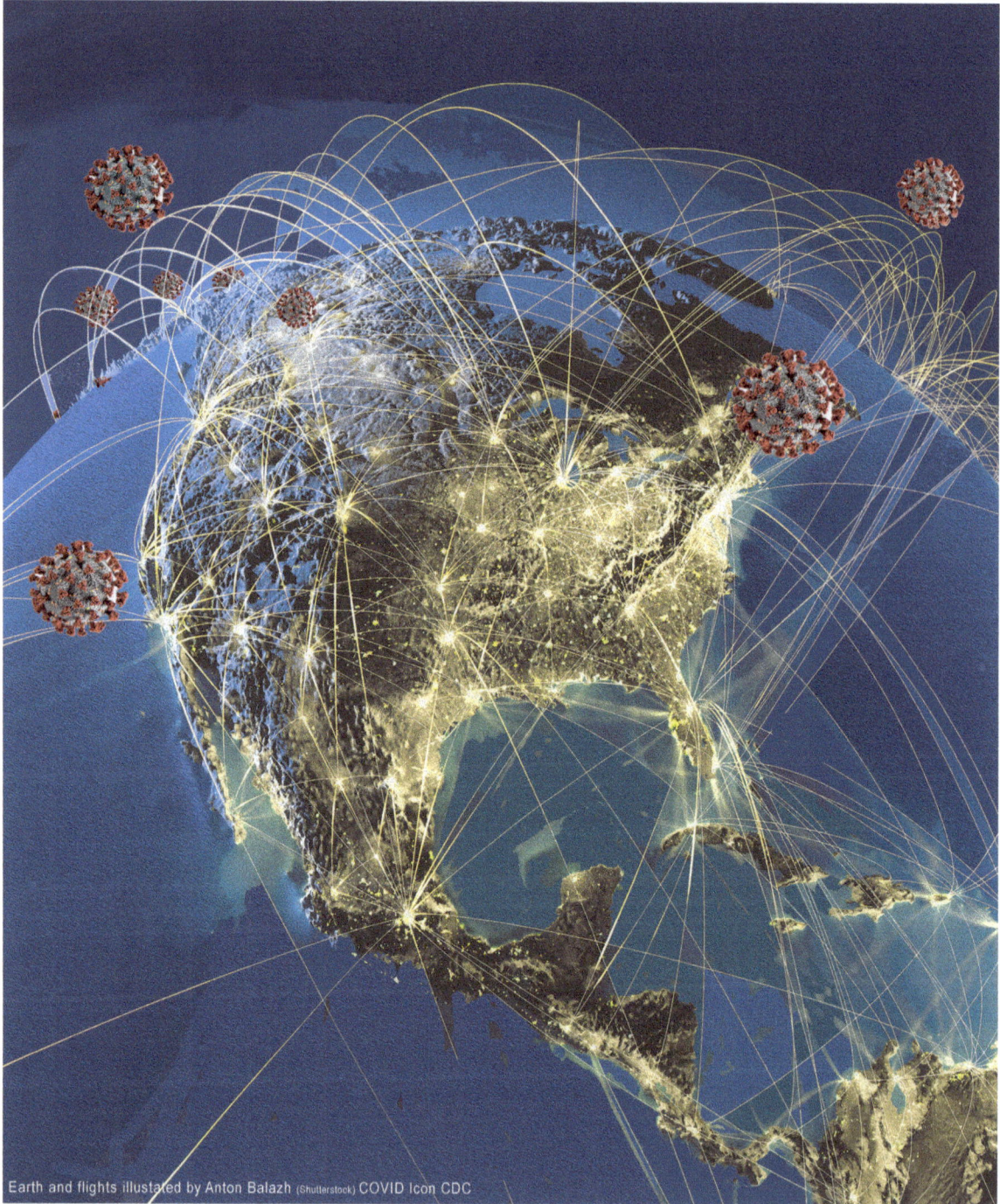

Earth and flights illustated by Anton Balazh (Shutterstock) COVID Icon CDC

*COVID Air*

# 2

# FLYING COVID AIR

*Most things which are urgent are not important,*
*and most things which are important are not urgent.[1]*
*— President Dwight D. Eisenhower*

**DECEMBER 2019**

We should have prepared when we had the time. Every day, over 200,000 people arrive from somewhere else to set foot in the United States of America. Travel on they do, mixing with every state, city, and hamlet.[1]

By Christmas, as nearly everyone opens presents, many celebrating the birth of Jesus Christ, many telling stories of Santa Claus to children, some gathering for family meals, others eating alone, the Hanukkah candles already having been lit a week before, the Winter Solstice having just passed, some doctors in Wuhan, unknown to us, notice something different about their pneumonia patients.

The unexplained pneumonia is filling hospitals in Wuhan. The first confirmed infection is on December 1st.[3] Officials are downplaying it, forbidding healthcare workers from discussing the rising caseload. By the end of the month, doctors know it spreads from person to person but are afraid to say, fearing retribution.

On the 25th, an unnamed Chinese technician posts an analysis of a sequence, writing it is a "subtype" of SARS (Severe Acute Respiratory Syndrome). SARS is a coronavirus that started in the province bordering Hong Kong in late 2002 with a fatality rate of 6% to 10%, spreading when in close contact with someone with symptoms. It was controlled by early 2004 by testing and isolating people with symptoms, holding worldwide fatalities to 774, with only 8 in America.[4] Although the new pneumonia shares the SARS heritage, the "contagious when symptomatic" rule is not the case —an important distinction.

The new virus has begun its worldwide travels, riding along roads, hitching a ride to the nearest airport — a virus travels. When it lands in the United States, the new pneumonia slips through customs unnoticed, hitching a ride on a passenger as they walk off an airplane.[5] It arrives as we are arguing over impeachment and immigration.[6]

Antibodies found in Red Cross donations from December are likely, but not provably, a response to a coronavirus infection.[7] In December 2019, 4,000 times across the country, "pneumonia" is written on a death certificate. This number is not out of the ordinary. No one is looking for a new virus as the next patient arrives — "pneumonia" is written on the chart.

This time, it is urgent and important because if COVID-19 isn't already in the United States, it soon will be.

| 1 | 5 | 7 | January 2020 | 13 | 15 | 17 —— 17 | 20 | 22 | 31 |

NEW YEARS Day

Chinese Health Offical publishes genetic sequence for SARS-CoV-2

CDC Incidence Team Formed

Thailand 1st case

Japan 1st case

MARTIN LUTHER KING Day

CDC screens passangers from 3 Wuhan Airports

First suspected US Case

CDC team to Seattle Case, sample from 18th

CDC confirms 1st US Case

WHO not an emergency

WHO confirms 1st human to human transmition

HHS health emergency

WHO declares emergency

*Aerial view of Wuhan skyline and Yangtze river*

Photo by
sleepingpanda (Sutterstock

# 3

# TRYING TIMES IN WUHAN

*It's clear that when the Government is telling us where to look,
they're also telling us where not to look.[1]*
— Nanfu Wang

## JANUARY 2020

Hamilton, Montana, is named not for a president but for a Copper King's man, James Hamilton, the man who laid out the town in 1894, a town of 4,827, a town that had the nerve to call itself the City of Hamilton when the population was but a quarter that size, a town platted in the center of the Bitterroot Valley, rooted in western Montana. The town is designated as the county seat of Ravalli County, a county whose western boundary scribes an imaginary line in the middle of the Selway-Bitterroot Wilderness, a wilderness whose mountainous wildlands have much to do with how people here live and much to do with what we argue about.

The Copper Kings stand as rough and ready giants in Montana's history. Just before the turn of the Twentieth Century, two men became rich in Butte, Montana, on a mother load of copper, tunneling, smelting, competing, litigating, bribing, fighting it out on the streets and in the mining tunnels under the streets.[2] Marcus Daly is our Copper King, the man who founded Hamilton, logged the hillsides to feed his refineries in Butte and Anaconda while creating one of the nation's leading thoroughbred stock farms east of town. His name and legacy permeate the streets and hillsides of Hamilton.[3]

Some event is usually held on Martin Luther King Day, often organized by the Montana Human Rights Network.[4] This year, there is little about the struggle and more about the underlying idea. The *Ravalli Republic* newspaper leads with a front page photo caption, "First Graders celebrate Martin Luther King Day, shouting 'Love and Kindness' at passing cars."[5] This is a precursor to the course I am preparing for the Bitterroot College, *Effective Grassroots Activism*.[6]

I have no idea how relevant this course will become in the coming year. The class is an expansion of a session I had given to graduate students at the University of Montana in Missoula a year ago, the idea being to share what I had learned since the 90s, having had the good fortune to meet and be mentored by some of the leading and well-known forest activists in the country. The course is about grassroots organizing, the framework of environmental law, how to wade into an Environmental Impact Statement (EIS), do press, and the like.

Activism has always been an avocation. I kept my day job. As I went along this path, the question became how to move away from anger as a motivation. As more activists around me began looking for a "positive" response, I found the answer in the works of Mahatma Gandhi,[7] whose ideas were adopted by the Southern Christian Leadership Conference, Reverend Martin Luther King, Junior being the most well-known spokesperson.[8] Still practiced today, it can be found in surprising quarters.

The practice is effective, even when the tactics don't include civil disobedience. The practice requires a surprising amount of self-discipline to respect and engage with the very people with whom you disagree — taking the opportunity when it presents itself when it is not a public meeting degenerating into slogans and shouting matches, to listen to those who have opposing views — not to persuade, but to understand. Of course, my passions and ego sometimes get the better of me.

When your "opposition" is arguing why they are right, don't argue with them; take notes. My mentors told me, "You don't understand the issue until you can present the opposition's arguments as well as they can."

In the course description, I assure everyone that we will strictly focus on the process, not the issue. I promise we will talk about at least one advocacy they disagree with. No matter what the issue, the principles of non-violence are the same.[9]

I am often labeled a liberal in Ravalli County. However, the opposite appellation would probably be applied in a county that voted 67% for Ms. Clinton instead of Mr. Trump in 2016.

Fourteen people sign up for the course; some are friends, some would be called liberal, some would be labeled conservative, some very mistrustful of the Government. I hoped for more, but it is too expensive for some, a conflict with working hours for others. Two are well-known active Republicans. I heard the Ravalli County Republican Central Committee paid their tuition.

The Bitterroot College is housed in Hamilton's old junior high school building. The hallways, the classrooms, and the texture bring to mind years past, though now computers, projectors, and screens have been added to bring it up to date. After hauling several boxes of notebooks for the class, I test my laptop's connection to the projector. I pause to close the shades to get rid of glare. Class members file in.

We begin. After introductions and "what is activism," the presentation moves to Gandhi's prin-

ciples, noting that it is firmly rooted in the "love your enemy" verse found in the Sermon on the Mount.[10] That is Gandhi's observation, not mine.

The discussion focuses on a video of the entire "I Have a Dream" speech, pausing to ask who King was addressing at that moment. The surprising observation is that King is often not addressing the throng of supporters gathered at the National Mall; instead, he addresses his oppressors, reminding them of irrefutable commonly held truths, from America's founding documents to widely held religious beliefs.

Good to my word, I follow with a video of a tree-sit to save the redwoods. Then, an antiabortion sit-in, both rooted firmly in the principles of non-violent action.[11] I note that if grace is to be found, it's not in the result but in the act.

Covid-19 has yet to interrupt our daily routines. The world goes slower on Third and Main than on First and Main. First Street is US Highway 93, a through-way that carries cars and trucks from Canada to Mexico. I take a break almost every day around 2:00 to have coffee and a few puffs of a cigar at Big Creek Coffee on Third and Main, about eight blocks from my house in Hamilton. I sit outside at one of the sidewalk ta-

*The Table*

bles, standing watch at the east end of the block. The watch post at the west end is a sidewalk table at the River Rising Bakery. Chapter One Bookstore sits catty-corner across the intersection, a shelf full of discount books posted on the sidewalk next to the door.

A cigar is like having a puppy; strangers stop and comment, sometimes leading to interesting conversations. Tim Binder began to show up almost every day some time ago. We talk of philosophy, religion, and science, from physics to the theory of language — and, of course, the latest conspiracy theory, a fair appellation because the narrative is usually about a conspiracy of some sort. Sometimes, someone has an alternate theory of Government — different from the one they teach in high school civics.

Tim, a traditional healer, naturopathic doctor, and philosopher, has quite a following, and some of his circle show up; sometimes strangers, sometimes people I know, will stop for a chat. Our rule is that respect is always required. When political discussions get hot, a commitment to a non-violent response is asked. The names are multiform: chinwag, gabfest, bull session, café society; I call it The Table.

It is primarily men who show, but not always. Tim introduced Hollie Rose Conger one day. Then her blonde husky shows up, well-behaved, as calm as the Buddha, so I call him Buddha Dog. Hollie is always there when we do a group I Ching cast, which happens every couple of weeks.

Hollie is from Winter Park, Colorado, but spent time in Denver. She is a health adviser, primarily online, so she can live anywhere. When she arrived, she was trying out different towns to see where she wanted to live. She settled outside of Corvallis.

Thirty, busy working, busy being single, Hollie will take time to have a cup with some old men once a week or so to talk philosophy, religion,

health, and counter-narratives. Counter-narrative is a fair appellation, as that is the phrase its adherents use. Hollie brings calm patience to a conversation, which is uncommon in American culture. However, it can come with a firmly held opinion. Not having navigated the '60s, there would come a day when we end up talking about how protests work, on a day when she was planning a trip to riot-filled Portland.

There is an ominous tone in the news. Something is going to happen. It begins like a movie trailer flashing dramatic highlights of the coming pandemic, the unseen monster in the basement. On January 1st, no one in the United States seemed to take notice of this headline as the year 2020 opens,

> Hong Kong Takes Emergency Steps as Mystery "Pneumonia" Infects 27 in Wuhan. — *South China Morning Post.*[12]

Since I subscribe to the online version of the *South China Morning Post*, I receive this news in my email inbox as the new year opens. A post in Hong Kong from the afternoon of December 31st that I saw shortly after midnight had me looking for it.[13] The mysterious virus may be coming to America if it is not here already. Is it another SARS? SARS was deadly but not much of a problem for the United States.

It is almost a week before this news shows up in America, and it is still not front page news: you have to turn to page 13 in the *New York Times*.[14] The first death is recorded in Wuhan on January 9th and reported on the 10th.[15]

I soon learn that Taiwan started checking and isolating symptomatic travelers from Wuhan on December 31st. Singapore, Thailand, and South Korea followed quickly, having been through the SARS outbreak.[16] It seems strange that no one else has travel restrictions. Even the World Health Organization (WHO) advises against travel restrictions,[17] even though the medical community is convinced it is new, a "novel coronavirus," in official parlance.

I am nearing the end of a four-year project. My book, *I Ching of a Thousand Doors*, has just been published. My first book reading is scheduled for March at the Chapter One Bookstore. I decided to do some merchandising and package the book with related goodies. I am looking for a vendor to make a custom box at Alibaba. By mid-month, things get strange.

Alibaba, a Chinese company, can be thought of as the "Amazon-factory-outlet" of Asia.[18] It has the benefit of a contract guarantee so that you don't have to navigate the various ways different cultures do business, including China. I have had success buying things like a book press and watch parts before when the price difference compared to other sources is too large to ignore.

It works by searching the Alibaba site for a product you want, then selecting one of the sellers, almost always a factory, which almost always displays their manufacturing process and products with photos or videos. You can send a message to a factory rep and expect a response within 24 hours, often female, with a headshot and name — maybe a pseudonym. The interaction is inevitably polite and specific.

So, in late December and into January, I am shopping around. By mid-January, I am almost ready to close on an offer when the factory rep does not respond. Strange — although Chinese New Year is coming, I expect some delay. As this thought comes to mind, Alibaba broadcasts a message saying that the Chinese New Year may cause delays. As expected, the subsequent response takes a few days: price agreeable. I send a message to close the deal.

Suddenly, radio silence. No response. Something is going on in China, worse than we are being told. This, as much as the news about the new virus, suggests something serious is happening in China. It's got to be the new pneumonia.

No one, including me, seems to foresee its impact in its full measure— the virus and the re-sponse, how it will affect how we live, how we will die, how we will see each other, how we will not see each other, how we will work, and how we will not work.

The new disease is a coronavirus. Officials in China are struggling to understand where it came from. The working assumption is based on statements by local Chinese officials, saying it originated from a "wet" fish and wildlife market in Wuhan, specifically the Huanan Market. But a paper from China will soon cast doubt that the Huanan seafood is the source of the outbreak in Wuhan:

> 41 admitted hospital patients... 27 patients had direct exposure to the Huanan seafood market... The... first patient infected... was Dec 1, 2019. None of his family members developed fever or respiratory symptoms. No epidemiological link was found between the first patient and later cases.[19]

The suggestion of a mammal-to-human transmission hangs in the air, like other novel diseases ranging from Ebola to Severe Acute Respiratory Syndrome (SARS). SARS is a coronavirus that crossed over from bats to civet cats to humans — so the bat becomes a prime suspect.[20]

*Bat Captured in Net*

Wuhan officials probably feel much political pressure to act; they close the market, call up the guards, and mobilize sanitation crews.

As the first death from the novel virus is reported, people in Wuhan are told not to worry — even if it is infectious, the risk is low, according to the Director of CDC Wuhan. But stories of sick people spread on social media faster than the censors can censor.[21]

Having been decoded by Yongzhen Zhang from a sample received on January 3rd at the

Shanghai Public Health Clinical Center, the genome sequence of the Novel Coronavirus is sent to the world on January 11th by an Australian colleague, Eddie Holmes, so scientists around the world can get to work. Critical parts of the sequence were quietly posted by Zhang on the 5th to the National Institutes of Health (NIH) data bank — quietly because the attitude of Chinese officials was not clear, as they were already shutting down social media critics. Authorities soon made themselves clear — do not post anything.

It takes a bit of cloak-and-dagger by a few scientists scattered across the globe to post the entire sequence on behalf of Zhang.[22] Zhang's post either went unnoticed or was handled with kid gloves by NIH. The NIH leadership announces the significance of Holmes' upload on the 11th. However, the post is 30,000 letters that need to be assessed by scientists who know what they are doing — and so they do worldwide.[23]

That same day, the 11th, a provincial health official in Wuhan says they do not see evidence of human-to-human transmission.[24] He notes that the medical staff who care for the patients have not come down with the disease. I'm not sure I buy it. I imagine healthy young people who could be infected while exhibiting no severe symptoms. The WHO quickly regrets a January 14th Tweet,

> Preliminary investigations conducted by the Chinese authorities have found no clear evidence of human-to-human transmission of the novel #Coronavirus (2019-NCoV) identified in # Wuhan, #China.[25]

The statement is literally accurate, misleadingly accurate, becoming misconstrued because "no clear evidence" doesn't mean "none."

The WHO's willingness to broadcast the official Chinese line during this fumbling by Chinese officials has its roots in the international criticism of China in similar circumstances during the SARS outbreak. The Chinese Communist Party (CCP) attempted to inoculate itself from a repeat performance by placing at least one politically motivated Chinese national in a key position in the WHO.[26] While this is not unusual for any major power, it is unfortunate that the WHO passed along statements without investigation.

By the 20th, it is clear this is tragically misleading. In a turn-about, Chinese officials announce a human-to-human transmission in Wuhan. An Associated Press (AP) news article notes that this "raises the possibility that it could spread more quickly and widely." The admission makes many wonder, "Can we trust statements emanating from Wuhan?" The follow-up question, "Is the WHO too cozy with China?"[27]

Now, people in Wuhan are worried. "It is probably worse than the flu but not as bad as SARS," says a taxi driver to a reporter who will later smuggle the video to *Al Jazeera*. There are now cases in Beijing.[28] But, these questions obscure an essential question for the United States. Should the United States shut down travel from Wuhan?

Too late. On the 20th, the first case is in the news from Washington State, noting that six people had already died from the disease worldwide.[29] The infected man had traveled to Seattle from Wuhan a week earlier, having had no symptoms when he arrived in the United States, only to develop respiratory symptoms later. Seeing the news about the symptoms of the new virus, he decides to tell his doctors where he had been. Alex Greninger, Assistant Director of the Clinical Virology Lab at the University of Washington Medical Center, says,

> Everything was so positive... The sample overnight to the CDC, [which]...turned around a test, which is crazy. They actually had a test for the virus when the genome had just gone online for 10 days.[30]

The Centers for Disease Control (CDC) flies a team out and track down the 16 contacts he recorded since his symptoms.

King County, home of Seattle, does not wait for the CDC test result reported on the 20th. The county has a plan. They had already isolated one patient five days earlier — as soon as they had his travel history. The media does not pick up on the fact that he reported nausea and vomiting two days before seeing the doctor.[31] Reports from China confirm that digestive issues can precede respiratory symptoms.[32]

Having tested positive for COVID-19, the patient deteriorates. His doctor gets authorization to try remdesivir. The patient recovers, but the doctor is unsure if it is the drug or the natural course of events.[33]

"This is the tip of the iceberg," says Doctor Todd Ellerin, Director of Infectious Diseases at South Shore Health, in South Weymouth, Massachusetts, to a reporter from the *Boston Herald* on the 21st — something many infectious disease experts are realizing.[34] However, their voices are buried among the news articles about the new coronavirus.

"We have it totally under control," says President Trump in an interview on the 22nd — almost in passing, as the interviewer at *CNBC* is more interested in talking about the economy:

> One person coming in from China. And we have it under control. It's going to be just fine.[35]

The next day, the first human-to-human transmission of COVID-19 within the confines of the United States is reported from Illinois.[36] The CDC directs airport screening, but this does not noticeably stem the tide.[37]

Asymptomatic passengers slip through.[38] An obvious empirical observation: the man in Washington with the first recorded case would have slipped through since he did not have symptoms when he landed. It should not be a surprise — asymptomatic infectious carriers of a disease go back to Typhoid Mary.[39]

On the 28th, the United States Department of Health and Human Services (HHS) holds a press conference — it is informative. HHS Secretary Alex Azar begins,

> As of today, the CDC has reported five cases of the novel coronavirus infection here in the United States...

Americans should know that this is a potentially very serious public health threat. But at this point, Americans should not worry for their own safety. This is a very fast-moving, constantly changing situation...

Part of the risk we face right now is that we don't yet know everything...

Finally, ... to determine whether there is an asymptomatic transmission... China has reported there may be evidence of asymptomatic transmission.

The playbook for responding is relatively simple and multi-tiered:
  ...identify cases, isolate people, diagnose them, and treat them.
  ...track down all the contacts
  ...the same with those people and the same with contacts of contacts if necessary.

The United States has the world's finest public health system... I'll give you a brief sense of the work being done,
  ...assessing the level of preparedness...
  ...research for vaccines, diagnostics and therapeutics...

The CDC director and Doctor Nancy Messonnier, Director of the CDC's National Center for Immunization and Respiratory Diseases, present more on contact tracing and isolation. The protocol they describe would work for SARS—wait for close contacts to exhibit symptoms—then test. However, this will prove to be a fatal flaw for SARS-CoV-2.[40]

At the end of the month, President Trump restricts travel from China for non-US Citizens, mandating quarantine for symptomatic passengers from China.[41] Public pronouncements are sporadic.

No one seems to know what is happening in Washington — at least not the public.

## BEHIND THE CURTAIN IN WASHINGTON, DC—

There is intelligence, and there are plans — Alex Azar is informed of the outbreak on the 3rd. The Office of National Intelligence and the CIA include coronavirus status in the President's

Daily Intelligence Briefings — by the end of the month, it is the dominant topic. The Senate Intelligence Committee, in closed session, is told in no uncertain terms by Doctor Robert Kadlec, HHS Assistant Secretary for Preparedness and Response (ASPR), and the CIA that the virus presents a serious threat — "Americans will need to take actions which could disrupt their daily lives." Yet, no course of action is recommended.

ASPR has a written "framework" for responding to a pandemic. Homeland Security has a book-long *National Incident Management System,* which the ASPR Framework fits into. FEMA created a *National Threat and Hazard Identification and Risk Assessment* notebook in 2019, which deals with all sorts of disasters — a pandemic is one. These plans are not dusted off, at least not right away.[42]

The Obama White House wrote a *Playbook* for responding to infectious disease outbreaks —a checklist and questions to be answered. The current White House appears to have left it unopened.[43] It will be a repeat — Presidents of both parties are more often than not too slow off the starting line to respond to emerging diseases — the problem is that this one is more contagious than anything seen since 1918.

The CDC quickly goes to *Community Mitigation Guidelines to Prevent Pandemic Influenza — United States, 2017,* for their starting point. This document is based on research and reviews the effectiveness of Non-Pharmaceutical Interventions (NPI) or flu outbreaks: school closures, social distancing, extra hygiene and cleaning, voluntary home isolation if infected — and selective use of masks, but not masking the general population.[44]

There was a scenario-driven exercise with HHS and several states for an infectious outbreak in 2019 called *Crimson Contagion 2019.* The lessons from *Crimson* are slow to take hold.[45]

When I look back on these plans and exercises, the bottom line is that *Crimson Contagion* is a decent exercise — a new strain of flu from China. It uncovered several issues, highlighting the overlapping responsibilities and confusion in coordinating HHS, Homeland Security/FEMA, and the States. Of course, it assumes a non-partisan political environment and transparent information to the public.

The plans are guidelines and outlines that aid in figuring out what various agencies do and a checklist of questions to ask. They assume that a specific plan for dealing with an emergency will be crafted by someone who knows what they are doing or someone who depends on a small cadre of people who know what they are doing.

There is little mention of a discipline known as "Public Health." Science, as in scientific research, is part of it, but not all of it; epidemiology is the foundation, but not all of it. Public Health is a well-known discipline, taught in universities at the master's level and practiced by every Public Health Department in every county and city in the United States. The country needs a Public Health response.

The problem is coordinating a coherent national plan based on public health and epidemiology principles: the United States will never get there. Instead, these plans will become partisan by May, being waved around at news conferences in a blame game in which both political parties are players.

The national response does not account for available critical information, the tricky nature of the disease, asymptomatic contagious people, and the importance of medical capacity. Secretary Azar urges a proactive response. The President demurs, downplaying the risk.

The 2020 campaign has started, and President Trump needs a replay of 2016. He runs downfield, the crowds cheering as he shakes off the Democrat's tackle, impeachment. He is gaining gaining yardage in the polls. Even though people are getting sick in the bleachers, a pandemic is not in the MAGA[46] campaign playbook.

The Democrat operatives need an error, a penalty, a mistake. As a result, anyone in the Government who tries to organize a rescue convoy for the increasing number of people falling ill in the grandstands finds themselves axle-deep in the muddy bog of Washington politics.[47]

Some people in HHS feel the need for urgent action. Doctor Rick Bright is one. By mid-month, Bright campaigns for a Disaster Leadership Group (DLG) meeting, which is called for in the ASPR Framework. According to Bright, DLG meetings worked well for Ebola, Zika, and flu outbreaks.[48]

Rick Bright is a deputy to Robert Kadlec with the title Director of the Biomedical Advanced Research and Development Authority (BARDA). BARDA is responsible for developing and procuring medical countermeasures and drugs during essential health emergencies and biological, chemical, and nuclear attacks. Of course, there is overlap with other organizations — a common problem in large, complex organizations.

It takes until the 23rd for Kadlec to call the DLG meeting. At a meeting the same day called by Secretary Azar, Bright raises "the need to get started immediately on commercial test development, drugs, and vaccines, and the need for money to do it."

Azar looks at Bright and says, "You need money?"

Referring to *Crimson,* Bright says, "Yes. We've exercised that it would take about $10 billion."

This declaration created a "shit storm" after the meeting — Bright's fiscal year budget of $1.6 billion is plenty, some say. It won't be. $10 billion is an underestimate.[49]

Managing the crisis through a DLG will not hold. Most Presidents want a crisis team that is

their own creature, and President Trump is no exception — though not yet because it is not a crisis in the President's mind. Despite the President's distaste for the United Nations, his administration seems to follow the WHO's lead. First, the WHO says it's not an emergency. The President and his leadership echo the message. Then, the WHO declares a health emergency on the 31st, as does the HHS that same day.[50]

Having just cut a trade deal with China, President Trump tells the American public he trusts President Xi Jinping, "I have a great relationship with him."[51] The problem is that we don't know what details President Xi Jinping shared nor how well the situation is understood in China.[52]

*The Emperor's New Clothes* comes to mind. The national public health response, clothed in assured, sanguine words, stands naked before the gaze of the child who has no need of favor from the King.[53] Doctor Fauci sees "troubling" reports from China that asymptomatic carriers can be contagious. However, he is not sure at this point.[54]

By the end of the month, a cabal of scientists inside and outside HHS "who have been in public health and emergency response for a long time" form an email chain called *Red Dawn*[55] — from the 1984 movie.

Red Dawn is trying to "assess the scope of the calamity and how to mitigate it," says Doctor Eva Lee, Research Director for the Georgia Institute of Technology — known for her models that predict the likely spread of disease. "When you see the first case, it means you have many cases you don't see." She thinks we are in for a pandemic. When she sees the first case in Seattle, she notes he has been moving around the city for four days before being tested. Based on his movement history, Lee estimates there would be about 2,000 contacts who would have needed to be tested from that one case.[56] Of course, there are few test kits. When Greninger is asked his thoughts about not having another case in Washington for a month after the first one, he says,

It was a total lie that there were no cases... lie is not the right word, but it was a delusion.[57]

It is a trade advisor, Peter Navarro, inside the White House, who, by mid-month, believes an aggressive response is needed. The balance of risk: if it turns out to be the seasonal flu, it won't cost that much, but if it is a pandemic, it is likely to cost "1-2 million souls" and "trillions in economic damage." If we act fast, $3 billion of supplemental funding could avoid much of that. He starts issuing sole source contracts to companies he has met for PPE. Navarro wrote several books about China's rise a decade ago; he advocates an immediate, total, and, if needed, lengthy travel ban from China.[58]

No one seems to know what is going on in China.

_____

It is an intelligence failure of the first order — I don't mean this in the sense of military or diplomatic intelligence, which is inherently adversarial — but it is the business of the CDC to monitor health issues around the world so that the United States can prepare.[59] Even though the Chinese Government keeps its own public in the dark, plenty of public information and social media chatter is available if anyone chooses to look and analyze — *passive open-source intelligence* in the parlance of the trade.[60]

Chinese broadcast TV gives a big hint on the 1st. New Year's Day celebrations are broadcast, some featuring the Chinese version of *America the Beautiful* — *My People, My Country*, celebrating a people's emotional bond with a beautiful landscape. The celebratory news is interrupted:

> Eight people were punished for spreading rumors about the new pneumonia. No one can get away with spreading rumors online.[61]

The most well-known of these is 32-year-old Doctor Wenliang Li, who had warned fellow doctors in a chat group to take precautions on December 30th, having seen something resembling

SARS. It went viral on social media. He is accused of two crimes, "spreading rumors" and "severely disturbing the social order." The police order him to appear at the police station a few days later to give him a written warning:

> We solemnly warn you: If you keep being stubborn, with such impertinence, and continue this illegal activity, you will be brought to justice - is that understood?[62]

Li pens, "Yes, I do," under the warning.

It is regrettable that Chinese officials did not investigate the validity of criticism before issuing a police warning. It was probably not a coincidence that on December 31st, Wuhan officials announced that 27 people were sick from a new virus.[63]

Is the number true? It would be nice to have a United States Government report on what happened in Wuhan in December 2019 and January 2020, but Washington seems to be busy with viral origin investigations. So we must rely on four sources that give a look inside Wuhan. One offered footage to the American press, including the *Washington Post* and the *New York Times*, to no avail; one had to be smuggled out of China to be posted on YouTube in March; one is available in January.

It is Nanfu Wang who offers the footage to American news media, footage that includes publicly available social media. It also includes footage shot by local camera operators, which involves some risk of arrest. The footage will become the movie *In the Same Breath*.[64]

Nanfu was raised in a rural farming village in Jiangxi Province, China. Her father died when she was 12, forcing her to leave school. She taught in primary school and studied English literature. A fellowship at Shanghai University led to a scholarship at Ohio University and New York University. It was only then that she saw her first documentary film. That became her passion. She married an American she met at school and settled in New Jersey. She has several award winners under her belt when she travels through Wuhan to see her family for the Chinese New Year on January 23rd. *In the Same Breath* is born.[65]

Two Chinese journalists from Beijing travel to Wuhan to report on the outbreak just before the shutdown. Their work is censored. Their film will be smuggled out of China and given to the Al Jazeera Investigative Unit. It is posted on YouTube under the title, *3 Days that Stopped the World*.[66] In addition, *Al Jazeera* posts at least one video from inside an overcrowded hospital as it happens.[67] A *Wall Street Journal* article, "How It All Started: China's Early Coronavirus Misstep," gives a retrospective into the Chinese response to the outbreak.[68]

The third source is a blog by Fang Fang (方方).[69] Anyone who can read Chinese and following @方方 is getting an idea of what a lockdown is like and the Chinese Government's pronouncements.

Fang Fang grew up in Wuhan and is now 65, old enough to remember the Cultural Revolution, old enough to have gone through the *Reform Opening-Up*, China's transition to managed capitalism. She has already written some best sellers.

Her blog is on Weibo,[70] a Chinese language micro-blog similar to Twitter, and WeChat, a popular messaging app. It quickly attracts millions of Chinese-speaking followers.[71] At its height, her posts get between 3 million and 10 million views in a few days.[72] If you don't know Chinese, you likely don't know about it.

The blog will be introduced to English readers in March when the *Los Angeles Times* and a few other English-language press outlets have an article about her blog.[73] The blog translated to English and collected in a book, *Wuhan Diary*, which will be in March 2020.

A fourth source is a Chinese magazine article that was quickly pulled down, but not until its was repeated and saved, sometimes in Brail, Morse Code, or Emojis.[74]

## BEHIND THE CURTAIN IN WUHAN

At the beginning of each year, each province holds its Provincial People's Congress, a time for leaders to give speeches and announce plans for the coming year. Wuhan hosts the Congress for the Hubei Province, in which Wuhan is the largest city. On January 6th, the national CDC of China quietly activates emergency measures. Chairman Xi Jinping discretely takes personal control.

The momentum of political tradition seems to cloud the leadership's judgment. The Congress commences that same day. The national color, red, is everywhere, with red ties adorning well-tailored suits, red badges, red drapery, and red carpets. It opens with China's national anthem, an anthem similar in theme to ours in that it celebrates a military triumph under trying circumstances. In China's case, it is *The March of the Volunteers*, a call to fight the Japanese invasion in the 1930s:

> Stand up! Those who refuse to be slaves!....
>
> We are billions of one heart,
> Braving the enemies' fire, March on!

The meetings go on for two weeks. Each day, the Government announces no new cases, assuring citizens that the disease is not dangerous.

The story is different on the streets. Chen, owner of a medical clinic a few blocks from the Huanan Market, knows that people from the market had been coming for symptom relief since mid-December. Her husband falls ill from treating them. On the 1st, they cannot find any hospital room. On the 2nd, they travel to the Red Cross Hospital and get assigned a bed. But, after a CT-scan, the doctor says they cannot treat him — even after Chen begs on her knees, they kick them out. After another taxi ride to the Tongji Hospital, they are told there is no treatment. The doctors know this because they had given up on a patient a week earlier and forwarded him to the Jinyintan

Hospital, which specializes in infectious diseases and has an isolation ward.

On January 7th, amid thunder and rain, an ambulance arrives; the medics lay Chen's husband on the gurney. He squeezes her hand and smiles as they lift him into the ambulance. She never sees him again — he dies.

Doctor Li is now having breathing problems of his own. He checks himself into his hospital. He sends posts wondering why officials are saying medical professionals aren't getting the disease when he is sick.

It is now a loud whisper — people are sick — where is Chairman Xi Jinping? The Congress ends with a massive Lunar New Year celebration. On the 20th, Mr. Xi appears on TV in a display of confidence and control. Officials announce that the new pneumonia is contagious.

Is it another SARS? As people attempt to go about their business, it is with a growing fear, a fear that grips the city of 11 million souls. Now, they might see a body on the street. If they try to get medical help, they end up in a crowded hospital hallway only to see doctors and nurses overwhelmed — but who keep working. Healthcare workers are now fully suited up, head to toe, with tape sealing seams where gloves meet sleeves and boots meet trousers. It's a hush-up campaign. Healthcare workers are forbidden to talk to the reporters. The Chinese People magazine reports that,

> Doctor Ai Fen received a virus test report for a patient with an unknown pneumonia. She circled the word "SARS coronavirus" in red... she took a picture of the report and circulated it. ... those who forwarded the report included the eight doctors who were disciplined by the police.
>
> Ai Fen... suffered an "unprecedented and severe reprimand"; it was said that she was acting unprofessionally by creating false rumors.[75]

Chinese social media has a forum where people upload their identification cards and X-rays,

hoping to get medical help. Lines at the hospital doors are blocks long. Some sleep in their car next to the hospital. An uncounted number die at home.

At 2:00 AM on the 23rd, authorities announce that Wuhan will lock down at 10:00 AM. 300,000 people flee the city. At 10:00, crews erect street barriers, public transportation is switched off, streets are quiet, and movement requires permission. Officials order travel restrictions in other cities in Hubie Province; excavators drop earth barricades at the entrances to towns.[76]

The Central Government arranges for 40,000 healthcare workers from other provinces to travel to Wuhan — white tent clinics sprout in parking lots like mushrooms after a rainstorm. It is not enough. An uncounted number of people die at home.

It is clear from her words that Fang believes in the governing system for the society in which she lives. Her blog begins on January 25th, a couple of days after the Government shuts down Wuhan. It is also day one of the Lunar New Year, Chinese New Year, when it is traditional for Chinese to celebrate, travel to see families, and return to their hometown.

Fang Fang's diary begins, in the first sentence, with Fang worrying if she will be censored. She is not sure if she can send anything out on her Weibo account because it had recently been shut down for criticizing "young nationalists" for using foul language to harangue people on the street, insisting that being a patriot is fine but does not excuse someone from acting like a "hooligan."[77]

Of course, the Chinese Communist Party (CCP) is heavily involved in setting the rules for social media and has had almost a decade of practice in keeping their social media mastiff on a leash, with opaque rules, using methods ranging from deleting posts to arrests. The Government is very explicit in the People's Republic of China (PRC) that they have the right to use personal data to maintain order; going even further, it should further socialist goals.

The pattern is to allow corruption scandals by low-level officials and expelled party members to go viral but immediately pull down any suggestion of collective action. Fang seems well practiced at navigating this maze — though before long, it is reported that her posts are getting deleted by Beijing within an hour — but not before they go viral as her followers repost them.[78]

Censorship exists in America as well. For example, movies were censored for decades because pressure groups and the Government at all levels turned the screws.[79] The Hay's Code came into its own in the early 1930s. It was not just about showing skin, but a moral code that promoted a sanitized view of America, insisting on specific themes, such as:

- Show religious ministers in a positive light.
- Criminals always get caught.
- Suppressing particular themes such as interracial love affairs.

By the late 1960s, Supreme Court decisions had eliminated most movie censorship. But, censorship did not go away. Today, the methods have changed, moving to state legislatures and school boards demanding specific ideas be taught or not taught in public schools.[80] Not to mention that private companies are taking it upon themselves to decide which post to delete.

During the lockdown of Wuhan, officials double down on censorship. While they can't keep up with social media, some of the most well-known reporters posting on social media are arrested and sent to prison. Anyone seen filming with a cell phone on the street is warned off by police or interrogated. Hospital staff and funeral home workers are forbidden to talk; cell phones are monitored.

Once the Chinese Government is forced to shut down Wuhan, it turns 180° from its silencing operation to putting on a media blitz. Camera crews arrive at hospitals to tell positive stories, people overcoming adversity — even some of Nanfu's camera people think that is needed:

> We're concerned that overseas agencies or individuals will take advantage of negative stories. The concern is that foreign hostile forces would report negatively on China.

I cannot disabuse them, given the penchant for China-baiting by some of our politicians.

Fang Fang's account is a preview of what will happen in the United States when faced with the same issues. Engaging in a cheering, jeering, no-holds-barred election, the United States has already prepared the ground for closed social media accounts, flagged tweets, and censored posts. Fang asks her readers to let her know if they saw her first post because she had learned that a Weibo user could be led to believe their post is live, but in reality, it could be tagged so no one could see it — it goes live. China's media mastiff slips its leash. Beijing deletes her posts, but not before her followers repost them.

Fang admires city officials for "making the hard choice that needed to be made."[81] She, like many others, assumes the shutdown will be over quickly. Fear subsides a few days after settling into the reality of a stay-at-home order. The streets are empty except for those who need to be there. Workers wearing white head-to-toe suits appear to enforce and implement the shutdown's logistics. They soon become known as Big Whites.[82]

If the shutdown and crisis had been over in a week or so, local officials would have escaped being blamed for missteps. But, hospitals were already at capacity when the shutdown started; it worsens daily for the first few weeks. A young man Fang knows is having trouble breathing but could not get diagnosed, recovering without ever getting seen by a doctor.[83]

Before the month is out, netizens[84] criticize the medical specialists who came to investigate the outbreak for saying it was not contagious and

Photo by Robert Way (Shutterstock) Cropped by James H. Olds

*"Big Whites," Street Scene Shutdown in Wuhan*

could be controlled. Fang waits in vain for someone to apologize; the delay in mobilizing for an infectious outbreak is destined to be investigated by the press, then by Chinese officials. Fang dubs it the "twenty-day lag."[85] It is a deadly mistake — a warning — that running out of medical capacity is as deadly as the disease.

Nanfu Wang, talking by phone to an anonymous funeral home worker at a Government-run crematorium, says,

> It was reported that Wuhan's COVID-19 death toll is 3,300. Is that number true?

He replies,

> Only a fool would believe that number.

The funeral home worker did the math: his one van, 10 trips a day to collect 10 bodies each — 10,000 to 20,000 for that one funeral home. There are eight funeral homes in the city. When Nanfu calls the people who posted x-rays on a social media forum, a family member says,

> Because there was no help, Mom didn't survive. My father and brother passed away... The virus got my whole family... We're just waiting to die.

A resident of a high-rise says,

> Do you know what the saddest moment was? It was when you suddenly heard the heartbreaking cry. We all knew that person must have just learned about the sudden death of a family member. The building was full with that cry. I'll never forget it.[86]

Fortunately, reports of symptoms being observed in Wuhan continue to be published. However, it is not clear if key details such as digestive symptoms and the distinguishing characteristics of a CT-scan are being used by hospitals worldwide.[87] Doctors, physicians, healers, airport workers, hospital workers, and people worldwide are now on the lookout for respiratory symptoms, a cough, a sneeze, a wheeze, or a fever. There is still be much to learn about how COVID-19 presents itself and that it often doesn't present any symptoms at all.

Scientists at the Rocky Mountain Laboratories (RML) in Hamilton, Montana, about six blocks from my house, are on the lookout for the new coronavirus, having studied other coronavirus outbreaks, particularly SARS in 2003 and Middle East Respiratory Syndrome (MERS) in 2013. "I don't think it came as a surprise to anyone here or anyone back at NIH that another pandemic was on the horizon," says Doctor Marshall Bloom, Associate Director of RML, in a later interview. Many of the scientists at RML work with colleagues around the world and heard reports on ProMED[88] "right around the end of December 2019." Doctors deWit and Munster have "their antennas up. They think it is a big deal." Getting approval for a series of studies on SARS-CoV-2 in early January, they start "as soon as they could get a sample from the CDC."[89] But, as scientists puzzle over the nature of the disease, an essential piece of the puzzle will be revealed from another "laboratory" first.

On January 25th, the cruise ship *Diamond Princess* arrives in Hong Kong five days into its travels, which began in Yokohama, Japan. An 80-year-old man who had been coughing for two days disembarks to go to a local hospital, a seemingly routine event. He has COVID-19. The passengers are not allowed off the ship.

The *Diamond Princess* becomes a deadly laboratory, giving the first definitive information on how the disease is spread in close quarters, at least for an older population. Many people test positive without having symptoms. Of 3,711 passengers and crew, 7 people die in the next month, all over 70. This gives a rough idea of what is coming. One analysis adjusts for the age distribution and the experience in China to estimate that about 0.5% of infections will be fatal. This will turn out to be higher than later results for the general population but will be a decent estimate for people in their 60s.[90]

It doesn't take long to ask the question in Hamilton, "Was it a lab leak?" The City of Wuhan

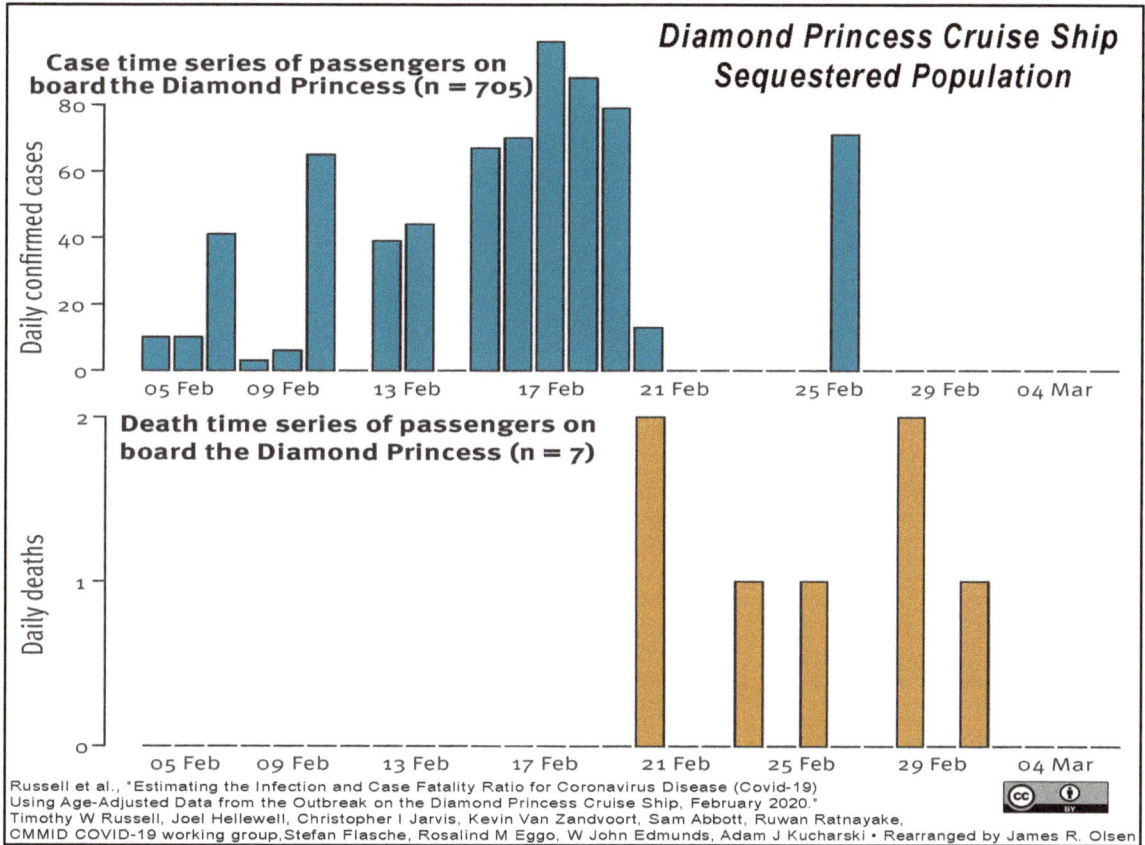

**Case time series of passengers on board the Diamond Princess (n = 705)**

*Diamond Princess Cruise Ship Sequestered Population*

**Death time series of passengers on board the Diamond Princess (n = 7)**

Russell et al., "Estimating the Infection and Case Fatality Ratio for Coronavirus Disease (Covid-19) Using Age-Adjusted Data from the Outbreak on the Diamond Princess Cruise Ship, February 2020." Timothy W Russell, Joel Hellewell, Christopher I Jarvis, Kevin Van Zandvoort, Sam Abbott, Ruwan Ratnayake, CMMID COVID-19 working group, Stefan Flasche, Rosalind M Eggo, W John Edmunds, Adam J Kucharski • Rearranged by James R. Olsen

*Diamond Princess Cases and Deaths*

hosts the Wuhan Institute of Virology. The institute hosts a BSL-4 lab, as does Rocky Mountain Laboratories (RML) in Hamilton.

The concern about BSL-4 labs and bioweapons is an old story. COVID-19 is the latest chapter for scientists, activists, and conspiracy theorists. Like gossip in a small town, where speculation over the back fence becomes gospel truth by the time it gets to the other side of town, it doesn't take long for speculation to become a truism: "the novel coronavirus was released from the BSL-4 lab in Wuhan."

The biosafety controversy was already in play when President George W. Bush initiated BioShield[91] after the 9/11 attack. Its purpose is to defend America against bioweapons. Biosafety was a concern in several locations around the United States as biolabs were expanded and upgraded. It came up right away in Ravalli County when we learned of the National Institute of Al-

lergy and Infectious Diseases (NIAID) plans to expand Rocky Mountain Laboratories to a BSL-4 lab in 2002. We raised concerns about biosafety in our scoping letters, the first to Doctor Anthony Fauci, director of NIAID. A local group quickly formed Citizens for a Safe Lab.[92]

A year later, having been appointed to the RML Advisory Committee, I gave a PowerPoint presentation on biolab safety, including a survey of the causes of pathogen leaks from biosafety labs worldwide. While the NIAID had touted their facility with HEPA filters and space suits, it turns out that pathogens escape in nearly every instance by hitching a ride on a staff member, sometimes an actual accident. But, more often, it is someone who ignores the safety protocols.[93] It happened at RML in the 1960s with Q-Fever, when the disease was carried out of the lab and passed around Hamilton.[94]

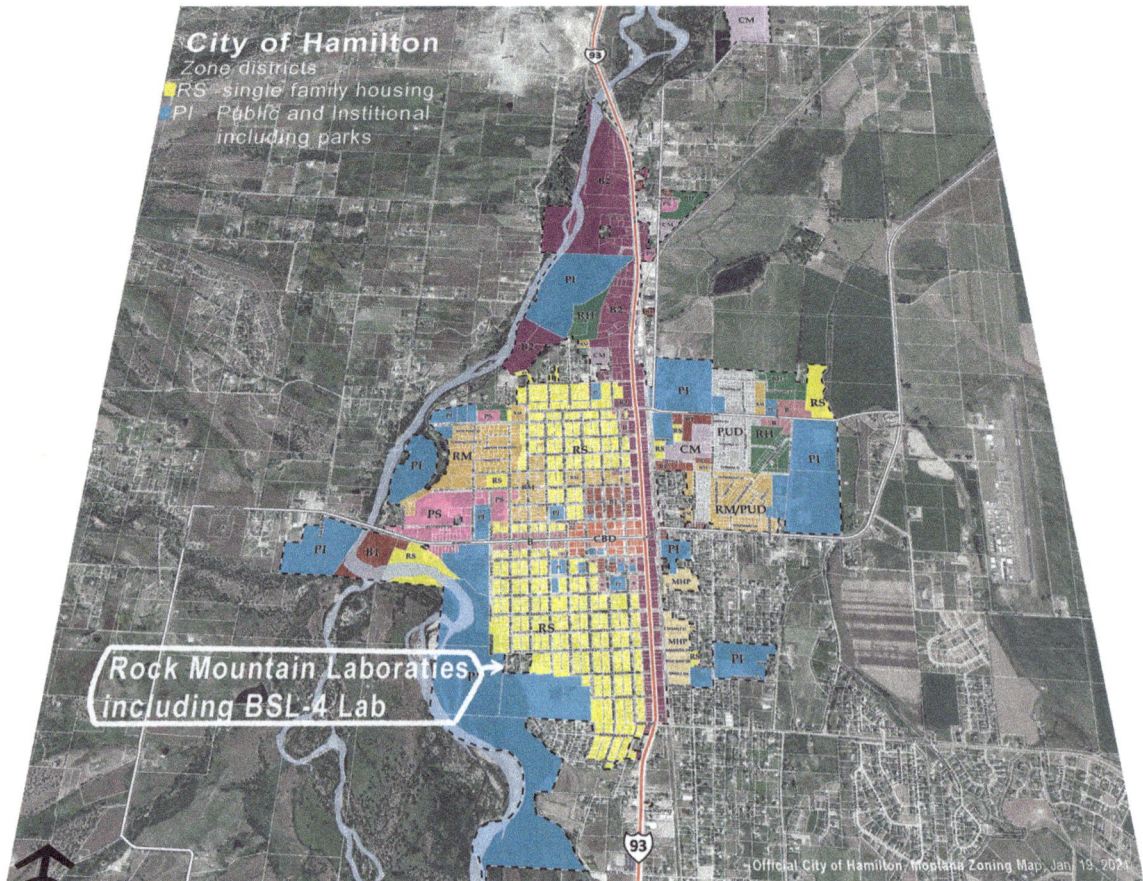

*Hamilton City Map, RML Location*

Lab leaks are more common than we would wish. The last vestige of the Spanish Flu[95] disappeared from the landscape by 1957. That is until it was released in 1976 or '77 from some lab, probably in the USSR,[96] killing 700,000 people.[97] At the time, several countries held stocks of the Spanish Flu. Labs were also working on Swine Flu, including in the United States.

> In 1976, H1N1 swine influenza virus struck Fort Dix, causing 13 hospitalizations and one death. The specter of a reprise of the deadly 1918 pandemic triggered an unprecedented effort to immunize all Americans.[98]

Biosafety regulation and enforcement in the United States is a patchwork, with NIH issuing guidelines and rules for high-risk experiments, the biosafety committee in each lab adjudicating the safety, the Federal Select Agent Program administering a list of 67 particularly hazardous agents, the FBI investing when one of them goes missing.[99]

In a court-ordered mediation, we extracted a commitment "not to weaponize a pathogen" at RML, which has led to assurances from NIAID that there have been no gain-of-function creations at RML.[100] The agreement also established relationships with the community and local public health officials. The deal was signed by Citizens for a Safe Lab, Friends of the Bitterroot, and Women's Voices for the Earth, countersigned by NIAID. RML got the go-ahead for its new BSL-4 lab in 2004.[101] I think we ended up better off. Is it perfect? No.

Our efforts to reduce the chances of an escape of a pathogen from the lab at RML had not done as much as hoped. Seven years later, a 2011 *Scientific American* headline reads,

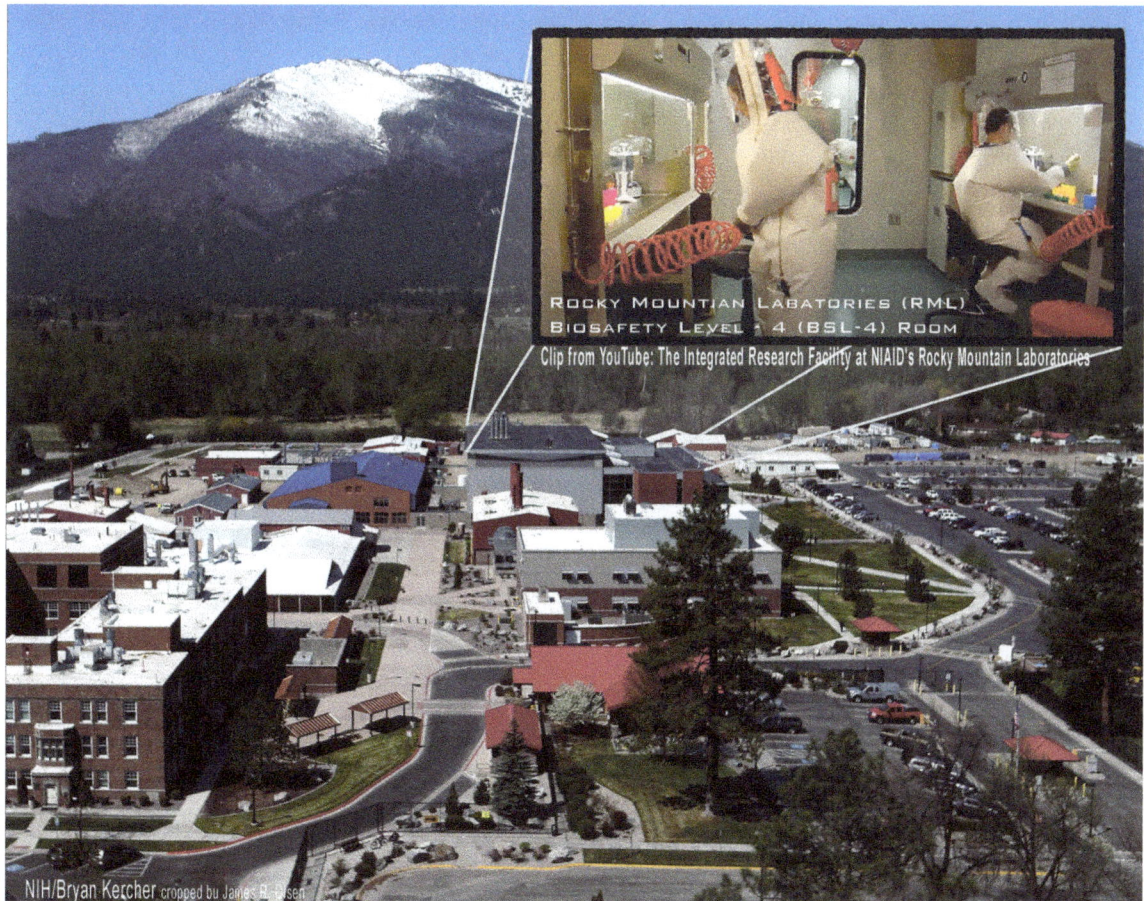

*RML BSL-4 Laboratory at Hamilton, Montana*

Nearly 400 Accidents with Dangerous Pathogens and Biotoxins Reported in U.S. Labs over 7 Years.[102]

Some of the accidents were at RML. In the end, safety and security are never perfect. Still, safety is much more rigorous in endeavors as diverse as commercial air travel and mining than at the CDC or NIH.[103] We had discussed this in RML Advisory Board meetings — *What is needed is safety that is disciplined, backed up by discipline*.

By 2014, reports of biolab accidents had infested the halls of Congress every summer, with hearings year after year. A Government Accounting Office (GAO)[104] Report in July 2014 noted that there were no national standards for biosafety, adding that the GAO had reported this to Congress in past reports for 12 years — the annual reports had not shown any progress.[105] The

2014 report includes a potential exposure of ten workers to Anthrax due to human error, adding, "A handful of smallpox vials were uncovered during a Food and Drug Administration (FDA) office move."

That same year, a story was picked up by the Associated Press (AP) that there were six security lapses in the BSL-4 lab at RML[106] — three of which were people lending out their badge to a coworker, or what is called piggybacking or tailgating through the door where people swipe their badge. Incidents at RML were described as "alarming" by Nicole Rowley, who had worked there and is a Missoula County Commissioner-elect.

There are known fatalities from biolab accidents. Two Chinese scientists had been exposed to SARS ten years earlier; one died.[107]

The biosafety investigation went on only slightly abated as another news article of a failure was reported by *USA Today* in 2016, "Newly disclosed CDC biolab failures read 'like a screenplay for a disaster movie'," at CDC headquarters in Atlanta in 2009. It took years for the CDC to respond to a follow-up FOIA (Freedom of Information Act).[108] Congress kept having hearings.

Today, 18 years later, Doctor Marshall Bloom summarizes biosafety. "You need 4 bites of an APPLE" for full lab safety to work in labs from high school level through biosafety level 4:

> A: Administration. Training, paperwork, etc.
> P: Personal protective equipment (PPE).
> PL: Laboratory procedures. Standard operating protocols.
> E: Engineering. Biosafety cabinets and labs that have protective features.[109]

The counter-narrative purveyors are ready with their shiny new sword, the Internet. Is the new virus a man-made bioweapon purposely released from the Wuhan lab? A January 5th Tweet begins,

> 18 years ago, #China killed nearly 300 #Hong-Kongers by unreporting #SARS cases, letting Chinese tourists travel around the world... to spread the virus with bad intention. Today the evil regime strikes again with a new virus.[110]

The blog in India, GreatGameIndia,[111] is quickly picked up by ZeroHedge[112] and others in the United States, spreading like a wind-blown prairie fire through social media, a network of blogs and sites with overlapping followings.[113] It quickly gets a counter-punch in the podcast Knowledge Fight.[114] *And The Beat Goes On*[115] in the "InfoWars."[116]

These sites are not new. COVID-19 is simply the new cause and counter cause for the InfoWars. The "mainstream" newspapers and members of Congress help give the idea wings, some almost by accident, others seeming to hope it is true.

Stories in mainstream media speculate, "US biosafety warned in 2017 that a SARS-like virus could 'escape'" from the Wuhan laboratory, "Coronavirus link to China biowarfare program possible, analyst says."[117] The idea of a man-made released pathogen flies across the Internet, breeding so many fire-breathing dragons that they are impossible to slay.

An accidental lab leak is more likely than a bioweapon. There is evidence that the Wuhan Lab was playing catchup on lab safety in 2019, coincident with a "flu-like" outbreak in October — but the evidence is circumstantial.[118]

While Senator Cotton stands on the Senate floor, urging closing the borders from China — which seems like a good idea to me — but he uses much of his time to cast doubt on the origin of the new virus, noting the existence of a BSL-4 lab in Wuhan, shifting the partisan focus from the origin of the virus to a growing list of purported Chinese misdeeds.[119]

## BEHIND THE CURTAIN IN THE VIRTUAL WORLD

The Chinese Government mounts a propaganda counteroffensive using fake social media accounts that will persist through a good part of the year to divert attention away from Wuhan. Attempts such as "US Army soldiers visiting Wuhan brought it" and "the virus was created at Fort Dix" falter. It doesn't work.

However, China's global assistance program gains traction in Asian countries in contrast to the America First policy.[120]

The Secretary of Defense issued a 2019 secret order to the military, raising the psyop competition with China to that of "active combat," allowing them to skip coordination with diplomats. Not to be outdone, using things like fake social media accounts, the United States military targets China's response to COVID-19 and then the effectiveness of China's vaccine even before they were complete. The misinformation campaign targets the Philippines[121] — and the Muslim world, claiming Chinese vaccines contain pork gelatin.

Several Pentagon officials will argue against the antivax propaganda, one saying, "We're stooping lower than the Chinese." It will go ahead anyway. It will go ahead anyway. Facebook will see through it, but the Pentagon tells it it is an anti-terrorism campaign. The Pentagon promises to pull down the COVID-related propaganda. They don't. It won't be until mid-2021 when Facebook meets Biden's National Security Council, that Asian antivax propaganda is put to a stop.[122]

D octor Li tests negative for the novel coronavirus several times, finally testing positive on the 30th. He posts a picture of himself on a ventilator on the 31st. By then, officials have agreed he was right. By then, the police have apologized. He will be dead within a week. Li will be honored by the Chinese Government as a "martyr," its highest honor.[123]

On the last day of January, looking from her apartment window in Wuhan, Fang Fang becomes an oracle, foreseeing the fate of the rest of the world:

> These days, there are very few cars and even fewer people.[124]

*31 Jan. 2020: Ø COVID-19 COVID-19 Related Deaths*

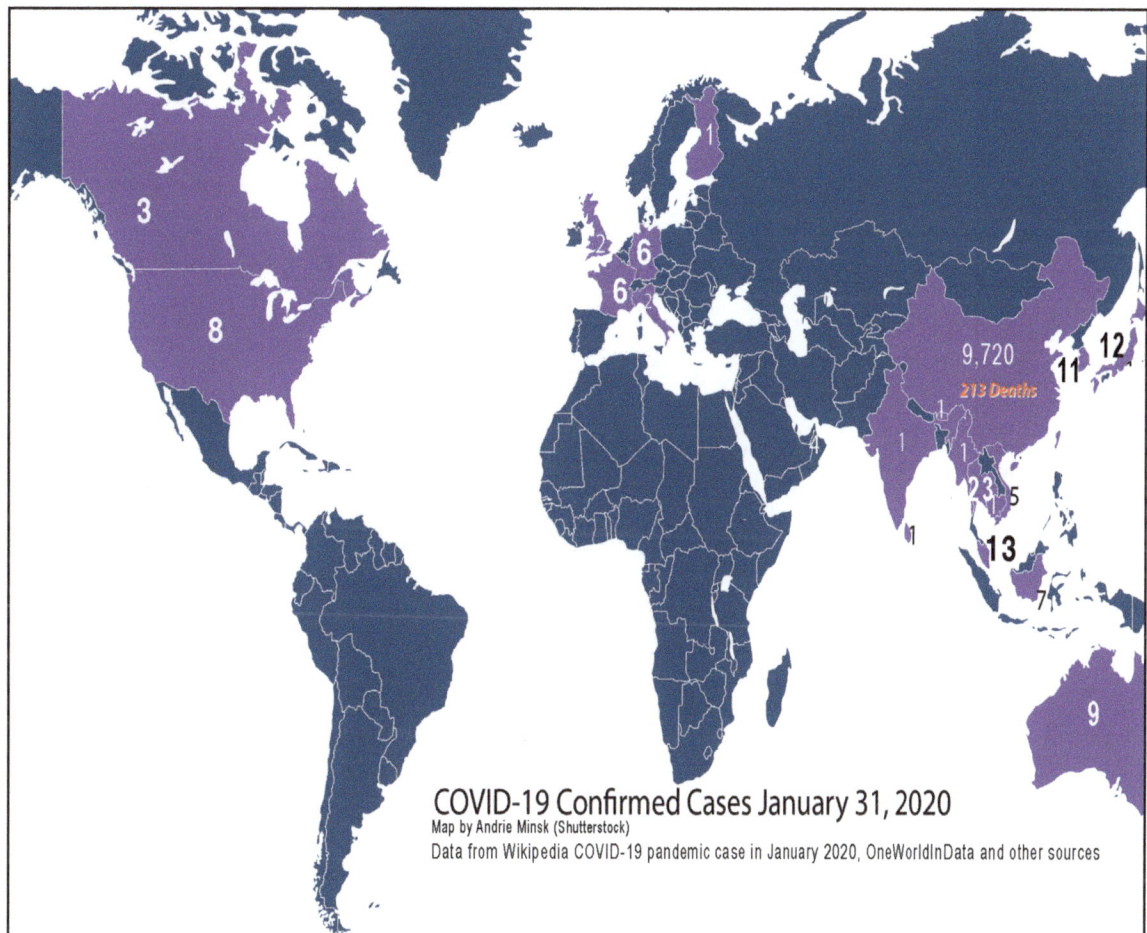

COVID-19 Confirmed Cases January 31, 2020
Map by Andrie Minsk (Shutterstock)
Data from Wikipedia COVID-19 pandemic case in January 2020, OneWorldInData and other sources

*World Map COVID-19 Cases January 2020*

| 4 | 8 | 11 | February 2020 | 17 | 23 | 26 | 29 |
|---|---|---|---|---|---|---|---|

FDA approves emergency CDC PCR Test 200 sent to states

"Untrustworthy results" PCR Test — East Manhatten, NYCity from CDC

WHO names disease COVID-19

P<small>RESIDENTS</small> Day

Italy 1st COVID-19 hotspot - nationwide lockdown

CDC Incident Manager, "disruption of everyday life may be severe"

CDC updates guides to allow testing if potentially exposed

FDA approves private companies to make tests

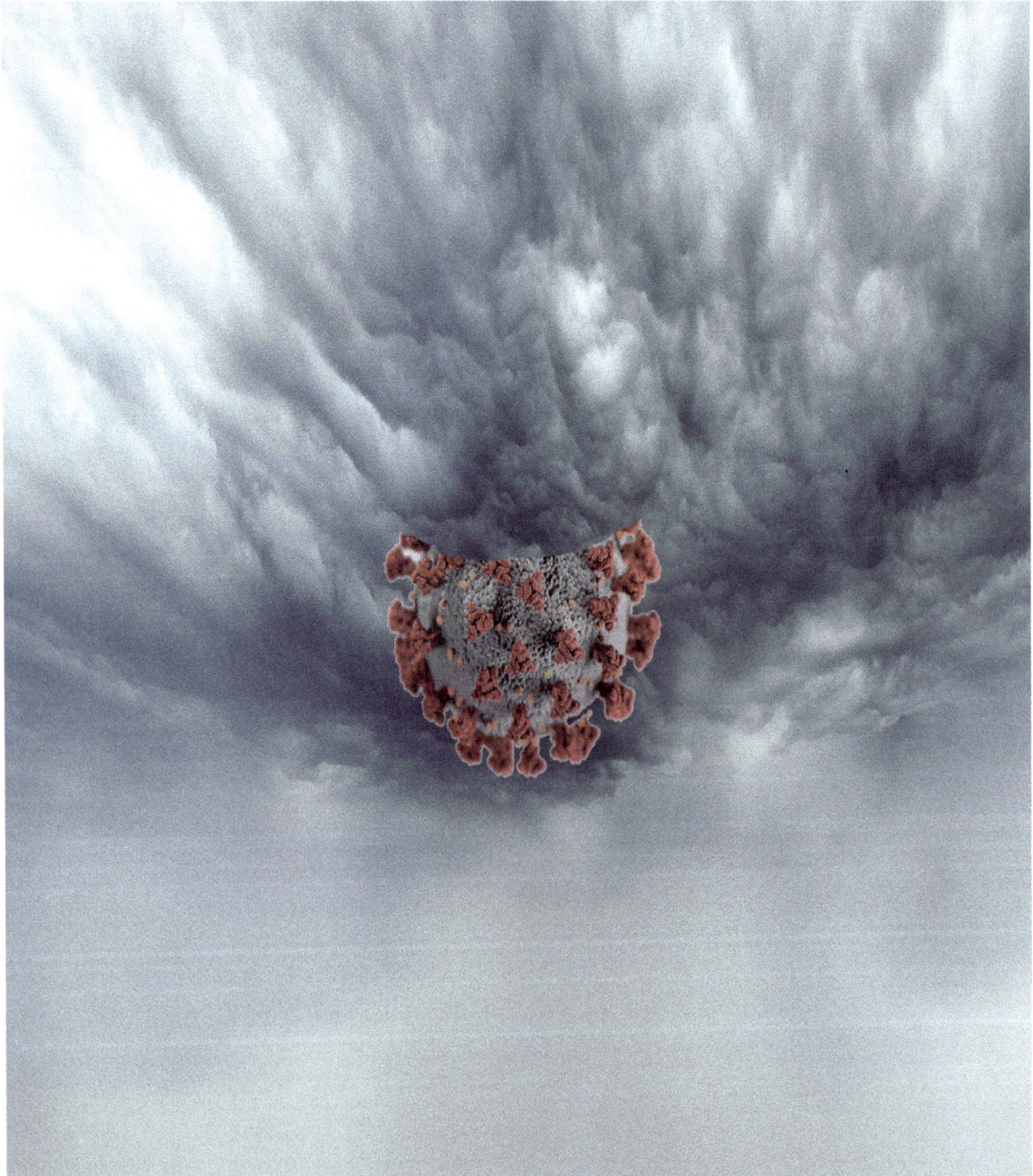

# 4

# STORM CLOUDS COMING FAST

You can't track what you can't see.[1]
— Michelle Mello

## FEBRUARY 2020

The virus is coming as inevitably as a bullet shot from a .45 Colt at high noon. Slow on the draw, all the political hacks can do is look for what is wrong about President Trump's China travel ban. It's a familiar meme for a President you didn't vote for — whatever he does must be wrong.

Even as US case numbers are rising fast, no one asks how to do travel restrictions better — even though it is obvious to anyone looking across the Pacific that travel restrictions being used in Asia are slowing the virus spread. Instead, before the first week of February is over, many commentators are searching for a way to call the President's restrictions on travel for non-citizens "racist." It is short-sighted, self-serving, and quickly back-fires.[2]

Everyone thinking it is another SARS is caught flatfooted. Many countries check for symptoms at ports of entry, knowing that SARS becomes contagious once symptoms appear. This convenient characteristic is not the case for COVID-19; people slip through the border because they can be infectious with COVID-19 before symptoms occur or have no symptoms at all.

Countries worldwide begin to impose various strategies for halting the spread of the virus at their borders, from banning travel from areas with outbreaks to mandatory quarantine for incoming travelers. A few countries, includ-

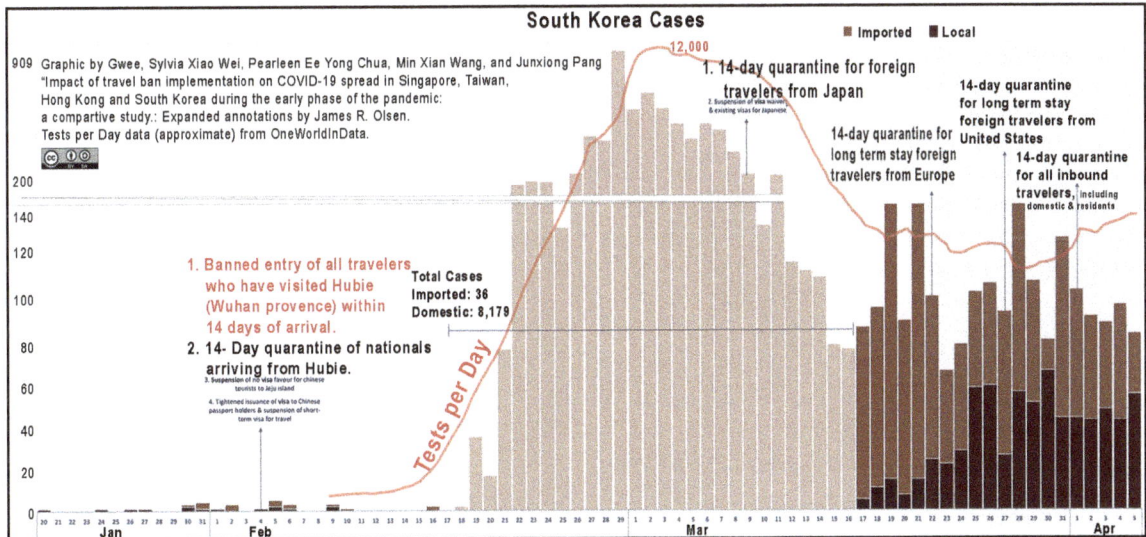

*South Korea Travel Restrictions and Cases*

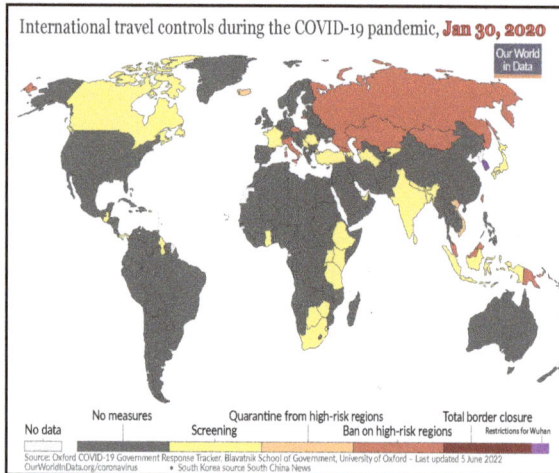

International Travel Restrictions Jan. 2020

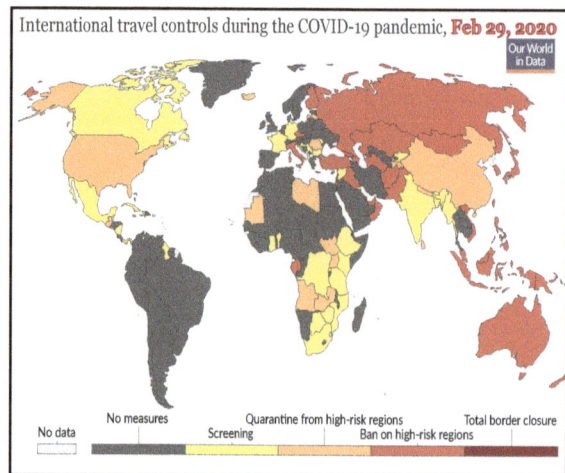

International Travel Restrictions Feb. 2020

ing South Korea, halt travel from Wuhan entirely— giving them some breathing room. Countries that are quick on the draw have the most success.[3] However, travel restrictions aren't enough to completely stop the virus. The case counts are "flattened," not driven to zero.

The question for public health is how to use the respite. Even those who slapped leather to get test kits ready must have the medical capacity and contact tracing teams ready to mount up. A few countries have thought it through, getting test kits, test analyzers, test stations, trained contact tracers, outpatient medical providers, hospital beds, ICU stations, and the doctors and nurses to staff them.

Knowledge exists down to the county level in the United States, but not the resources. Ravalli County Public Health knows how to test and contact trace, for instance, with a period Whooping Cough outbreak in the schools. Still, there is no plan to marshal the resources for a pandemic here or anywhere else in the United States.

It's a horse race. In the United States, partisans are too busy hobbling each other's horses for the country to get quickly out of the starting gate.

Now that the SARS-CoV-2 genome is known, the race is on to produce a test that will recognize a unique part of the virus RNA, a Reverse Transcription - Polymerase Chain Reaction test, abbreviated RT-PCR test, or simply PCR test.[4]

In South Korea, Chun Jon-yoon had seen enough. He owns the biotech company Seegene, headquartered in Seoul. In early January, he believed it was only a matter of time before the first case of COVID-19 would come to South Korea. Calling his team together on January 16th, he told them to start working on a test. "Even if nobody is asking us to, we are a molecular diagnosis company."[5] It was predictable, not prescient; the first case in South Korea was detected four days after his speech.[6] By January 24th, he had ordered the raw material for the test kits. They used a supercomputer to design the test, setting up its framework from a previously designed test for inflammation of the urethra — saving weeks." The test kit design only took a small team.

On the 27th, South Korean health officials summoned 20 medical companies to a meeting at a conference room at a train station, telling them "they are cleared for take off"; emergency use authorization is forthcoming.[7] Kogene Biotech is the first to have an approved test on the 4th. The Korean CDC signs off on the test a week later after verifying it themselves using patient samples. Just to be sure, samples are crosschecked with multiple tests to see that they get the same results.[8]

## South Korea's Response to COVID-19

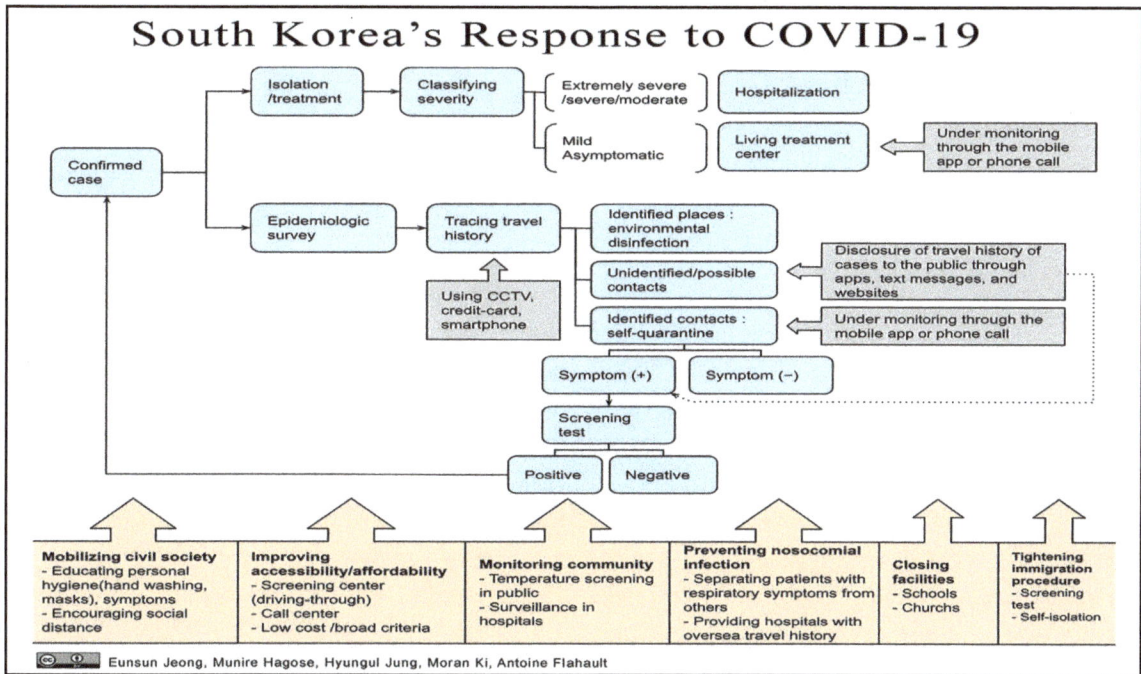

Eunsun Jeong, Munire Hagose, Hyungul Jung, Moran Ki, Antoine Flahault

*South Korea's Response to COVID-19*

The Seegene team has a PCR test for SARS-CoV-2 ready on February 5th. The kits are available in time as cases spike in South Korea, driven by a large crowd of worshipers in a mega-church.[9]

President Moon Jae-in raises the crisis level to its highest as he announces,

> We're now at a watershed moment... the next few days will be very critical. We need to identify the infected people as soon as possible and prevent the virus from spreading further.[10]

That day, Chun put his entire 395 staff to work making test kits, saying, "Emergency operation means all divisions, you have to change your job." That means PhD microbiologists dropping their research and taking a shift on the assembly line. By month's end, Seegene is producing 1 million tests per week for $20 each. The word gets out. Orders come in from Italy and Germany.

The government deploys 118 test stations, including drive-through. Samples are analyzed automatically. The Health Minister summarizes the policy,

> Detecting patients at an early stage is very important. South Korea is an open

society and would like to protect the freedom of people moving around and traveling. That is why we're conducting a mass amount of tests.[11]

In Germany, a similar story emerges. A small, high-tech company in Berlin led by Olfert Landt reports, "My wife and I have been working 16 hours a day, seven days a week." Landt's company ships 1.4 million tests to the WHO in February for distribution worldwide. Why didn't the United States request some of the WHO test kits? The CDC will take care of America.[12]

A test's actual real-world effectiveness in controlling the spread of the disease depends on how it fits into the public health response. South Korea implements a contact tracing and quarantine regime, as illustrated above.

Techniques in the gray boxes require a democracy to protect privacy, which South Korea does under their laws — mostly — some activists in South Korea accuse authorities of going beyond the law in tracking movements by geolocating cell phones and following credit card use without leave from a court.[13] It remains to be seen whether or not this will be enough.

## BEHIND THE CURTAIN IN ATLANTA

Three weeks lost. That is the *Wall Street Journal's* judgment for America.[14] The rules dictate that the CDC produces the first test kit for a new disease — in the name of quality control. Thus, the design of the test is the job of the Respiratory and Virus Diagnostic Lab in the CDC Laboratory in Atlanta.

It is a cascade of mistakes, egos, and errors.

Stephen Lindstrom runs the diagnostic lab.[15] He has successfully created diagnostic tests for flus. Lindstrom starts right away. Reluctant to go through the procurement delays of having a private company provide the material, he tasks another CDC lab, called the "Core Lab," to produce a list of raw materials. He does not have a human sample, so, using the genome, he synthesizes the strands needed for the test verification. Even as WHO announces the recipe for the test, the one developed by Landt's in Germany, the CDC has its own formulation nearly complete.

By January 20th, Lindstrom's lab was able to confirm the first case from Washington.[16] As the only facility in the United States allowed to test, more and more test requests come flowing in from all over the country. The Food and Drug Administration (FDA) has yet to approve the test for distribution.

While Lindstrom waits for approval, he kicks off the production of test kits. He verifies the formulation with his synthesized viral stand-in, then packages it in vials. On February 3rd, the vials are tested by exposing them to his synthesized SARS-CoV-2 strand to ensure it returns a positive result. Then expose vials to distilled water to ensure it yields a negative result. One of the two lots shows a false-positive for the water sample. This calls for a root cause analysis to determine why.

Lindstrom's CDC management puts him under tremendous pressure to get the test kits out despite being told about the test failure. It de-

volves into table-pounding meetings — get the tests out. Lindstrom relents. The flawed lot is set aside, and the vials for other lots are sent to public health labs across the county. While they are in transit, more lots are made — the verification tests again give inconsistent results.

A California lab email on February 8th complains that the kit shows positive for a water sample. "There is likely a widespread issue that will need to be addressed immediately," is the feedback from the field.[17] The same problem shows up in a New York lab. Doctor Nancy Messonnier, director of the CDC's National Center for Immunization and Respiratory Diseases, tells the public,

> Contamination is one possible explanation, but there are others, and I can't really comment on what is an ongoing investigation.[18]

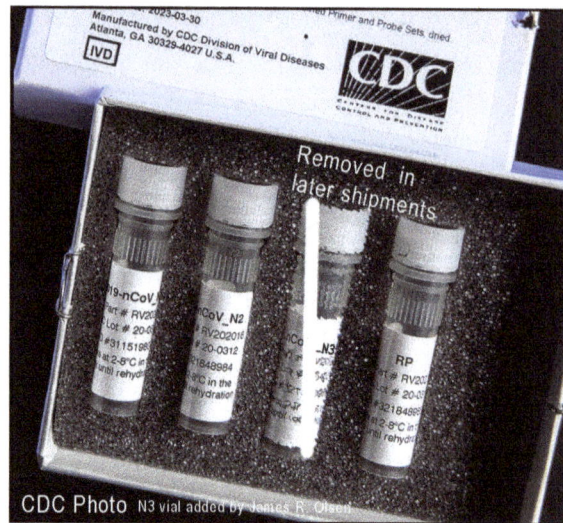

Vials: SARS-CoV-2 CDC PCR Test Kit

The rest of the story is destined to remain uncovered for eight months. Labs around the country start using a workaround. Despite the workaround, only a fraction of the 160,000 kits produced are used, 203 in state labs and 3,125 at the CDC. The source of the contamination is never sorted out. Lindstrom believes it was from the Core Lab.[19] But he is the one who signed off on it and will lose his position as the whole thing becomes political.

The inertia of the FDA is surprising. Some are starting to wonder if it is driven by numbers — 15 cases, and it's going down, says the President as he measures his performance in front of the American People in terms of low case counts.

Of course, the virus spreads across the country, remaining undetected, invisible, until someone gets really sick. We know from the *Diamond Princess* that if you flood the nation with tests that work, the invisible becomes visible, and case numbers go up.[20] The number of reported cases triples overnight on board the vessel when most people are tested instead of just those with symptoms.[21]

You can't just count people with symptoms; you must test people without symptoms. Public health is stuck with a dearth of test kits, not knowing how many people are infected, not even a good guess.

There are fewer than 100 tests per day being done in the United States. Public and private labs know how to build a test kit — it's not black magic. Some call the FDA, begging for permission.[22] But they are not allowed to.

The CDC test kit contains four vials — N1 and N2 specific to different areas of SARS-CoV-2 (labeled 2019-nCov in the photo). N3 detects another area but may also detect other kinds of coronavirus. The RP vial detects human mucus to ensure the sample was taken correctly. It is N3 that is often failing.[23]

So, CDC test kits that could be used but are not approved are sitting around the country — simply don't use the N3 vial. It takes until the 26th for the FDA to update the authorization to allow this.[24] Finally, on the 29th, other labs are released

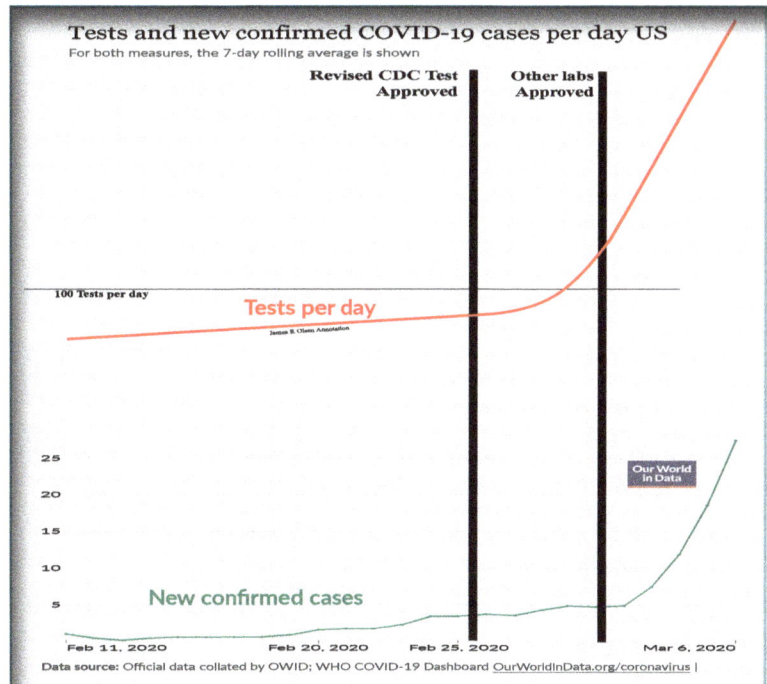

COVID Tests and Cases Feb 2020

to build test kits — LabCorp will announce one in a week and a half.[25]

Three weeks lost. The domino effect will last the rest of the year with severe consequences. The lack of kits results in triage, like a Battalion Aid Station ignoring the walking wounded until they can stop the bleeding of the stretcher bound. The public is subject to a series of ever-changing triage rules for tests restricted to severe symptoms, then any symptoms, then close contacts. This leaves many people carrying the virus with mild or no symptoms undetected.[26]

The critical characteristics of the new virus are being reported. America's testing regime is a chaotic, frustrating process. Most cases are asymptomatic or mild; many people don't know they have it, don't ask for a test, and pass it along without knowing it.[27] Then there are errors: test too early, the first week or so after exposure, and the viral markers have not made it to the upper nose; test too late, and the body's immune system has often beaten down the infectious virus while still hosting the noninfectious viral parts the test is designed to find. The new virus is a lot trickier than SARS because it can be passed on prior to symptoms — a feature in common with the common cold.[28]

Sure enough, the first community infection of unknown origin is confirmed on February 27th.[29] The next shoe doesn't take long. It is a leap year, so February 29th will be the day the news reports the first fatality in the United States of COVID-19. He is a 50-year-old man in a long-term care facility in Washington State, one of three who had tested positive at that location, making the worldwide reported fatality count 2,937, all but 102 from China.

As an insightful afterthought, Washington State officials attribute rising case counts to the availability of more test kits.[30]

Doctor Nancy Messonnier says,

> Our hearts go out to the family of the patient who died, as well as the families of the people who are caught up in this outbreak. The health of the residents, staff, and community of this skilled nursing facility are a top priority.[31]

When a gunslinger in a black hat strolls down Main Street, everyone knows something deadly has arrived. This time it is SARS-CoV-2. When a closer look is taken at previous mysterious deaths, a woman in San Jose, Santa Clara County, California, found dead by her daughter in the kitchen after having "flu" symptoms on February 6th, will be declared a COVID-19 fatality based on results from a tissue sample analyzed by the CDC.[32] It will take a year for a few doctors from California, Oklahoma, Alabama, Wisconsin, and Kansas to revisit and amend death certificates from January to include COVID-19 — though the forensic evidence in these cases is unclear. he earliest is on January 9th, when a woman in Leavenworth, Kansas, home of a military base — certainly had COVID-19 symptoms.[33]

Any hope that the CDC can focus on only one county is short-lived; the problem is about to multiply itself by the 3,143 counties, parishes,

and boroughs in the United States, of which Ravalli County is one. If you don't know who is infected, you cannot control its spread. You can't, in a phrase that has now entered the English language, "flatten the curve."

By now, it's clear that the virus has not been contained; lockdowns have become inevitable in the United States.

As far as anyone knows, the first novel corocnavirus arrives in Ravalli County in a FedEx box, the standard method for shipping pathogens to the Rocky Mountain Laboratories.[34] However, the first infection in the valley is something that no one will ever know.[35]

The virus is placed under an electron microscope at the lab, turned into a picture and colorized to become a media star in its own right.[36] Rocky Mountain Laboratories researchers aren't just taking pictures. They performed experiments to determine how long the COVID-19 virus could survive on a surface.[37] Their study shows half-lives, the interpretation being that SARS-CoV-2 lasts up to 72 hours on most surfaces. So, we all scrubbed with cleaners — Clorox Company stockholders will make over 30% by Christmas.

Scientists, doctors, and nurses around the world are beginning to understand how this new disease works and how our body responds. Since a virus does not carry the material to reproduce on its own, it must have a way to enter a cell so that it can hijack the cell and its contents. This is where the spike protein comes in.

For scientists, the spike protein is the primary candidate for how the new coronavirus enters the cell, being the method used by SARS and MERS.[38] A team led by a University of Texas researcher publishes a preprint (prior to a peer review) on the 15th, which then gets published in *Science*. The paper describes the spike protein for SARS-CoV-2. A Chinese team, also working on the problem, publishes an explanation of the biology in *Science*: the protein enters the human cell

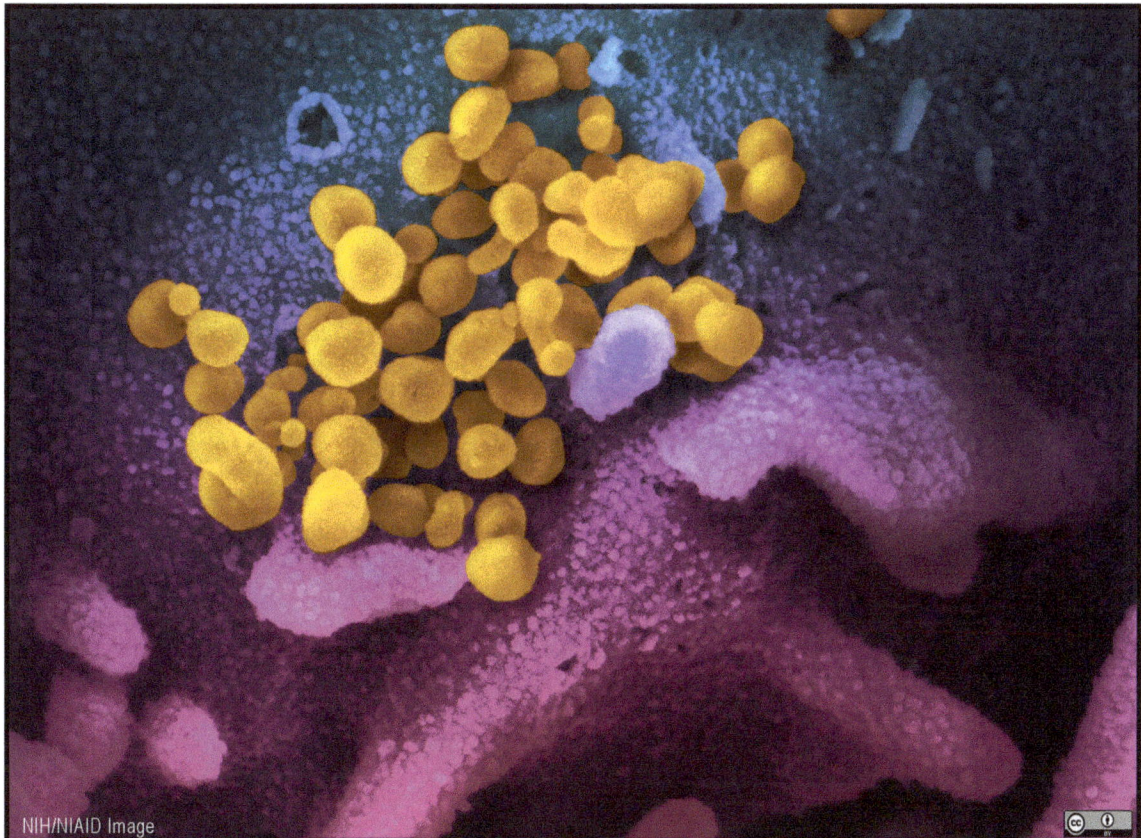

*This scanning electron microscope image shows SARS-CoV-2 (yellow)—also known as 2019-nCoV, the virus that causes COVID-19—isolated from a patient in the U.S., emerging from the surface of cells (blue/pink) cultured in the lab. Image captured and colorized at NIAID's Rocky Mountain Laboratories (RML) in Hamilton, Montana. Credit: NIAID RML*

*RML Colorized Electron Microscope Image of SARS-Cov-2*

by locking onto a complex of molecules embedded in the cell membrane (surface).

The entry point on the cell surface is an enzyme, Angiotensin Converting Enzyme-2 (ACE-2).[39] The digestive symptoms reported early on from Wuhan are starting to make more sense, as ACE-2 is present in digestive, circulatory, and respiratory cells.[40] While there are many illustrations of how the SARS-CoV-2 works, one of the best series is by the Coronavirus Structural Task Force.[41]

In the biological world, it's not just chemical bonding that counts, but the shape and geometry that the long strings of molecules take on. So, when the coronavirus is floating around, its spike proteins, sticking out like spines of a blowfish, brush against ACE-2 — they fit like a puzzle piece with the added force that some of the molecules are negatively or positively charged, lining up with the opposite charge on the same principle as a magnet.

Then, interactions with the body's naturally occurring enzymes (TMPRSS2 and furin[42]) snip the end off the spike at a cleavage in the spike protein, the cleavage being a feature that appears to make entry into the cell easier.[43] The remaining spike protein is cocked like a spring-loaded needle and penetrates the cell membrane. As multiple spikes latch the virus onto the human cell, they act like scissor-jacks to pull it through the membrane.

Once inside, the cell becomes a factory. The virus RNA becomes the factory floor supervisor, directing a process that synthesizes and collects viral parts using the cell's own machinery to assemble more SARS-CoV-2 viruses, which

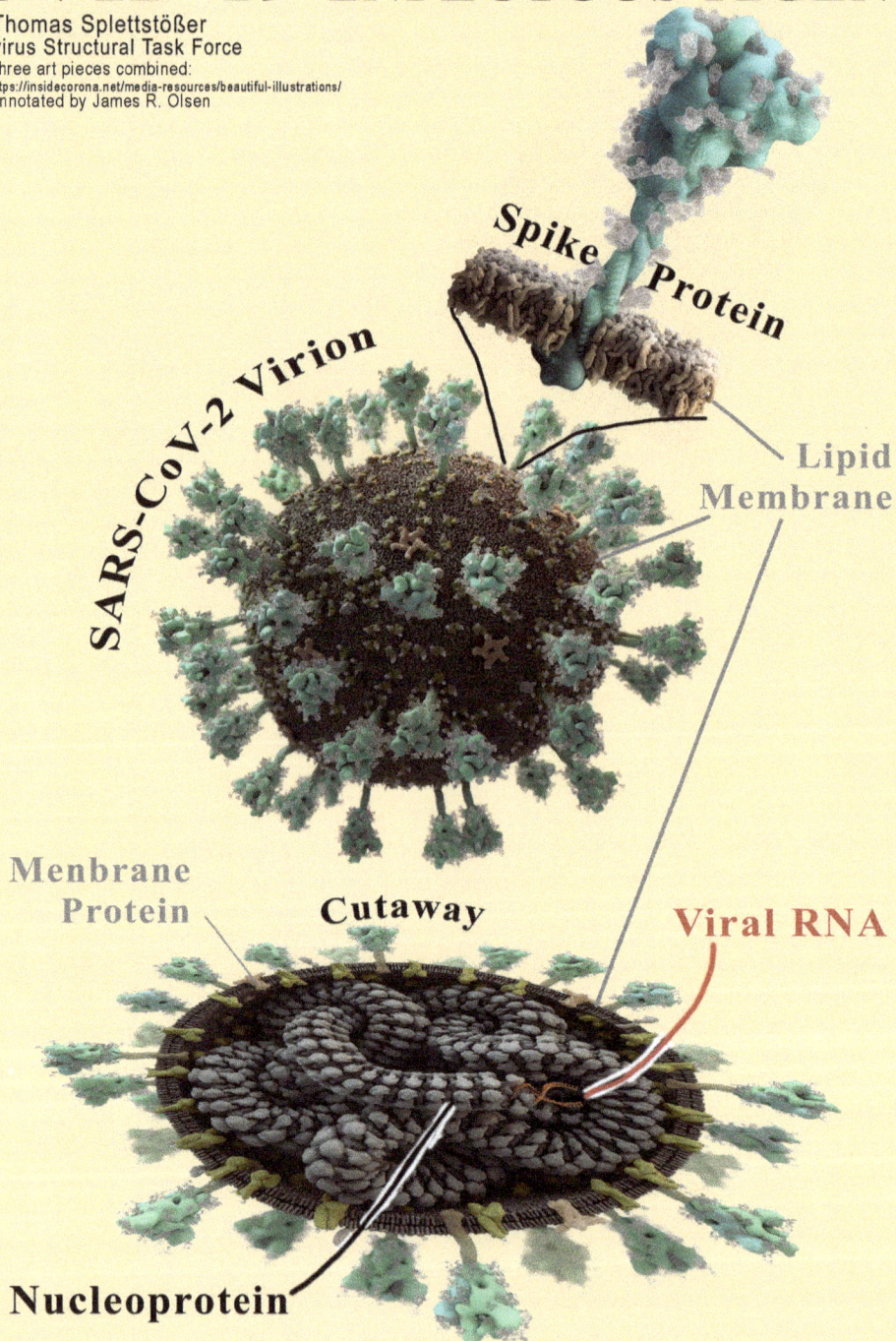

*SARS-CoV-2 Virus Illustration*

are pushed out into the body by a process called "budding." Finally, the cell bursts, scattering viruses by the thousands or tens of thousands, each ready to latch onto another ACE-2.[44] As this process multiplies, the body becomes host to hundreds of thousands of viruses, called a "viral load." The intensity and duration depend on several factors, including the virus variant and the body's immune response.[45]

Of course, each new virus must have a copy of the virus RNA. The virus hijacks the cell to make the viral RNA copies, but many copies contain mutations or copy errors. Most are "fatal" because the flawed replicate cannot reproduce itself. Some make no difference regarding virulence to humans; some make it less virulent; some more so because either they are more deadly or more contagious; some make the virus infectious to other species. If two viruses are at work in the same cell, parts of their RNA can combine. In the process of reproduction, mistakes leave detritus — many noninfectious RNA parts floating around.[46]

We wouldn't see this as anything other than an intellectual exercise if SARS-CoV-2 didn't make us sick. In making us sick, the virus runs into our immune system. In very oversimplified terms, the orthodox understanding is:

1. The virus must enter our body. With SARS-CoV-2 and other well-known respiratory diseases, such as the common cold and the flu, the most common way is to breathe in airborne viruses. Viruses are also called virions[47] when outside an invaded cell, intact and infectious. A virion must run a gauntlet of general-purpose immune layers before finding a suitable cell.[48] Our airway lining is covered with mucus, trapping virions, dust, and bacteria.[49]

2. If the virus makes it past these barriers, our general-purpose immune response then quickly attempts to slow the viral reproduction process down, even killing the hijacked cells. The cells in the tissues react by releasing interferons made in the cells themselves. Interferons slow down viral reproduction by interfering with the operation of the virus, putting the factory "on strike," sometimes making it possible for the cell to rid itself of the virus and survive. Interferons also activate "natural killer" cells already circulating in the blood. These stop the virus by killing the infected cell, depriving it of its factory. Unfortunately, SARS-CoV-2 has some mechanisms to avoid the full effect of interferons.[50]

3. A recent infection with the most common virus that causes the common cold, a rhinovirus, however, primes the upper respiratory tract with interferons, creating a faster, more effective response, one study suggesting it reduces the risk of severe symptoms.[51] However, lab experiments suggest that past infection with the less common cause of the common cold, a coronavirus, may slow antibody production specific to the SARS-CoV-2 spike protein, increasing the risk.[52]

4. The interferon and natural killer cell response often do not finish the job. Getting rid of the virus needs a final step.[53]

5. The final step is a specific response where adaptive immune cells recognize and respond to the particular virus. T-cells specifically kill infected cells and also help B-cells to produce specific antibodies against the virus.

6. An antibody locks onto a part of the virus. The antibody chooses a SARS-CoV-2 "antigen" from among all others like a key to a lock:

    An antibody (Ab) is a large, Y-shaped protein used by the immune system to identify and neutralize foreign objects such as ...viruses. The antibody recognizes a unique molecule of the pathogen, called an antigen. Each tip of the "Y" of an antibody contains a paratope (analogous to a lock) that is specific for one particular epitope (analogous to a key) on an antigen, allowing these two structures to bind together with precision. Using this binding mechanism, an antibody can tag a microbe or an infected cell for attack by other parts of the immune system, or can neutralize it directly (for example, by blocking a part of a virus that is essential for its invasion).[54]

The adaptive response takes some time to do its work — making and priming of T-cells to adapt to the specific invader. When they are primed and ready, T-cells, which come in several

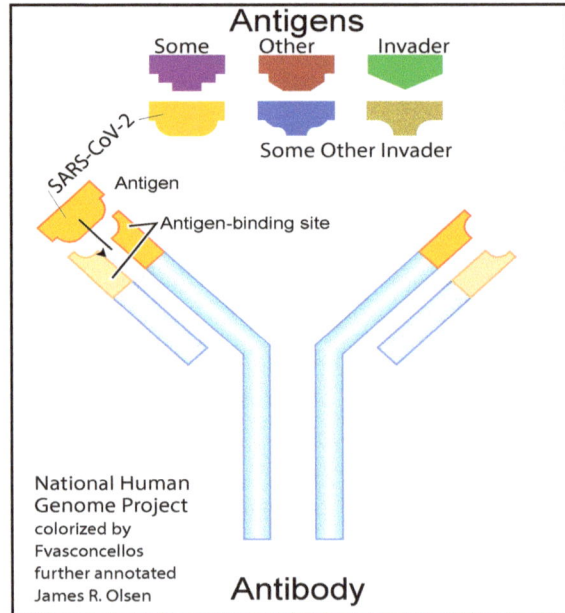

*Antibody mechanism*

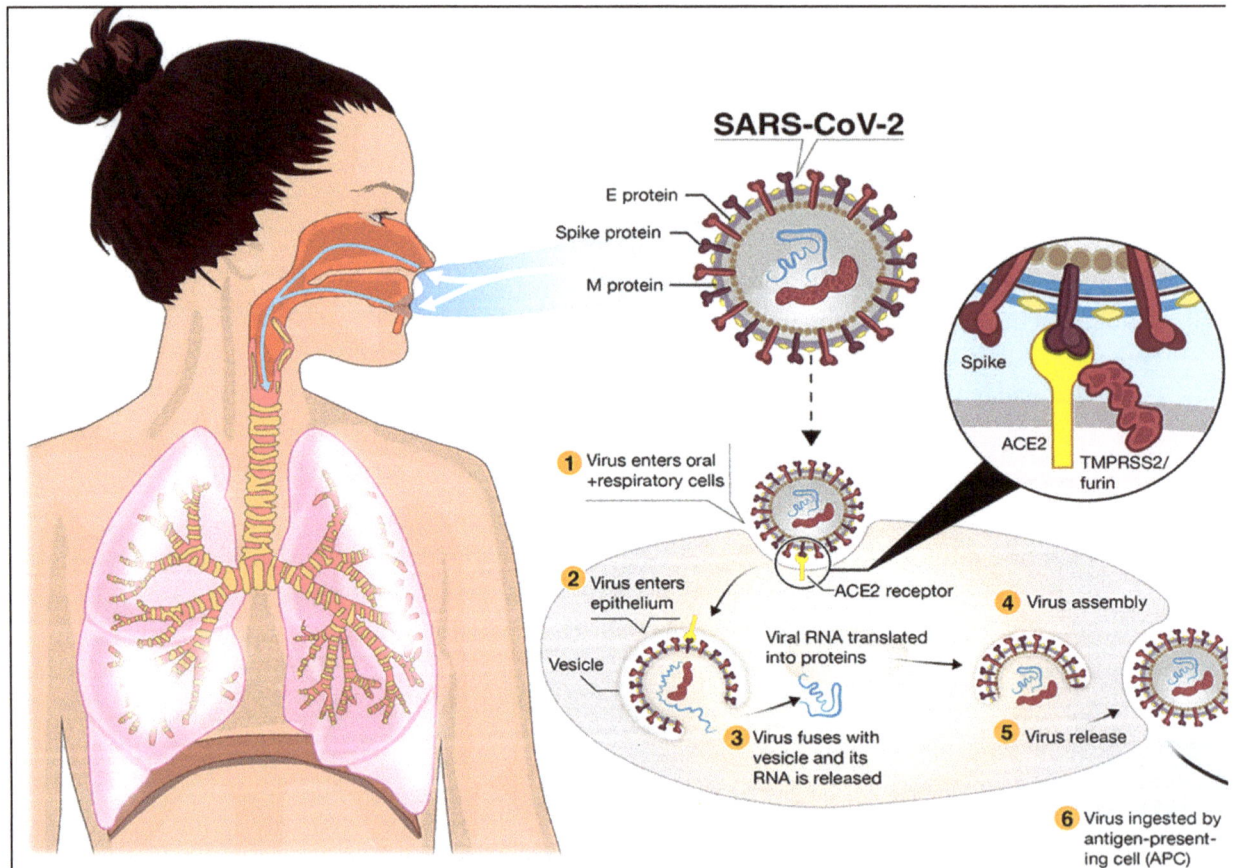

Funk CD, Laferrière C and Ardakani A (2020) A Snapshot of the Global Race for Vaccines Targeting SARS-CoV-2 and the COVID-19 Pandemic. Front. Pharmacol. 11:

*Immune Response to SARS-CoV-2*

varieties, as well as B-cells, which make antibodies, go to work. The bottom line is that the SARS-CoV-2 specific antigen is recognized and remembered to produce specific antibodies.

So, it's not whether a virus finds you but how many find you and how fast it multiplies. It's a race between the multiplying viruses and how fast the immune system fights it off. It is likely that the higher and more frequent the dose of virions one breathes in, the worse the symptoms.[55] The literature calls it the "infectious dose." One of the MDs on the Health Board noted that this is what they were taught about the infectious dose in medical school. There is little agreement on what dose is contagious — it could be as few as 100 virions, but more likely a thousand or thousands for COVID-19.

Virions do not travel alone. Breathe on a mirror, and it becomes apparent that the virions come in water vapor when breathed into the air. The size of vapor particles varies depending on whether a person is breathing, speaking, or coughing.[56]

A trial, or clinical trial, is a research study on human participants designed to answer specific questions about new treatments or interventions that involve carefully selecting control groups, one a placebo or no exposure. It won't be until 2022 before anyone purposely gives subjects COVID-19 to discover the infectious dose in a clinical trial. Still, it was done with the flu in 2014.[57] The flu trial shows what COVID-19 studies are still struggling with.

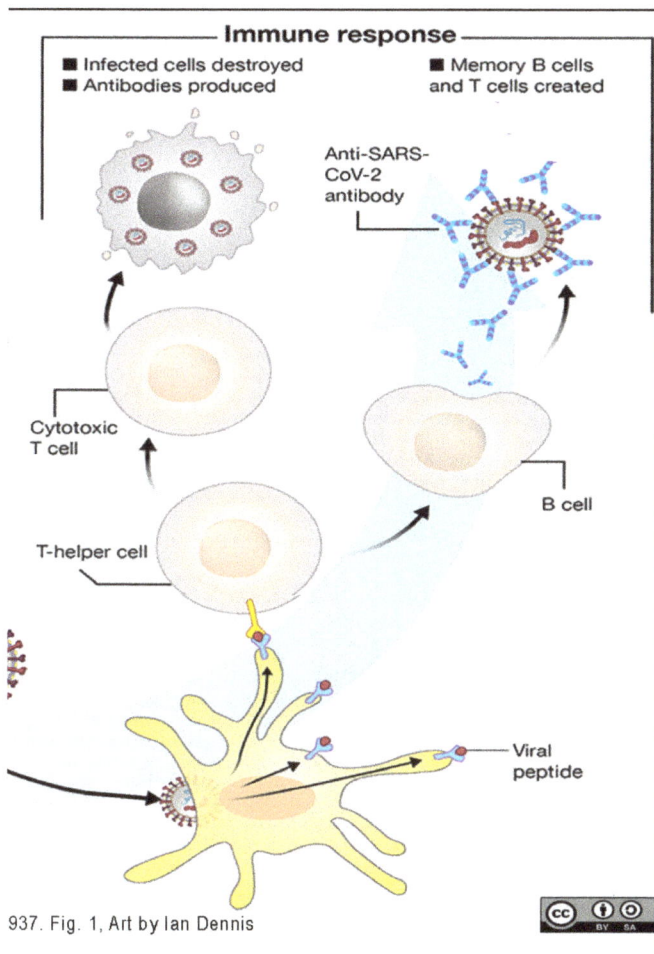

937. Fig. 1, Art by Ian Dennis

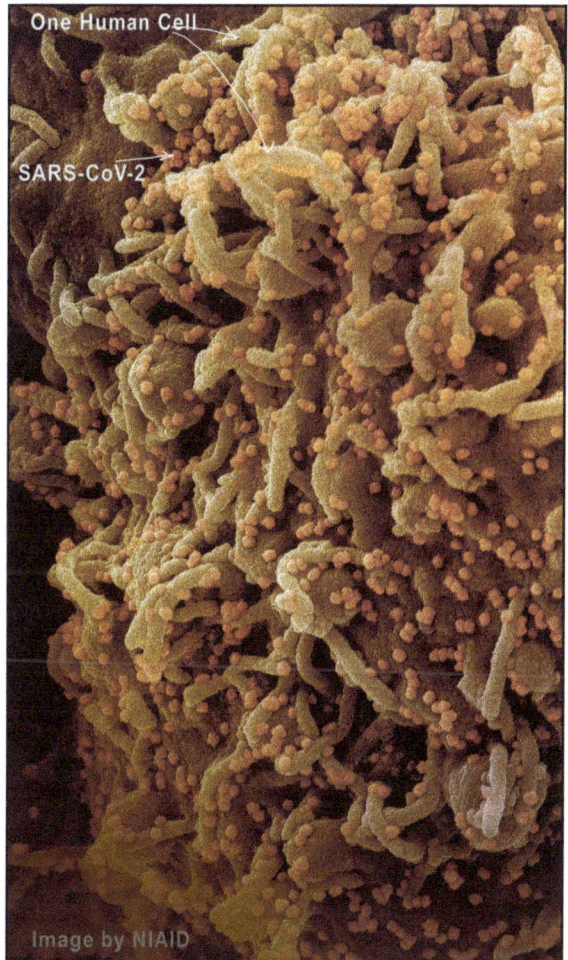

*SARS-Cov-2 Virus on Human Cell*

# Immune Response to Mild to Moderate COVID-19 ILLUSTRATIVE
## Timing and Levels Are Uncertain and Vary from Paper to Paper and Person to Person

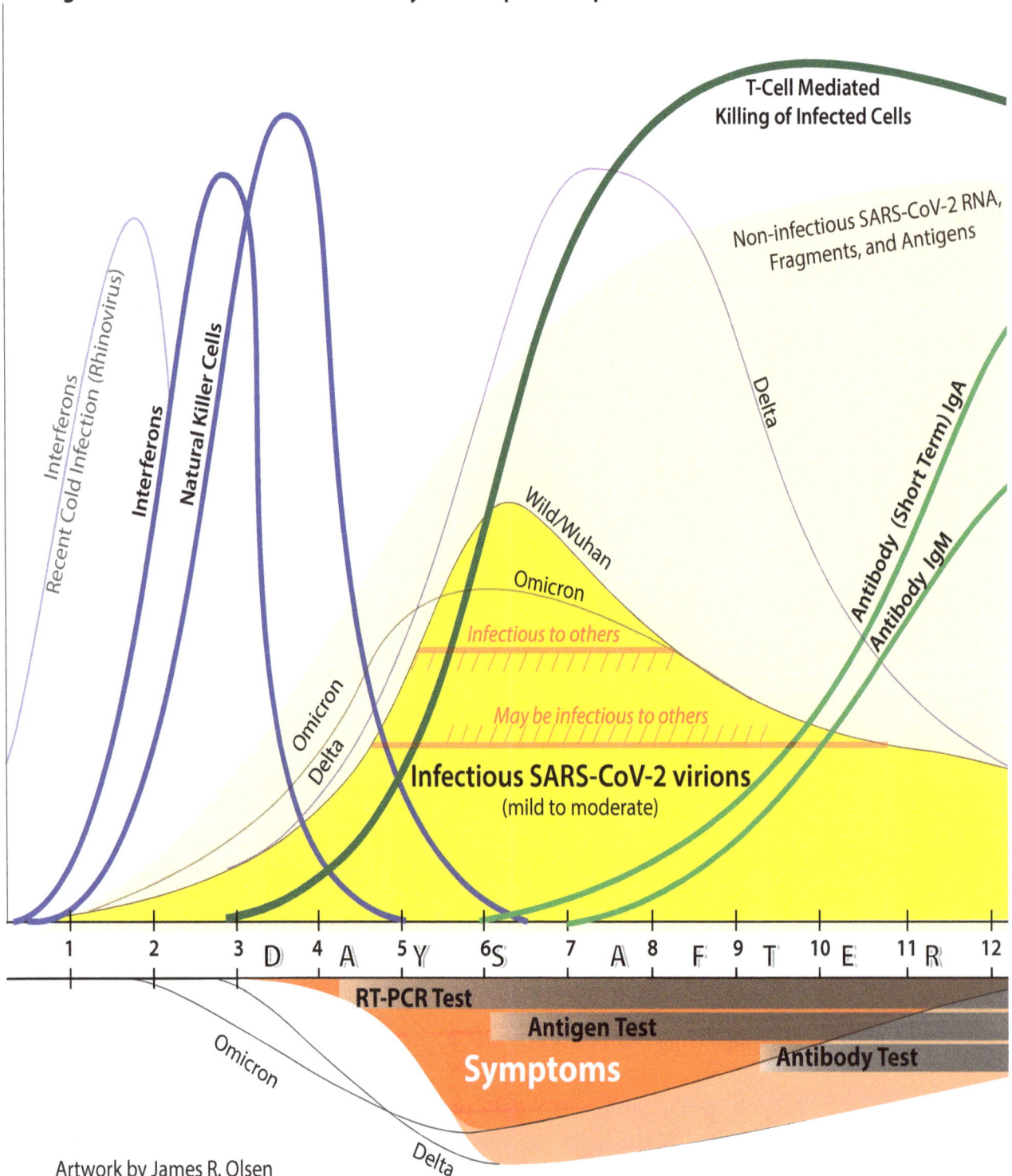

Artwork by James R. Olsen

*Immune Response to Virus Time Graph*

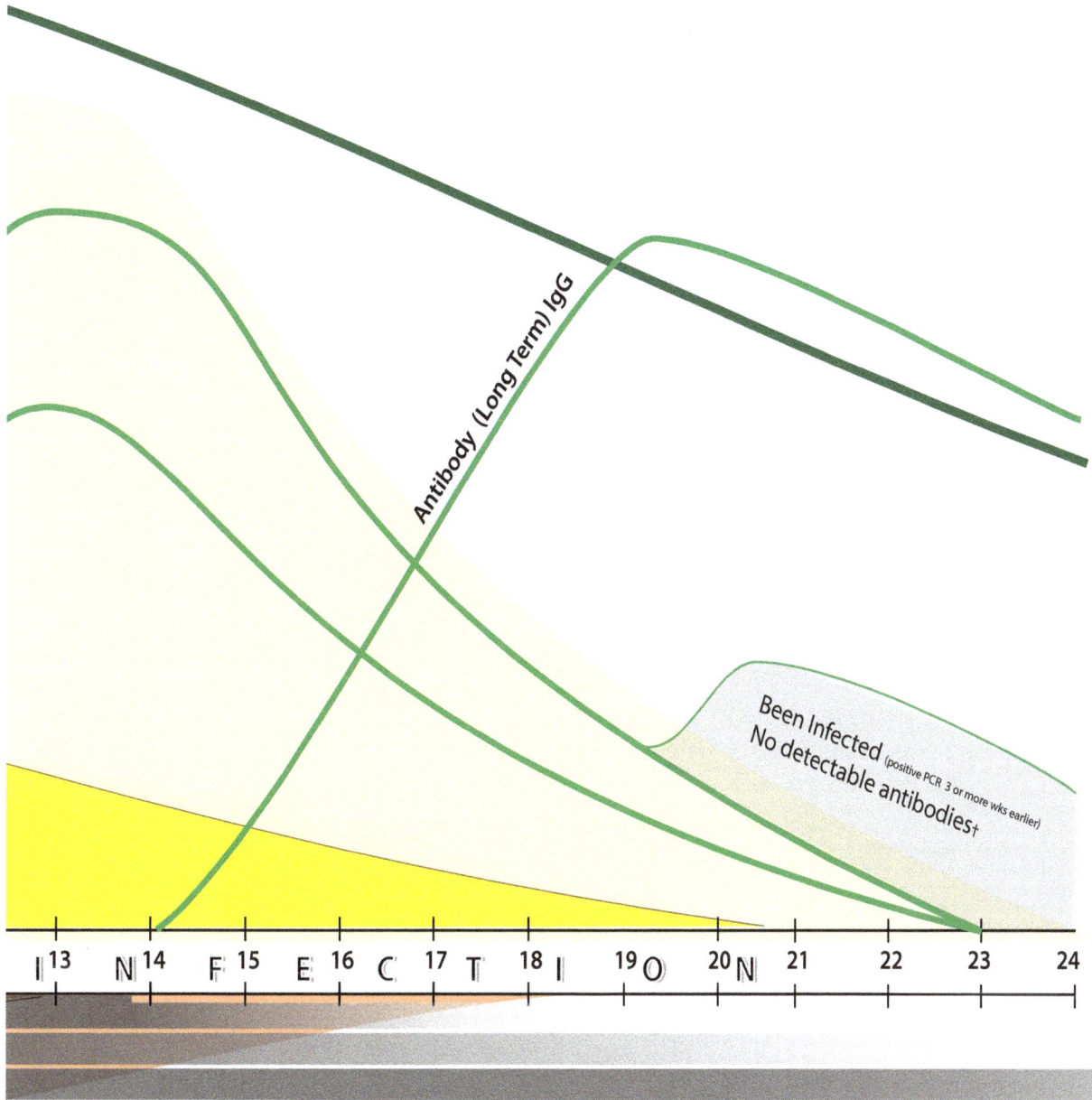

Antibody (Long Term) IgG

Been Infected (positive PCR 3 or more wks earlier)
No detectable antibodies†

| I 13 | N 14 | F 15 | E 16 | C 17 | T 18 | I 19 | O 20 | N 21 | 22 | 23 | 24 |

**Data from:** Janeway, *Immuno Biology*, p. 83; Murray, *Medical Microbiology*, pp. 86-91; Puhach, "SARS-COV-2 Viral Load & Shedding Kinetics";
Karita, "Trajectory of Viral RNA Load Among Persons With Incident SARS-CoV-2 G614 Infection (Wuhan Strain) in Association With COVID-19...";
Jeong, "Revisiting the Guidelines for Ending Isolation for Covid-19 Patients."; Wu, "Incubation Period of Covid-19 Caused by Unique SARS-COV-2 Strains.";
Cheemarla, "Dynamic Innate Immune..."; Owusu, "Persistent SARS-COV-2 RNA Shedding without Evidence of Infectiousness: ..."; Khatun, "'Convalescent Plasma...";
Wei, "Presymptomatic Transmission of SARS-COV-2 — Singapore, January 23–March 16, 2020."; Drain, "Rapid Diagnostic Testing for SARS-COV-2";
†Liu, "Predictors of Nonseroconversion after SARS-CoV-2 Infection."; Luo, "Infection with ... Delta Variant Is .. Higher Infectious Virus Loads.";
Sterlin, "Iga Dominates the Early Neutralizing Antibody Response to SARS-COV-2."; Zhou, "Viral Emissions into the Air and Environment after SARS-COV-2 Human Challeng"

A higher level of exposure for a longer time may make it more likely to get symptoms.[58]

Once a person gets infected, they can become infectious, that is contagious, when they produce or "shed" enough virions by breathing, coughing, sneezing, or talking to infect someone else. COVID-19 spreads so readily because this viral shedding happens a day or three before symptoms appear — before people realize they are sick.

## Flu Dose Versus Symptoms

*46 Healthy volunteers received nasal doses of the H1N1 Flu*

| Approx # Virions (2) | % Who Got Symtoms | % Mild to Moderate Flu | Viral Shedding (Infectious) |
|---|---|---|---|
| 1,000 | 40% | 0% | 0% |
| 10,000 | 50% | 0% | 0% |
| 100,000 | 80% | 20% | 20% |
| 1,000,000 | 89% | 47% | 47% |
| 10,000,000 | 85% | 69% | 77% |
| Total 46 | 78% | 41% | 43% |

(1) Memoli et al., "Validation of the Wild-Type Influenza A Human Challenge Model H1N1PDMIST: An A(H1N1)PDM09 Dose-Finding Investigational New Drug Study."
(2) Racaniello, "Virus Watch: Counting Viruses," Virus Watch (Video), Jun. 2, 2016
Memoli uses TCID50 for Dose given. Table assumes one to one infectious virion

*Flu Dose Human Challenge*

In the 18 to 30-year-old healthy subjects who were purposely infected with the original SARS-CoV-2 variant in 2022, the period of infectiousness, mainly emanating through the nose, lasts for 1 to 3 days for most but not all — 7% occurred before the first symptom.[59] Several other studies show that people have given other people COVID-19 before they realize symptoms appear.

Studies also show that people can test positive for COVID-19 and never have symptoms. Can someone who never has symptoms give someone COVID-19? The literature is not definitive; it seems likely since the viral load in a cohort of infected children who tested positive for COVID-19 is the same regardless of symptoms.[60]

If you are infected, what symptoms will you get? First, you may not get any at all. You are asymptomatic even though you are infected. Estimates of how many people get infected but do not have symptoms range from 30% to 86%, with 50-70% being a good summary of the literature.

It would not be surprising if the asymptomatic percentage varies by age and changes with some SARS-CoV-2 variants.[61]

Being "symptomatic" means a person has one or more symptoms. It's a bit tricky when it gets down to it because a "symptom" is something the person observes — muscle aches, sneezing, coughing, loss of smell. When a doctor attempts to determine the cause, they look at "signs and symptoms." A sign is observed or measured by someone else, such as blood test results.[62] A fever can be either felt by the patient or measured by someone taking a temperature.

If a mild case only manifests for a day or two, some people will notice it, and some will not. Of course, if you are around someone who cannot afford to get COVID-19, it would be wise to pay close attention to subtle changes — not to rush to the doctor, but to take extra precautions for a day or so.

I am in my mid-70s. Giving my immune system every chance to work is a good idea. Specific vitamins and minerals are essential for the immune system's biology and come from my food. If I get a hint of any COVID-19 symptoms, am near someone with symptoms, or am in some crowded, poorly ventilated space when the disease is spreading, I help my immune system. Since the biology of the immune system depends on vitamins and minerals, I don't want to be deficient — I don't take mega doses, but I will sometimes take the daily requirement or so of vitamin B complex, vitamin C, vitamin D3, selenium, and zinc — and abstain from even a sip of alcohol. The studies of a vitamin's effect on viral diseases are suggestive rather than delivering a magic bullet. I mustn't overdose — some can have serious health consequences in large doses.[63]

What are the symptoms of COVID-19? First, they do vary by variant.[64] Many initial symptoms are caused by the immune response to the virus: a fever, a scratchy throat, a sore throat, a cough, a runny nose, a sneeze, fatigue, and muscle aches.

## Infected: Symptoms and No Symptoms

**313 Nursing Home Residents**
UK Outbreak
Ave age 83
Graham, "SARS-COV-2
Infection, …
in United Kingdom
Nursing Homes."
*There were
23 deaths*

Tested Positive
43% Symptomatic
57% Asymptomatic

PCR Test Results

Positive
9% Symptoms 2 weeks before test
13% Symptomatic
17% Asymptomatic

Negative
61% Asymptomatic

**215 Women**
New York City Outbreak
Ave age 29
Admitted for Birth
Sutton, "Universal Screening for
SARS-COV-2 in Women Admit-
ted for Delivery. *Observed
for symptoms for two
days after RT-PCR test*

Tested Positive
21% Symptomatic
79% Asymptomatic

PCR Test Results

Positive
12% Symptomatic
Asymptomatic

Negative
85% Asymptomatic

**382 Children**
Durham, NC
0 - 20 years old
Contact with infected persons
Hurst, "Severe Acute
Respiratory Syndrome
Coronavirus 2
Infections
among
Children
.. Respir-
atory
Virus
-Expos-
ed Kids"

*Selection
may be biased
toward symptom-
atic children*

Tested Positive
39%
61% Symptomatic

PCR Test Results

Positive
47% Symptomatic
30% Asymptomatic

Negative
23% Asymptomatic

Artwork by James R. Olsen

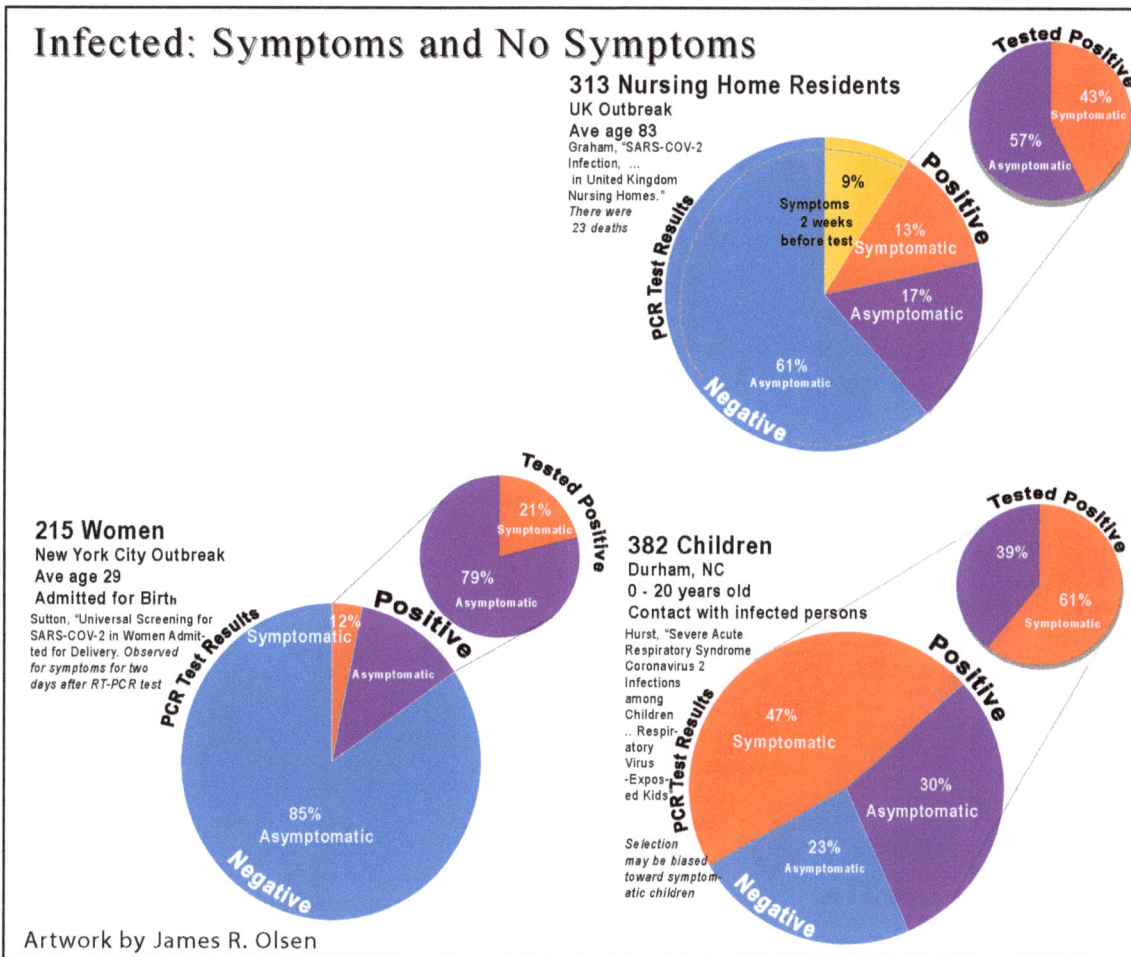

*Asymptomatic Infections with PCR Tests*

Apparently, the loss of smell is also due to the immune response, the invasion of too many T-cells in the olfactory cleft in the upper nose. The immune response can get out of hand, such as a runaway inflammatory response, which can rarely, but seriously, cause blood clots.[65]

Other symptoms are caused by the degradation of organ function by the virus. Digestive symptoms are an example which may be an early symptom or occur later in the course of the disease.[66] Viral machinations in the lungs and bloodstream can lead to dangerously low oxygen levels that can occur early and may initially go unnoticed.[67] There are various possible symptoms and severe outcomes driven by which organ or organs are involved, including lungs, heart, brain, and kidneys.[68]

Comorbidity is having another disease, usually chronic, when you get sick with COVID. The exact list of comorbidities that increase the risk for COVID-19 is hard to pin down. The work is incomplete, but the CDC, Johns Hopkins, and Mayo Clinic lists are nearly consistent.[69] The chart on the next page is a summary.

The risk of a severe outcome is related to age — it is quite different for a healthy 25-year-old and someone 75 years old. This difference can lead to a different perception of risk and the importance of government interventions — but even a healthy 25-year-old can get a dose that will have severe consequences.[70] For older age groups, the rising risk in relation to COVID-19 is not unique — the risk of dementia and strokes have a similar age correlation. According to a survey from Wuhan, the risk of infection is also related to blood type (This will be verified by a large-scale study in Canada in 2021).[71]

Figure 4.1 suggests that a rational personal choice occurs in the context of the overall risk as one ages. For example, high COVID-19 fatality counts for people over 85 is an add-on to an already higher fatality rate. But, when the risk is looked at as "% Deaths due to COVID" by age group, it becomes a cautionary tale for people in their late 40s, 50s, and 60s as well as older people — the risk .[72]

The world is going to start tracking deaths as a percentage of the population, called death rate, deaths per capita, or fatality rate, often presenting it as deaths per million people. The charts in Figure 4.1 are from CDC data collected state-by-state, county-by-county, and doctor-by-doctor filling out death certificates, entering from a list of "ICD 10-codes,"[73] often for patients with multiple health issues. The errors induced are already becoming controversial.[74] Even so, the risk or relative risk by age seem to be good guides.

With no approved medicines, doctors are reduced to treating symptoms and trying something off-label. Two favorites are hydroxychloroquine (a malaria treatment) and azithromycin (an antibiotic for bacteria). Doctors also prescribe a steroid for two-thirds of the ventilated patients.[75]

Someone needs to marshal the talent and resources to figure out the nature of the beast and tame it.

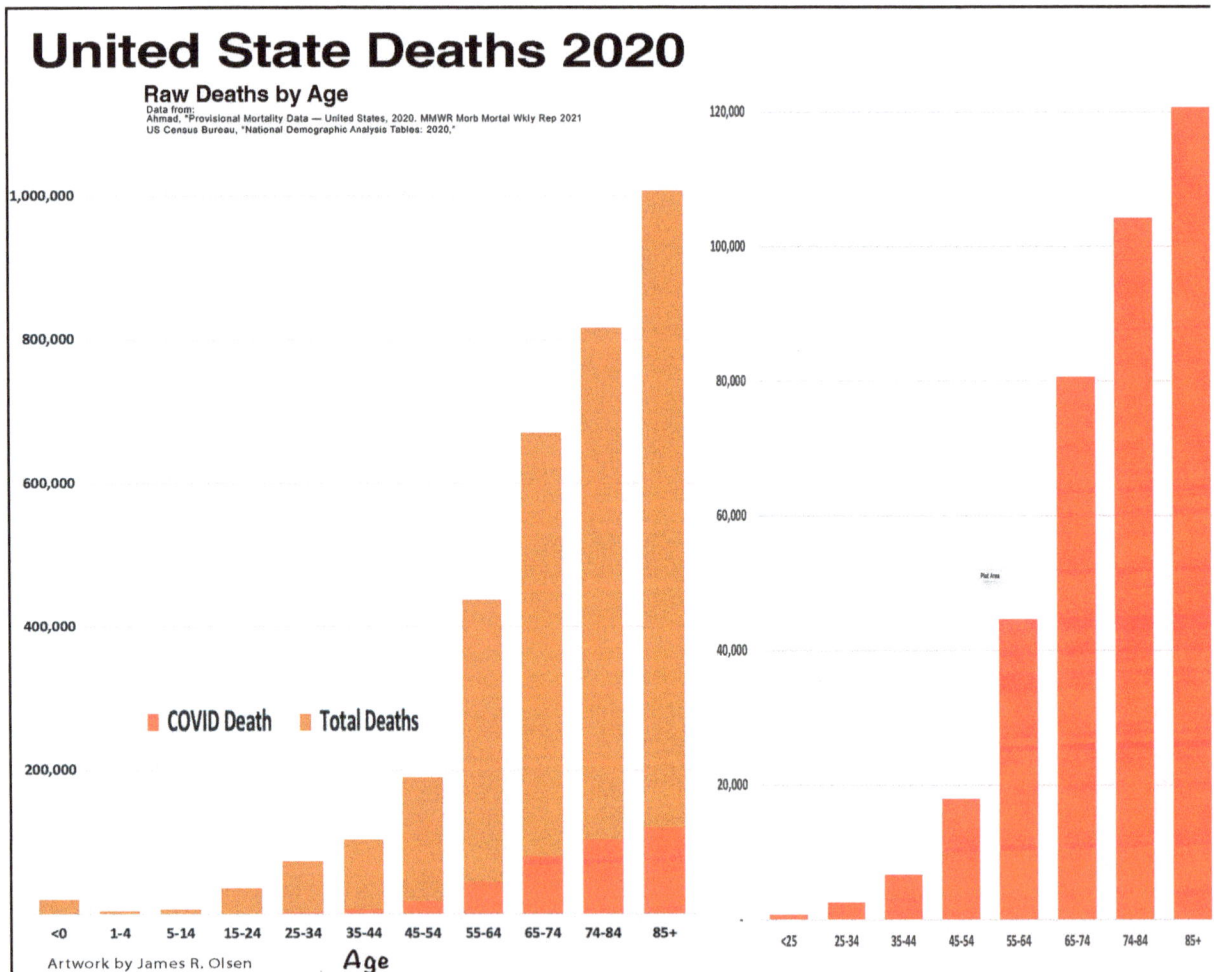

**Figure 4.1** - Fatalities by Age and cause, severity by blood type

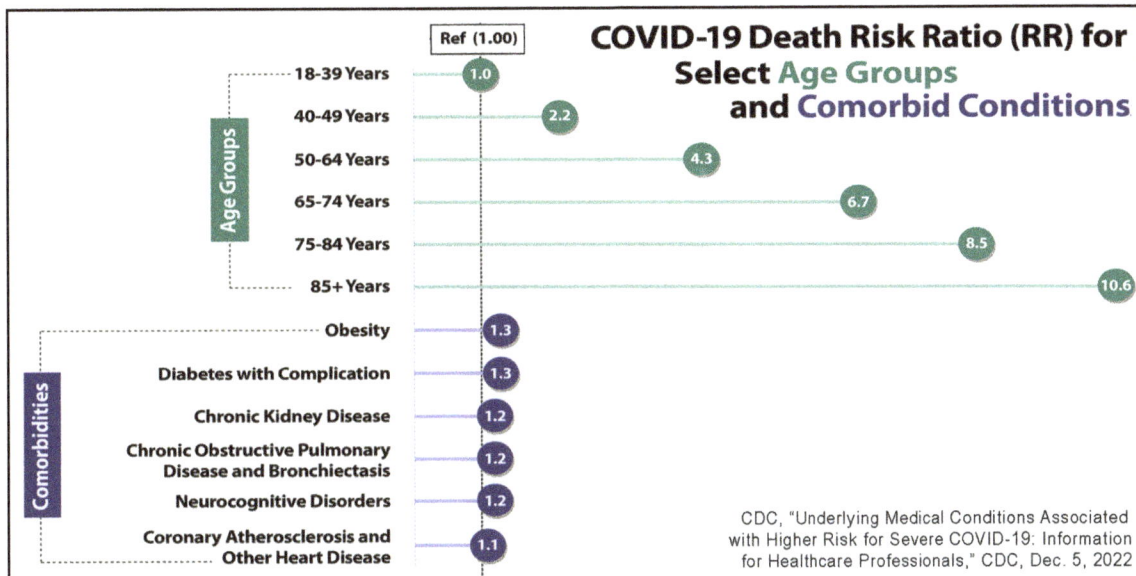

*COVID Risk Factors Illustration*

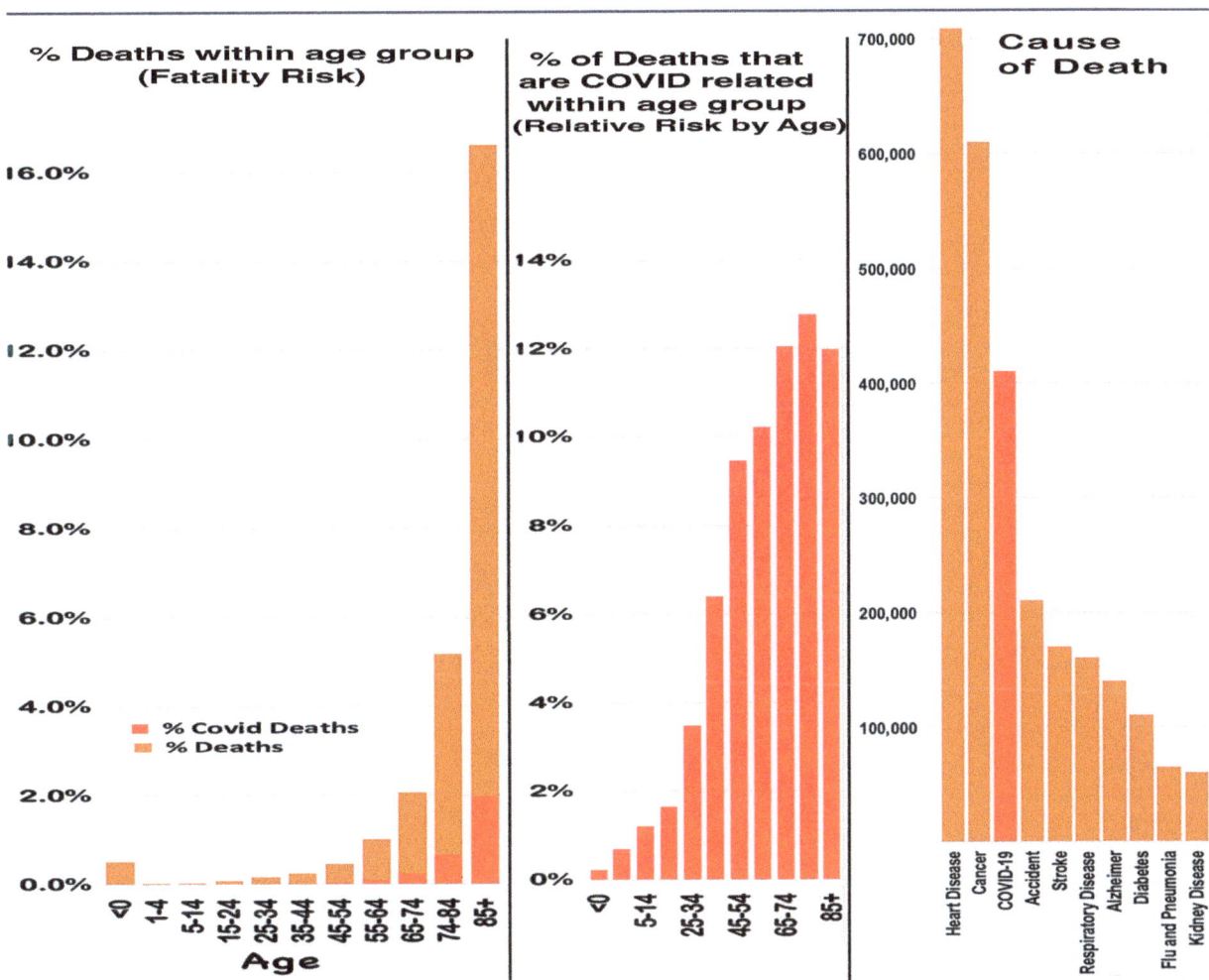

Doctor Anthony S. Fauci has dealt with a new controversial disease before. He has been the Director of the NIAID, situated in Bethesda, Maryland, for over three decades, with an impressive research record, publishing his work in peer-reviewed journals since at least 1976.[76] He was once in charge of research on HIV/AIDS (Human Immunodeficiency Virus/Acquired Immunodeficiency Syndrome).[77] The people who were getting HIV/AIDS were nearly all gay men. A *San Fransisco Examiner* article on Sunday, June 26, 1988, is not off the mark for the times,

> The past seven years, San Francisco gay men had suffered... No gay home has remained unscathed. No family or circle or club has remained unscathed... In the eyes of some, the plight of gay men is nothing more than a case of just desserts. Some would add, "They are suffering for their sins."[78]

Doctor Fauci is no stranger to politics. It was a time when there was a test, but no treatment, for HIV, leading some activists to say men should not get tested because they would be fired, evicted, or disowned. In a famous open letter to Dr. Fauci, Larry Kramer, a gay activist, called Fauci a murderer for not coming up with a cure after five years, adding,

> The gay community has, for five years, told the NIH which drugs to test because we know and we hear first what is working on some of us... How many years ago did we tell you about aerosol pentamidine, Tony?[79]

There was also a bioweapon story going around about HIV — with no social media, it appeared only occasionally in the news:

> There is even a theory, which arose first out of the Soviet Union, that says the HIV virus was man-made by the CIA for biological warfare.[80]

In this case, the fake news outfit *was* a Soviet Union government operation.[81] No cure has been found for HIV, though the disease, AIDS, can now be managed. Pentamidine is among the drugs that are now sometimes recommended for

Courtesy: National Institute of Allergy and Infectious Diseases

*Doctor Anthony Fauci*

AIDS to prevent pneumonia caused by a fungus.[82] Over time, the two protagonists came to respect each other.[83]

An impressive work ethic is the product of upbringing and education. Anthony's grandparents were part of the great immigration wave from Europe around the turn of the Twentieth Century from southern Italy and Sicily. He was raised Catholic. His formative years were spent under the tutelage of Jesuits in High School and while pursuing his college degree in the Greek classics, with pre-med thrown in, at the College of the Holy Cross in Manhattan. His Jesuit teachers, according to Anthony, "combine intense intellectualism with discipline" — with the motto of service to others.[84] Working hard "energizes him."[85] He takes pride in his three adult daughters.

> Broadly and generically, I am not a regular church-attender. I have evolved into less of a Roman Catholic religion person to someone who tries to keep a degree of spirituality about them. I look upon

myself as a humanist. I have faith in the goodness of mankind."[86]

As he mobilizes the NIAID to study and understand the SARS-CoV-2 virus, develop effective treatments, and kick-start vaccine development, a 2015 paper on work funded by his organization diverts some of his attention.

## Behind the curtain in Bethesda

In February, neither I nor the public knew of these details. They will be released in mid-2021 in response to FOIA Requests.

Dr. Fauci, "Tony," for those addressing him informally, starts the first day of the month answering emails, the daily toll now reaching into the hundreds.

An invitation to a WHO conference in Geneva is forwarded to Cliff, "Whom should I nominate."[87]

Another, from Greg Folkers, forwarding an article titled, "Exclusive, New FDA chief plans for 'the most serious scenarios' of coronavirus outbreak,"[88]

We should make a similar slide for upcoming talks.

Offers of help and thanks,

I am writing to say that your continuing leadership and dedication to our nation's health is deeply appreciated by me and millions of others."[89]

Doctor Fauci also responds to one of the early let's-hope-this-will-be-easy thought from ABC News,

Trump discussed Coronavirus briefly: He said he had a long talk with President Xi two nights ago, and the Chinese president told him "The heat generally kills this kind of virus."

Fauci responds with his usual diplomatic tone,

Coronaviruses... tend to circulate in the winter season...

Remember, however, that when it is summer here, it is winter in Australia and Argentina. This may be the genesis of the information given to President Xi and President Trump that the virus may "go away" in the warm weather. However, this is a brand new virus with which we have no prior experience...[90]

Doctor Fauci has two issues hanging over his head. The first is to determine the origin of the new virus, if possible.

Some scientists are initially suspicious of the SARS-CoV-2 genome, noting in a January 31 email to Fauci from Kristian Andersen that they find the,

Eddie, Bob, Mike, and myself all find the genome inconsistent with the expectations of evolutionary theory. But we have to look at this much more closely and there is much analysis to be done, so those opinions could change.[91]

Doctor Fauci arranges an international conference call, the details of which are still confidential, to get scientists thinking about the origin of the virus and whether it was lab-made. Fauci noting in an email that he wants to keep an open mind. After three days, Anderson changes his mind.[92]

Andersen and the others will publish a paper trying to "prove" natural origin, "The Proximal Origin of SARS-COV-2."[93] Still, they are relegated to an analysis of the genome, a well-forged blade— but not sufficient to nail down the origin of SARS-CoV-2 nor definitively prove natural origins.

Is it politics or science that changed the minds of the researchers? There is a case for both. It quickly gets political. But, "genome inconsistencies" do appear in nature in rare combinations, which are reproduced when highly adapted.[94]

We don't know the origin and are unlikely ever to know.[95] It's still being argued.

So, I will leave it there — you could write a whole book about it — in fact, David Quamenn did just that: *Breathless*.

The second issue has Doctor Fauci anticipating political inquisitions and conspiracy-theory narratives. One email is a copy of an article that he thought was helpful. He forwarded it to several people, including Bob Kaldec, Assistant Secretary of Preparedness for NIH.[96]

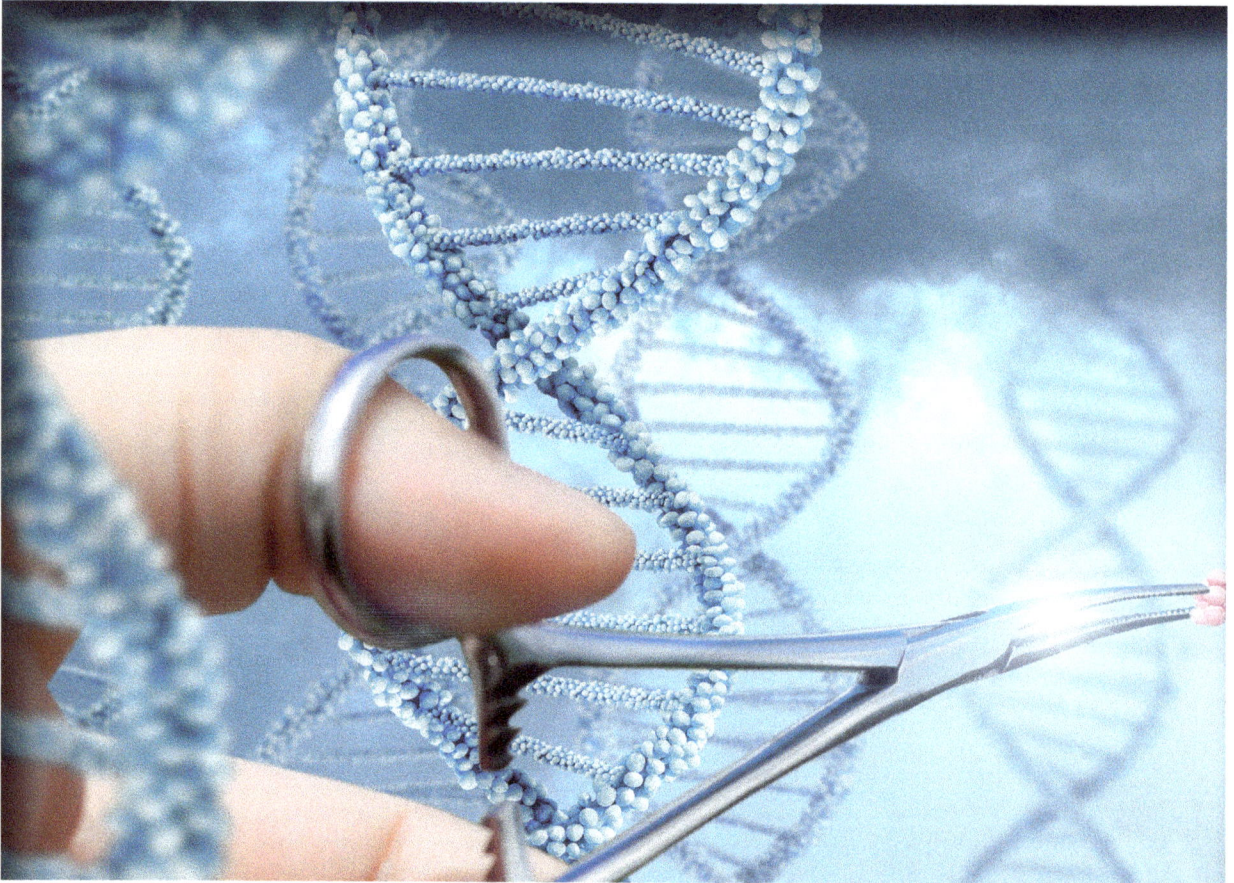

*Genetic manipulation concept*

Fauci sees the *Science* article as "balanced," which says that the role of the Huanan market is murky, adding that activity at the market accelerated the outbreak. The article noted that the SARS-CoV-2 genome is 96% similar to a bat virus and 80% with SARS. However, an article written in 2015 will generate a firestorm:

From: Fauci, Anthony (NIH/NIAID)
To: Auchincloss, Hugh (NIH/NIAID)
Sent: Sat, 1 Feb 2020 12:29:01
Subject: IMPORTANT
Attachments: Baric, Shi, et al. - Nature Medicine - SARS Gain of Function

Hugh:

It is essential we speak this AM. Keep your cell phone on... Read this paper as well as the e-mail I will forward to you now. You will have tasks today that must be done.

Thanks, Tony

From: Auchincloss, Hugh (NIH/NIAID)
Sent: 1 Feb 2020. 17:51:31
To: Fauci, Anthony (NIH/NIAID)

Subject: RE: Continued

The paper you sent me says the experiments were performed before the gain of function pause but have since been reviewed and approved by NIH. Not sure what that means since Emily is sure that no Coronavirus work has gone through the P3[97] framework. She will try to determine if we have any distant ties to this work abroad.

From: Fauci, Anthony (NIH/NIAID)
To: Auchincloss, Hugh (NIH/HIAID)
Sent: Sat, 1 Feb 2020 17:51:31
Subject: RE: Continued

OK. Stay tuned.

The attachment is not in the FOIA release. It is:

*Nature Medicine*, 9 November 2015, "A SARS-like cluster of circulating bat coronaviruses shows potential for human emergence," by Menachery, ... Zheng-li-Li[98] Shi, Ralph S. Baric.[99]

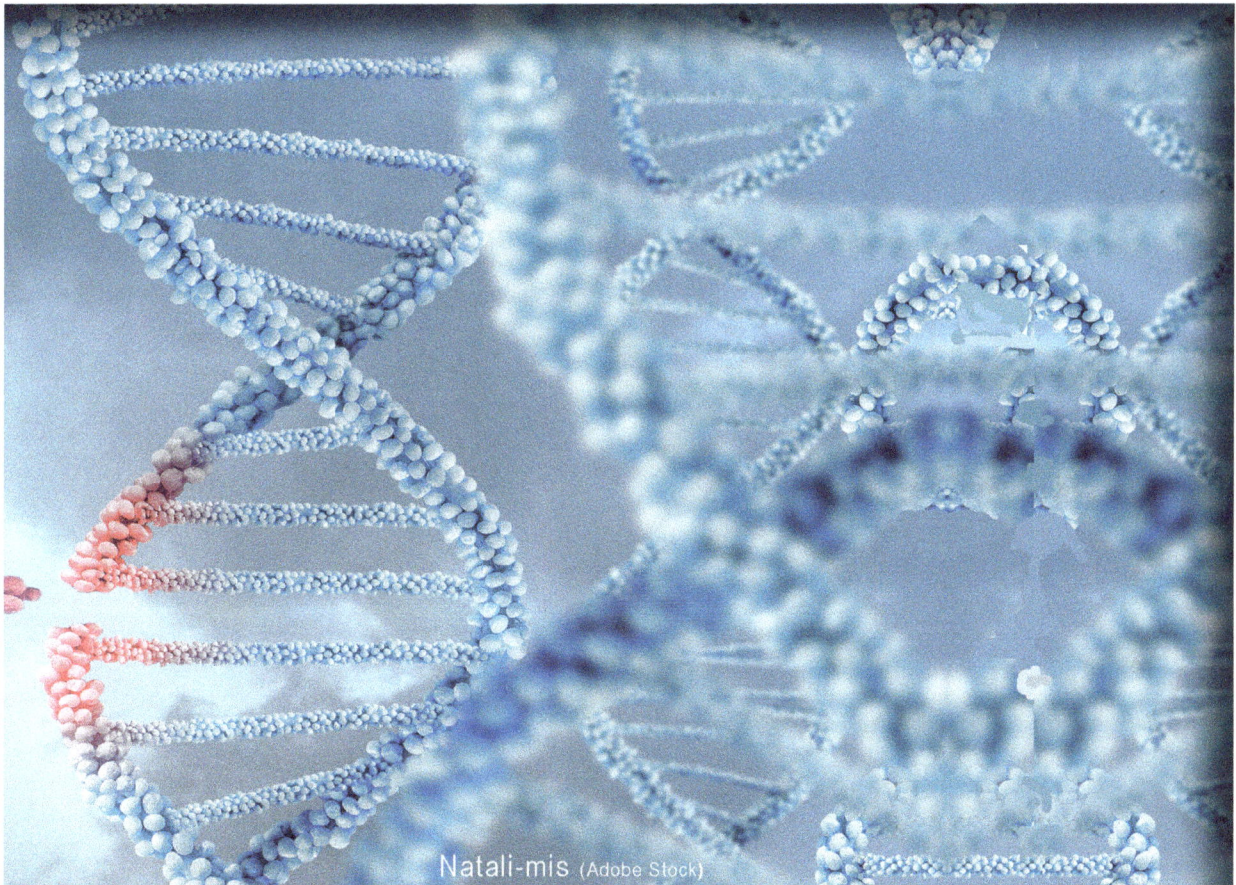

Natali-mis (Adobe Stock)

P3 is shorthand for Potential Pandemic Pathogens. Enhanced P3 is more popularly known as gain-of-function research. Even though a gain-of-function research is usually benign and happens in through natural mutations, *gain-of-function* is used in political dialog.[100]

Emily's belief that no coronavirus work met the definition of gain-of-function research will be questioned, examined, and publicly and loudly contradicted. When the term is used by researchers, gain-of-function can apply to a disease causing agent for any living thing. Enhanced P3, on the other hand, limits scope to gain-of-function risk the humans.[101] *This section uses the term as it applies to Enhanced P3, that is enhancing the risk of disease to humans.*

If gain-of-function was funded, in this case for a coronavirus, it should have been reported to the Office of Science and Technology Policy in the White House.

Gain-of-function research is the manipulation of genes to change or enhance the germ's abilities, such as allowing a virus to enter a cell more easily. It is like Frankenstein's Monster, a creation in a lab is harmless unless it leaks into the world, either because the guard at the gate is asleep or its creator releases it as a bioweapon. Lab leak and bioweapon conspiracy theories multiply almost as fast as the novel coronavirus.

This is an old controversy that is now being resurrected. In 2011, two labs were concerned about the mutation that they had designed. After creating the mutation, they tested it on ferrets, a lab-generated "gain-in-function," the gain being the virus could now infect mammals.[102]

It took three years for the virtually unregulated gain-of-function research to come to a head. NIH paused new funding for gain-of-function research in October 2014, requesting everyone halt such work.

This is not just a biosafety issue, however. Another concern is known as "Dual Use" technology, a term popularized after 9/11 — a technology that can be used for peaceful purposes, such as the know-how, engineering materials, and refined uranium for a nuclear power plant that could also be used to make an atomic bomb.

It is not a new concern, given that we and other countries developed bioweapons during the Cold War.[103] The concern is that some lab-made gain-of-function pathogens could be effective bioweapons, though non-biology issues would also be at play in the real world, such as deployment, delivery, security, harming an enemy but not you, political consequences, and morality — and its effectiveness compared to other means.[104]

Even so, the gain-of-function research rule was rescinded in 2017 with some additional guidance. Gain-of-function research can be funded by NIH today, as long it is not to create a weapon and as long as safety rules are followed.[105] However, no one at NIH checks each project to ensure the rules are followed. Privately funded labs have few restrictions on gain-of-function research.

Doctor Fauci's first problem with the *Nature Medicine* article is that Ralph S. Baric is an author.[106] He has been funded by NIAID and is known to have experimented with genetic modifications to understand better how to deal with emerging diseases. Of course, some people outside the research community will see his work in their own way, some claiming he made a bioweapon.[107]

The second problem is Zhengli Shi, also an author of the gain-of-function *Nature Medicine* article. She works in the Virology Lab in Wuhan.[108] Shi is having her own problems in China. The *South China Morning News* announces, "Pi-

oneering Bat Scientist Wins Both Praise and On-line Vitriol."[109] Shi's team also published a paper in 2017, "Discovery of a rich gene pool of bat SARS-related coronaviruses provides new insights into the origin of SARS coronavirus," listing NIAID as one of the funders.[110]

Shi had scouted the mountains of China for bat caves, bringing back samples of droppings to the Wuhan Laboratory to create the most extensive catalog of bat viruses in the world — which is invaluable to scientists assessing new diseases. An article reports,

> Her team determined that the new coronavirus is a direct descendant of a wild strain she and her team "culled from the droppings of a fruit bat in Yunnan province, sharing 96 percent of genes.[111]

Internet searches for her name explode by a factor of 2,000; most comments are negative, some calling her "mother of the devil," more saying that the virus killing people escaped her lab. As the pressure grows, she announces to her WeChat friends: "I swear with my life, [the virus] has nothing to do with the lab."[112]

What is lost in Shi's attempts is the service she performed. If you put thoroughbred stallions and mares together in a pen for long enough, you should not be surprised if nature takes its course and a foal that will take the racing circuit by storm is borne. The caves she is tromping through are breeding farms of a genetic pool of coronaviruses that she discusses in her study — as nature takes its course, the mixing of genes may well create a deadly, infectious virus — no one seems to read that part of the paper anymore.[113] Later studies suggest that an ant-eating Pangolin traded in the market is the suspect, only to be replaced by a Raccoon Dog also sold in the market. The wildlife had been removed before the market was swabbed for SARS-CoV-2, unlike for SARS.

The origin of COVID-19 is still a mystery.[114]

If Shi had stopped at the genetic analysis and not gone with lab-made variations, she might

*Wuhan Institute of Virology entrance*

have been a national hero. Instead, with only wild speculation about a lab leak, the social media people hounding her have made themselves trolls.

Ralph Baric, at the University of North Carolina at Chapel Hill in the United States, and Zhengli Shi in the Wuhan Laboratory in China are working together — a connection any good anti-globalist would find nefarious. The Wuhan Laboratory includes a BSL-4 lab to make the story even juicier. A BSL-4 lab is the kind of lab where one weaponizes viruses by genetic manipulation. People connect the dots — the NIAID funds Baric and Shi. And, for the grand finale, Doctor Fauci runs the outfit. A conspiracy theory is borne or uncovered. Take your pick.

The crux of Fauci's "leave your cell phone on" problem on February 1st is the *Nature Medicine* paper has a benign-sounding title until you read it and decode the technical language:[115]

> Experiments ... recombinant viruses were ... performed before the GOF [gain of function] research funding pause and have since been reviewed and approved .. by the NIH.[116]

Shi and Baric are two of the authors.[117]

Doctor Fauci's calendar shows a meeting with Baric at 2:30 on the 11th.[118] We don't know what was said.

The problem is that to show the potential for the virus to infect humans, the researchers made something new to see if it would make mice sick and replicate in a human lung tissue culture:

> We built a chimeric virus [fused from two different species]
> encoding a novel, [new]
> zoonotic [infectious from animals to humans]
> CoV [Coronavirus]
> spike protein [the protein that breaks through the cell membrane]
> — from the RsSHC014-CoV sequence that was isolated from Chinese horseshoe bats.[119]

In other words, they took existing bat viruses that didn't seem very efficient at infecting humans. They made something new to understand how it *might* happen in nature. Though Fauci will insist it was legal, he seems to anticipate it will "go viral."

It did. The journal would add this statement in a month. Editors' note, March 2020:

> We are aware that this article is being used as the basis for unverified theories that the novel coronavirus causing COVID-19 was engineered. There is no evidence that this is true.[120]

The grant was awarded in May 2014, a few months before the gain-of-function pause, and ended after the pause was over at the end of 2017. The grant was to EcoHealth Alliance, which showed that it would subcontract Shi's lab and a Chinese university. Both the original grant and later report to the NIAID say,

> We will then use receptor-mutant pseudovirus binding assays, in vitro studies in bat, primate, human and other species cell lines, and with humanized mice.... [121]

The question of whether or not this was gain-of-function research is defined in a document: *Framework for Guiding Funding Decisions about Proposed Research Involving Enhanced Potential Pandemic Pathogens* (HHS P3CO Framework). Potential Pandemic Pathogens is shorthanded to P3 in Fauci's emails. In this framework, an Enhanced PPP must be man-made, not occur in nature, and be,

> a pathogen that satisfies both,
>
> 1. It is likely highly transmissible and likely capable of wide and uncontrollable spread in human populations and
>
> 2. It is likely highly virulent and likely to cause significant morbidity and/or mortality in humans.[122]

This HHS definition became known as "gain-in-function" even though researchers may manipulated a genome to gain function in animal tissue and not humans.

The grant application says that a mutant spike protein will be embedded in a pseudovirus. The pseudovirus is then introduced to human cells in vitro (in a test tube or flask) and into mice "humanized" with human ACE2. It presumably is not "highly transmissible" because a pseudovirus should not replicate in normal tissues.[123] The grant does not propose gain-of-function research.

On the other hand, the paper highlights,

> ...the ability of a chimeric virus containing the SHC014 spike [from a bat] in a SARS-CoV backbone to cause robust infection in both human airway cultures and in mice.

NIH grant administrators noted this change in 2016. They accepted a rationale by the head of EcoHealth, Peter Daszak, that EcoHealth would notify NIH and stop the experiment if the chimeric virus replicated faster than the original strain by more than 10 times. In 2018, it was clear that the enhanced virus replicated 10,000 times faster. EcoHealth allowed the work to proceed despite the addendum. David Morens, Senior Advisor to NIAID, tried to run interference for EcoHealth, attempting to bypass FOIA only to find himself on the hot-seat.[124]

Being genetically very different from SARS-COV-2, the created virus is unlikely to be the source COVID-19, but seems to fall within the P3CO criteria. Whether this hybrid would be contagious in the real world is unanswered. Nevertheless, the researchers went beyond NIAID's approval, and NIH took unnecessary risks.[125]

While "BSL-4 lab" is often implied or mentioned when discussing lab leaks, experiments with pseudovirus and chimeric viruses frequently occur in BSL-3 and BSL-2 labs. So, was SARS-CoV-2 created in a lab? Did it leak from a lab?

We don't know.

The controversy will go from scientific journals to news and alt-narrative sites back to scientific journals. Genome experts discuss the implications of the furin cleavage site in the spike protein, suggesting that it is so perfectly adequate that it seems to be a gain-of-function product of a lab. However, cleavage sites do happen in nature.[126]

This controversy will end up at Senate hearings where Senator Rand Paul notes the existence of an NIH-funded project that describes gain-of-function research. He asks if NIH had funded gain-of-function research in China. Fauci says that they had never funded gain-of-function research at the Wuhan lab — which appears to be accurate when one reads the grant rather than the paper. Was Fauci made aware of the 2016 Eco-Health arrangement for the Wuhan grant before the hearings? He says not.[127]

However, Senator Paul had asked the question in connection with the paper. The nuances are lost in the heat of battle. This leads to a long-running interaction between Senator Rand Paul and Doctor Anthony Fauci that can only be characterized as a feud. When I watch their interactions, there is the classic inside-the-beltway talking past each other to the point where it is hard to sort out. The slippery meaning of gain-of-function is at play. Fauci later insisted that he was referring to Enhanced PPP when he used the term.[128]

So, I will leave it there — you could write a book about it — in fact, Senator Rand Paul did just that: *Deception, The Great COVID Cover-Up*.[129]

The international relationship with China devolves as the issue explodes into the U.S. Senate. Doctor Fauci is accused of being connected to the release of the coronavirus from the lab in Wuhan.[130]

Despite the political coup-counting, gain-of-function research is still allowed, inside and outside the HHS P3CO Framework, waiting for a deadly combination of events, the enhanced pathogen that leaves the lab. While the risk may or may not be worth it for significant breakthroughs when taking extra care to contain the experiment and destroy all of the byproducts when finished, in 2022, researchers at Boston University couldn't help themselves. They took the Omicron spike protein that made the virus more contagious by tacking it onto a more virulent version of SARS-CoV-2 in a BSL-3 lab.[131]

When asked by *Forbes Magazine* why they created the new virus, the reply is that it is to,

> Show that the pathogenicity of the Covid virus is determined primarily by something other than the Spike protein.[132]

Pathogenicity is "the power of the organism to produce a disease."[133] Numerous papers had already been written on the subject, some describing the biological mechanism that makes SARS-CoV-2 pathogenic, many examining the pathogenicity with methods that do not involve creating a new and more deadly organism.[134] The paper has no review of existing work on pathogenicity. Thus, it is a mystery how this will further the science. I would like to see the Boston University Biosafety Committee minutes on this one. The biosafety risk versus the scientific reward for this gain-of-function experiment is flawed.[135] It seems as useful as running a snowplow on a hot summer day.

What is the benefit of making a more deadly virus other than that you can?[136] While it hasn't nearly been as politically controversial as Shi's work, the balance of evidence is that she took much more care to make her research meaningful than at Boston University — though the safety protocols in Wuhan Institute of Virology had been questioned in 2018. The good words, P3CO Framework, and processes HHS put in place are not doing the job. The regulations for septic tanks in Ravalli County have more teeth than the frameworks, guidance, and policies than laboratory-enhanced pathogens.[137]

However, geopolitics is quickly becoming the driving force behind the growth of Wuhan Lab Leak theories. The Congressional House Select Committee report claiming a single source as key evidence is unpersuasive as it relies on a news source.[138] Nearly every intelligence agency and

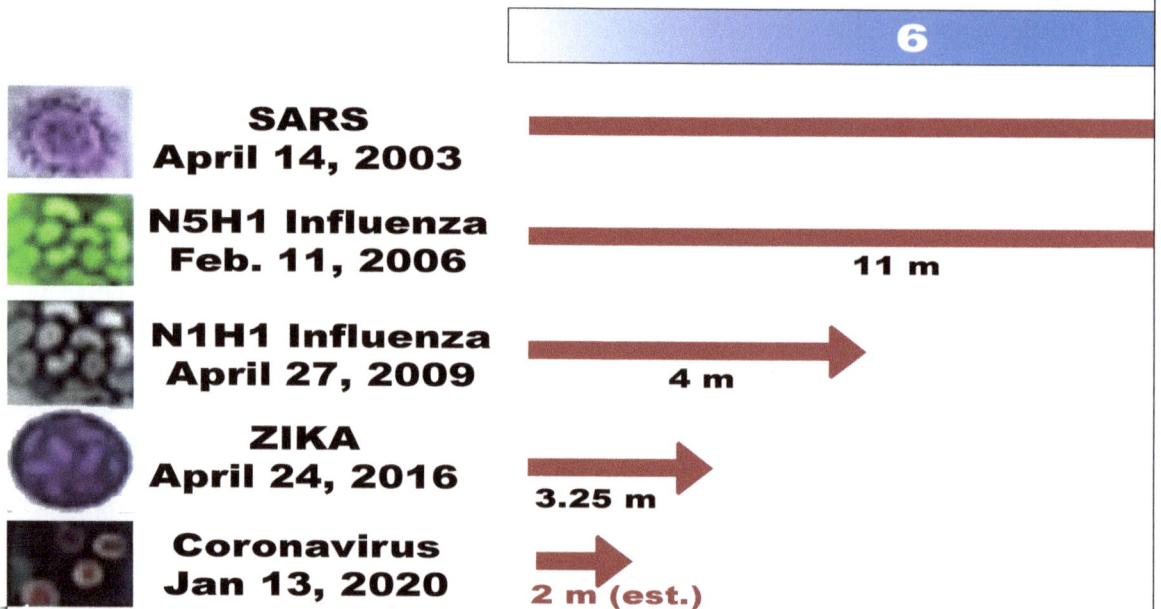

*Fauci Briefing to President Trump*

the FBI will investigate the question — lab leak or nature? It will remain speculative, political, and unresolved. FBI Director Christopher Wray will use double-speak about the origin,

> The FBI has for quite some time now assessed that the origins of the pandemic are most likely a potential lab incident in Wuhan.[139]

Meanwhile, the President is optimistic, citing low case counts, Doctor Fauci and others give interviews, saying that people do not need to worry about the novel coronavirus now — but with caution — it could change quickly. If the world starts seeing community spread, it will likely come to the United States. It could be a serious problem.[140]

## BEHIND THE CURTAIN IN WASHINGTON, DC——

This subtle back and forth between optimism and cautionary words explodes. On the 22nd, the HHS top leadership meets, among them Azar,

Fauci, and Messonnier. Though the restricted testing cannot determine how many people are infected, models predict thousands of undetected cases. It's agreed that Secretary Azar needs to brief the President before making this "right turn on messaging" and tell the President that "we have to be more aggressive about telling the American public that this where we're headed."

But before Azar gets the chance to meet, the President flies off in Air Force One to India the next day. It is "not the kind of discussion for a phone call," so Azar waits until his return.

When the briefing is put off, apparently, no one tells Nancy Messonier — and she is the one who gives the telecom briefings to the press. On the 25th, she delivers the Full Monte to the press,

> Ultimately, we expect we will see community spread in this country.
>
> It's not so much question of if... but ...

## h Center:
## g Diseases

**tion**

**Months**

| 12 | 18 |

**20 m**

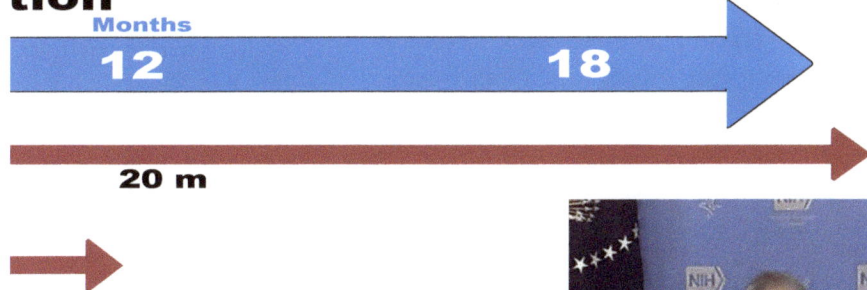

when... and how many people in this country will have severe illness.

Based on what we know now, we would implement these NPI measures in a very aggressive, proactive way...

Continuing, Doctor Messonier says these interventions include voluntary home quarantine, closing schools and remote learning, cleaning surfaces, canceling mass gatherings, and encouraging businesses to replace face-to-face meetings with video conferences.[141]

When the President hears about this on his flight back, he is livid — Messonier scared the American people. Presidents, no matter which party, will not abide such public discord.[142]

At a press conference the next day, President Trump, with Messonier now absent from the stage, notes the best people in the world have briefed him: the risk is "very low." "We are ready to adapt." Only 15 infections. Johns Hopkins rates the US number one for preparing for the pandemic. Congress is ready to authorize $2.5 billion.[143]

Now, President Trump needs a team that is his own creature. He appoints Vice President Pence to lead a task force to combat the disease. Doctor Fauci was on the team. Pence quickly appoints Deborah Birx as the White House Coronavirus Response Coordinator, placing her in the White House.[144]

Fauci finishes the month preparing for a televised briefing with President Trump at the NIAID Bethesda campus. Fauci displays confidence and pride in his outfit. He shows a timeline for when the first injection will happen — when human trials begin. The end of the year is the goal for an FDA Emergency Use Authorization (EUA).[145]

February is a month of waiting and speculating. I get sick, a weird version of the flu. It begins with aches in my muscles, worse than the flu. I notice a dry cough. I take my temperature: 98.8° — a degree higher than I usually run. There is no sense in going to the doctor; they only send tests to the CDC for people with "real" fevers. Exhausted for no reason, my body seems to want to stop working. I stay away from my wife. We sleep on separate floors. I lie down. I sleep. And, I lie there. I sleep. I start to recover by day four and back to my old self by the week's end.

Something is going around; a couple of people mention at The Table that it is being passed around the almost-homeless, a society that makes similar rounds, which church is serving free meals today, the hours at the library to get some Internet time. Is it COVID? I don't know. Maybe I am one of those undetected cases.

Given the news coming out of Wuhan, will the economy come to a halt? My modest stock market portfolio had been doing well. Will there be a crash? Seems likely. Though I am an amateur at this, it would be an excellent time to run to cash. The *Wall Street Journal* sounds a warning:

> Some U.S. companies say they could lose as much as half their annual revenue from China if the coronavirus epidemic extends through the summer.[146]

Mormons I know believe they should have a year's worth of supplies in the name of prudence, just in case. A few go beyond Mormon prudence — there are even a few bomb shelters in the valley. After reading the news and a few medical papers, it seems just-in-case might be coming. I am a laggard, as we have more than our share of preppers in Ravalli County. I begin by ordering some N-95 masks for my family and a couple of full-face respirators. Then, water storage in five-gallon containers, emergency food, gasoline, oil, first aid, compass, water filters, candles, lanterns, rope, alcohol stove, and radios. For sanitizer, I order 99% alcohol and mix it with 20% to 30% water. I organize the food by expiration date, combining freeze-dried, bulk, and canned. I finally decide to get an emergency generator, even though the house has solar power for short periods when the power goes out.

I am riding just ahead of a stampede. These supplies are getting harder to find just after I buy them. However, I forget about two American penchants for hoarding: ammo, which I have enough of, and toilet paper, which becomes a close-run thing.

At first, we sanitize surfaces and our hands religiously at home. After a while, the information I am getting indicates that airborne transmission is how it usually spreads. I learn the term infectious dose. Speculating that you cannot get a sufficient infectious dose by touching surfaces in the ordinary course of events, the rigor of sanitizing surfaces and hands abated in our household unless someone had symptoms.[147]

I don't know if shortages will happen or how the economy will fare. Still, by the end of the month, one thing seems inevitable: Pestilence is coming.

No one seems to notice a report from Wuhan that COVID-19 patients with Acute Respiratory Distress Syndrome (ARDS), lying on their back, hooked to a ventilator to force them to breathe; 19 of 22 died. People who were prone, on their bellies, have better odds, 2 out of 4.[148]

*29 Feb. 2020: 1 COVID-19 COVID-19 Related Deaths Reported*

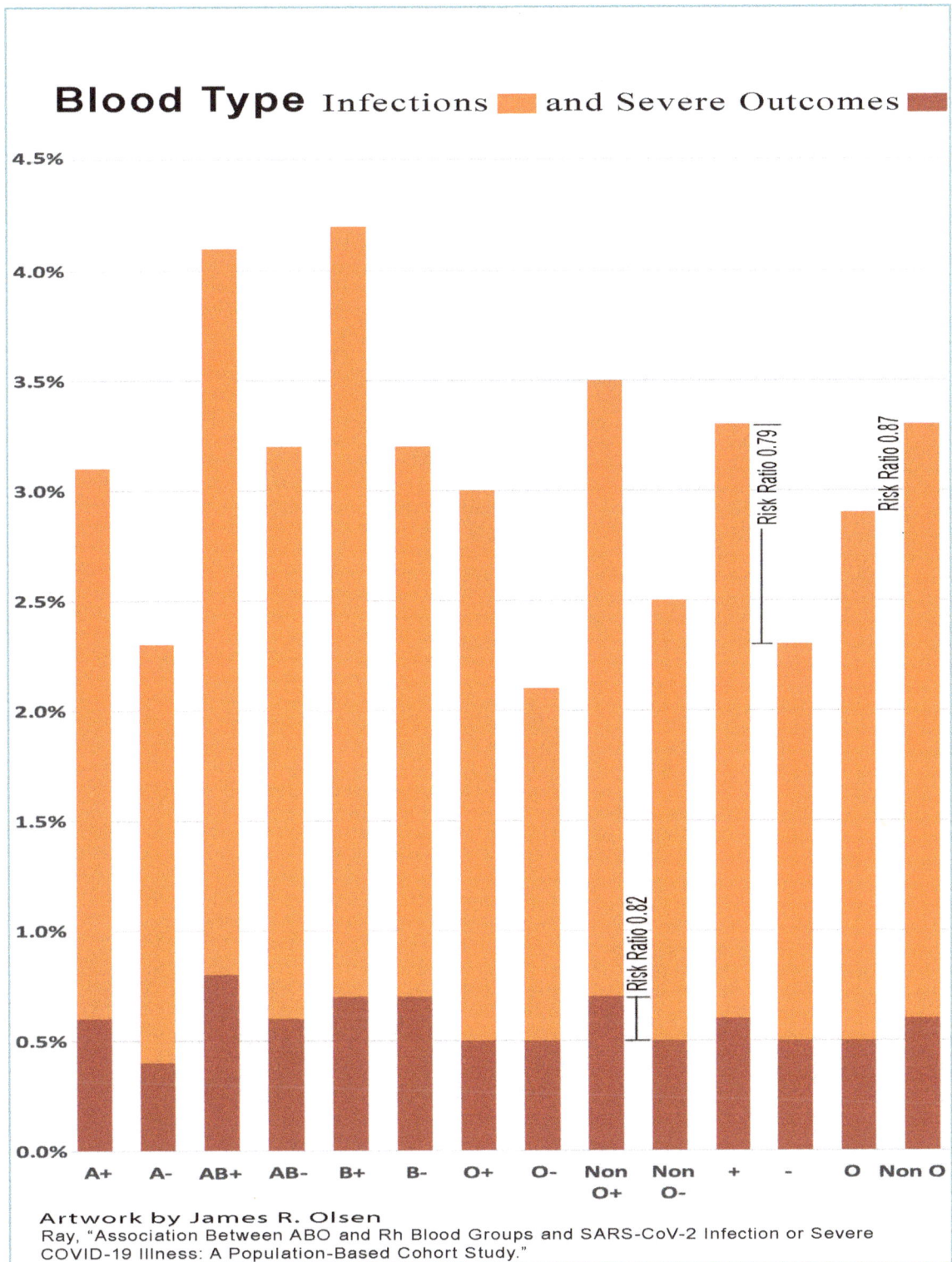

*COVID-19 Infection and Severe Outcomes by Bloodtype*

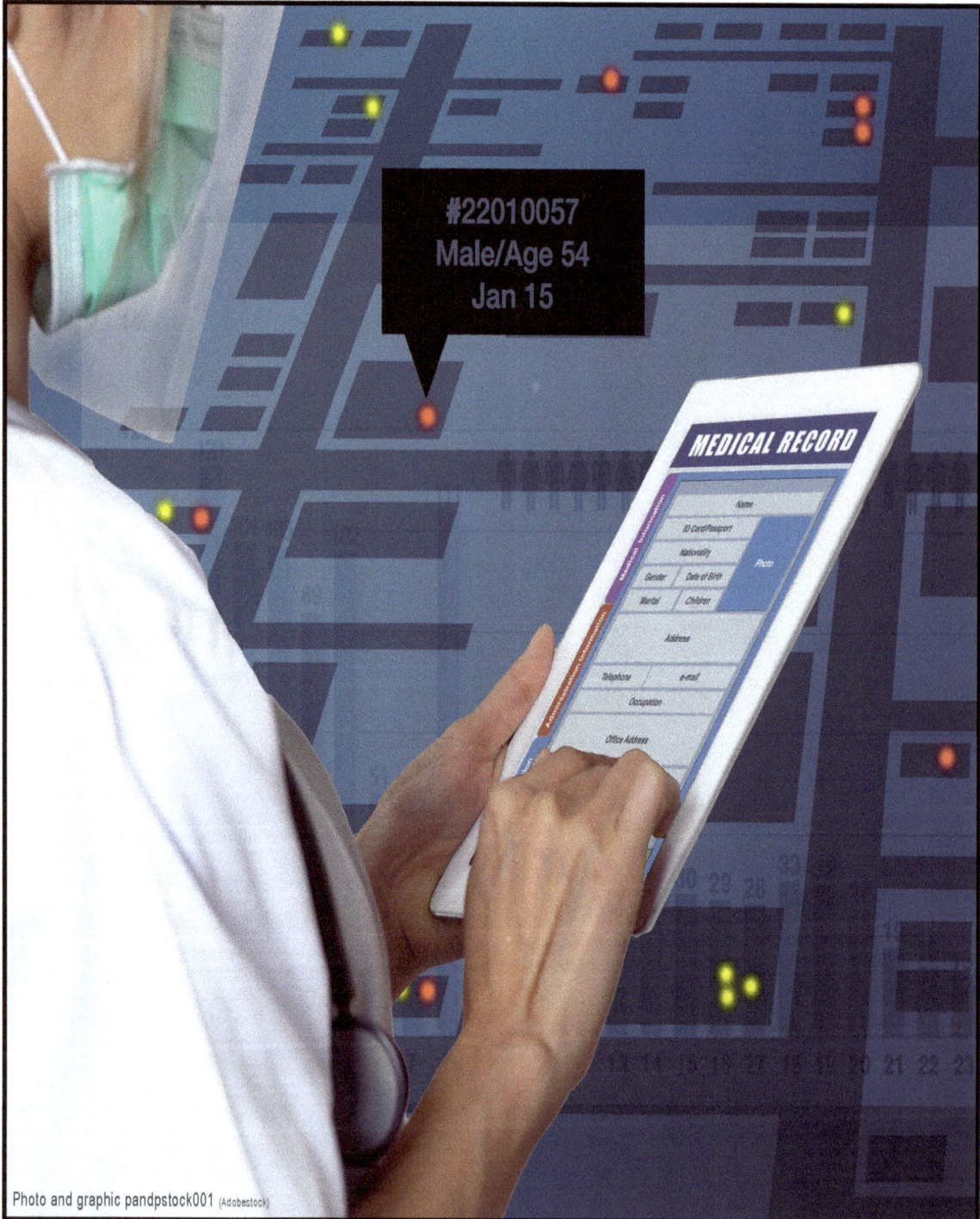

#22010057
Male/Age 54
Jan 15

MEDICAL RECORD

Photo and graphic pandpstock001 (Adobestock)

*Mapping a Case*

# 5

# EPIDEMIOLOGY

## SCIENCE AND PUBLIC HEALTH

### by Pamela L. C. Small, PhD.

*This text is based on an interview with Pamela Small who has reviewed it. Since this book contains a variety of viewpoints, her contribution does not imply agreement or disagreement with any other part of this book. Further, its presence does not imply agreement or disagreement by any other contributor.*

Pam Small has a PhD in Microbiology from Stanford. After working at RML as a Department Head, she was a Professor at the University of Tennessee, teaching infectious diseases and microbiology. Pam Small still has ties to the research community in Ravalli County, often returning for conferences. Now retired, living in the Bitterroot Valley, she remains in touch with her colleagues, having just finished a two-day conference at RML attended by researchers from around the world when we talked about this chapter. She follows the data.

Epidemiology is the study of disease distribution in space. You map where people are infected. You are looking at the disease in its native habitat. That allows you to focus on where most people are infected. In our case, it is people who live close to water. Then you ask, what's in common for the infected people? Then, look at what's in the water. What comes out of epidemiology can move into experimental science in the laboratory, looking at how biology works and developing treatments.

Having lived in India for many years, I saw that infectious disease was a really major limiter of mortality. If somebody lived to five, they would go on to have a life similar to ours. But a lot of kids died of infectious disease. I decided I wanted to be a microbiologist to have an impact on the developing world. I ended up knowing researchers from around the world.

While the CDC and the NIAID have become the subject of political controversy, the people I met from the CDC and who I worked with in the NIAID are sincere people who want to do a good job.

Imagining I would end up back in India, I ended up in Africa, doing case-control for Buruliosis, the Buruli Ulcer (*Mycobacterium ulcerans*), a known disease that has been "neglected" by the research community. It's a major cause of morbidity because when people get infected, there wasn't a good treatment for it. Their arm would fall off, or a leg would fall off because the disease causes open lesions, which, in poor countries, are often left untreated for too long because

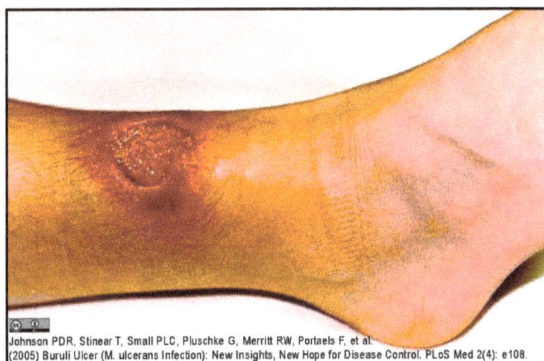

Johnson PDR, Stinear T, Small PLC, Pluschke G, Merritt RW, Portaels F, et al. (2005) Buruli Ulcer (M. ulcerans Infection): New Insights, New Hope for Disease Control. PLoS Med 2(4): e108.

*Buruli Ulcer*

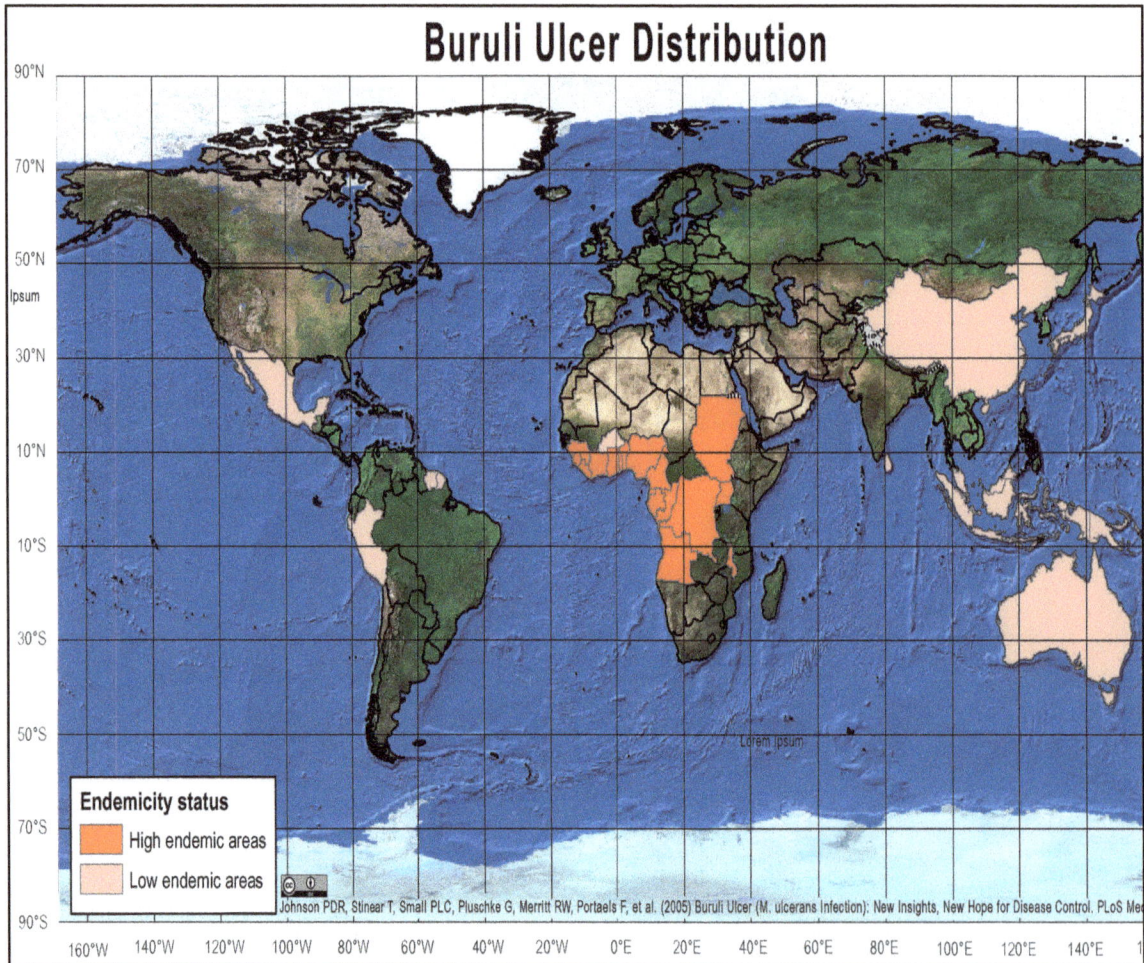

## Buruli Ulcer Distribution

**Endemicity status**
- High endemic areas
- Low endemic areas

Johnson PDR, Stinear T, Small PLC, Pluschke G, Merritt RW, Portaels F, et al. (2005) Buruli Ulcer (M. ulcerans infection): New Insights, New Hope for Disease Control. PLoS Med

*Buruli Ulcer Distribution*

they are usually painless. It didn't kill them, but it disabled them. If they were disabled, nobody would feed them — they'd go way down on that chain, and they wouldn't live as long.

Nobody really knew where cases were occurring, nor what caused it. We only knew it had something to do with proximity to water. We'd go to a village, go to different houses, and ask if anyone was sick. We also took water samples from the environment, analyzing them and finding it could live in the water by itself; it had to be in some organism, like fish, which turned out to be the case.

We did mapping to correlate the water samples with instances of the disease. We found that it occurs in patches, where, for example, 25% of the people in one village would get the disease,

but no one in the village on the other side of the river would get it. The cause was isolated, a bacteria, *Mycobacterium ulcerans*, whose relatives in the genus *Mycobacterium* include the species that causes Tuberculosis.[1]

We examined several possible avenues once we isolated the bacteria as the cause. The added benefit of understanding how it works might give insight into how other similar bacteria work. Since the bacteria affected cell growth, it could also lead to the use of therapies for cancer.

There was a surprising result when we identified the toxin made by the bacteria. Toxins are almost always a protein, so that's where the search started. The idea was that maybe someone could make an antitoxin. But it wasn't a protein; it was a lipid (a fatty acid). When we described it, Abbot Laboratories, which delves into the chemistry of natural products — drugs and chemicals derived from plants and insects — took one look and knew what it was. It inhibits the immune response. So, that's why it doesn't get inflamed or painful. It didn't turn out the way I thought it would. That's science. That's why you do it.

You can't make an antitoxin for a lipid. Sometimes, low-tech solutions are best, especially in poor countries. While expensive high-tech research may yield results, if a disease is associated with some plant, insect, or area, simply telling people to stay away can prevent the disease. The key to controlling Buruli Ulcer is thorough contact tracing, early treatment, and antibiotics to keep secondary infection down. The people who live there are still doing it today.

When the source of a disease is found in an epidemiological study, the nature of the germ and the toxins it may create can be followed up in the lab. When I am doing genetics, I can look for the characteristics of the people more likely to get it: For example, males over 50. This gives a hint as to which genes a toxin is targeting. In trying to figure out which genes are the target for a toxin, you make random mutations in the host in a lab culture. A mutation is simply changing the genes in the genome of the host. You then see if the toxin works on the mutated host. By process of elimination, when the toxin does not do its dirty work, you have found the genes that are targeted by the toxin. Knowing this, you can begin to develop effective therapies.

Epidemiology and experimental can work together to create effective therapies for neglected or emerging diseases — COVID-19 being an emerging disease. In the case of the Buruli Ulcer

project, the effort included the NIAID, researchers from other groups, fieldwork, and a relationship with public health officials where the disease is present. The benefit of publishing and sharing results with the research community is also evident.

At Rocky Mountain Laboratories, the work is funded by Federal Grants targeted to what they want studied. Of course, you must have some experience in the subject to win the grant. The grants can run into the millions of dollars, usually funding a team, not just one individual.

A researcher can strike out on something new if they are well known and well respected. Since things change between the time you apply and the time you get the grant, it is possible for the team to set out on a new research path. Once, I called a well-known colleague and said, "Do I have to really do what I said in the grant?" "You should at least start out that way," was the advice. Obviously, you are dealing with the agency giving the grant, so they have to go along with it.

In thinking about the NIAID trials for COVID-19 treatments and vaccines, it would be an advantage if the United States had a national health program like that in the United Kingdom because it would make it easier to design and recruit for large-scale trials. But everybody here is so afraid of a public health system.

In thinking about our response to any disease outbreak, you must know the number of cases to make good decisions. You don't have to test everybody, but a randomly selected, good-size sample of the population would tell you — it's expensive, though. It is a basic approach to epidemiology:

*You have to know the number of cases in order to make good decisions.*

| 1 | 5 | 11 | 12 | 13 | 14 | 15 | 17 | March 2020 | 27 | 28 | 30 |
|---|---|----|----|----|----|----|----|------------|----|----|----|

CDC creates COVID-NET monitor hospitalization for COVID-19

First test arrives in Montana

WHO declares pandemic

CDC confirm COVID tests results

Montana (MT) no longer required to

First confirmed case Montana

President Trump (MT) Emergency Declared

CDC no-sail order all cruise ships

States begin shutdown

President Trump calls National Emergency

Moderna: 1st human vaccine trial

President Trump signs CARES Act

MT Lockdown - Stay at Home

Chloriquine self-adminstering fatality

EUA Hydroxychloriquine

MT Evictions Suspended

MT Traveler Quantine

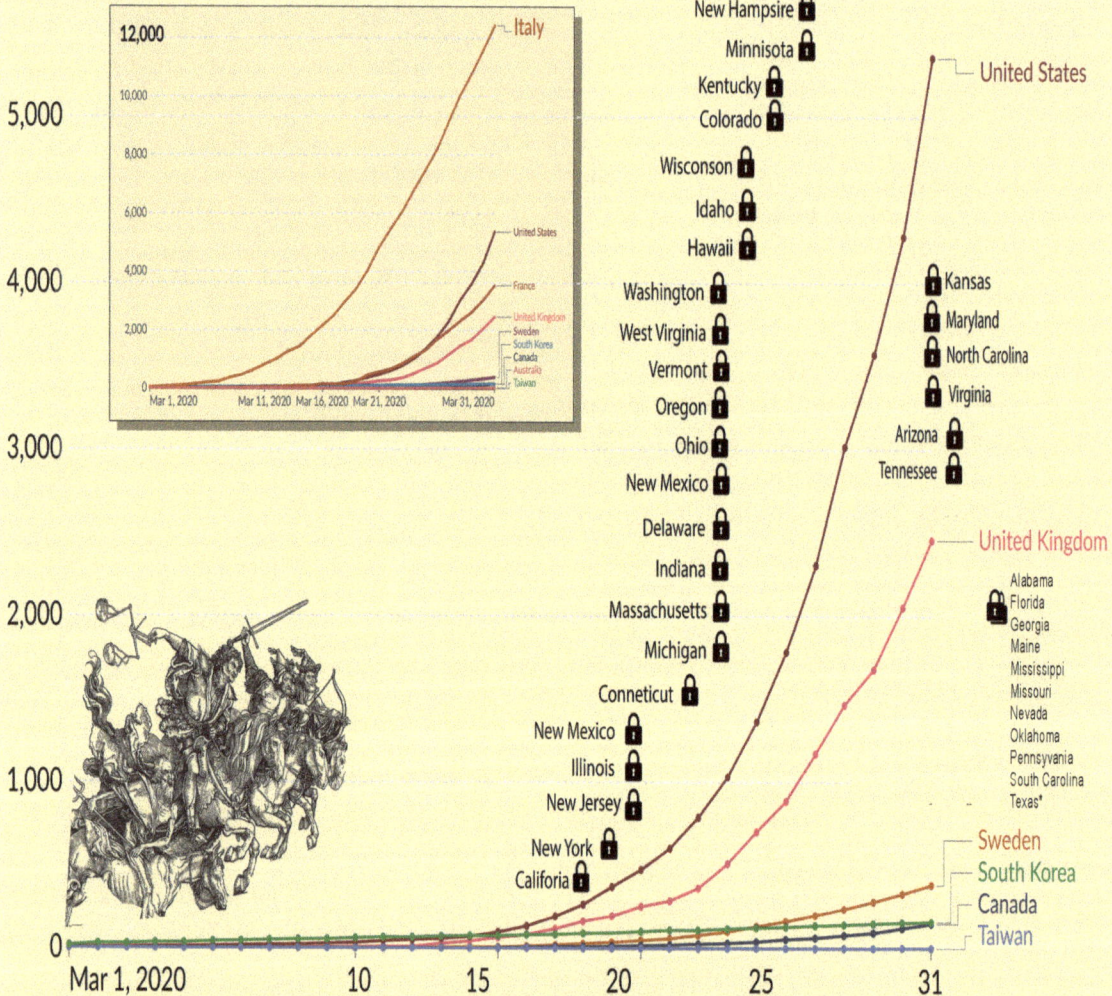

# Cumulative confirmed COVID-19 deaths 🔒 = Stay-at-home order

Due to limited testing and challenges in the attribution of the cause of death, confirmed deaths can be lower than the true number of deaths.

Source: Johns Hopkins University CSSE COVID-19 Data

Four Horsemen illustrated by TPYXA Illustration. Lock Icon by svyatoslavius (Shutterstock)

* Alabama, Florida, Georgia, Maine, Mississippi, Missouri, Nevada. Okalahoma, Pennsylvania, South Carolina, and Texas issues stay-at-home orders in the 1st week of April.

Arkansas, Iowa, Nebraska, North and South Dakota, Utah, and Wyoming did not shut down.

*Fatalities & State by State Lockdowns, March 2020*

# 6

# THE FIFTH HORSEMAN IS FEAR[1]

*And I looked, and behold, a pale horse — pestilence! And its rider's name was Death...*
*— based on Revelation 6:8 ESV[2]*

## MARCH 2020

If *The Ghost Riders in the Sky* appear anywhere, it's in the big skies of Montana. Emerging from clouds glowing like the coals in a hot forge, four riders on fire-snorting horses, forever trying to catch the devil's herd.[3] Many a night glows from their flaming eyes.

But not tonight, the first of March, a still frozen night blanketing a pitch-dark sky. Finally, as the clock swings past midnight, a sliver of moon rises, a glimmering beacon. And there they are, the riders in the sky.

But, there is no herd to chase this night. They are *the* four horsemen; the fourth, the pale horse,[4] pestilence — coming for us. Out of the glow of shadowy fog that hugs the ground this night, a fifth horseman appears at the edge of town, a herald trumpeting what is to come; the fifth horseman is fear.

Inundated with advice about staying healthy, the daily count of coronavirus cases marches across TV sets and news feeds. People are dying in Italy, hospitals overflowing, total lockdown, pictures of empty streets.[5] Almost everyone I know is watching for news of the coronavirus. Jargon previously reserved for *Lab Rats*[6] enters everyday conversation: spike protein, PCR, N-95, PPE, genome.[7] Are we marked to get really sick from the virus?[8] The question is, "How bad will it be, health-wise, job-wise, money-wise?"

American government institutions seem unsure how to respond to the unwanted immigrant's arrival. On March 2nd, the Surgeon General begins an interview on *Fox and Friends* with a report of the second death in Washington, a man in his 70s who "did have underlying health conditions... the outbreak growing to at least 79 cases in the United States." The Surgeon General, wearing a crisp uniform styled after the US Navy, Vice Admiral Jerome Adams, says,

> One of the things they shouldn't be doing, the general public, is going out and buying masks. It does not help.[9]

He goes on to say that masks have not been proven to prevent the spread of coronavirus among the general public — in fact, people who do not know how to put them on correctly can actually "increase your risk if you are not a health care provider." His message is to wash your hands and stay home if you are sick — not just for the novel coronavirus but for the flu, which could be much worse.[10] Is this the point? Or, is the main point to get people to stop buying masks because of the worldwide shortage — a point he made a few days earlier in a Twitter tweet?[11]

A few days later, Doctor Fauci gives pretty much the same pitch on masks on *60 Minutes*,

> The masks are important for someone who is infected to protect them from infecting someone else.

After talking about all of the pictures in Asia where most people are wearing masks, he adds,

> Right now, people in the United States should not be walking around wearing masks... When you are in the middle of a pandemic, wearing a mask might make people feel a bit better. It might even block a droplet, but it is not providing the perfect protection that people think it is.[12]

Fauci is providing an accurate account of the medical literature for flu and SARS —and the CDC's conclusion in their review of interventions for novel flu outbreaks.[13] The federal administration's medical leadership appears to be on a public persuasion campaign to deal with a problem — the lack of masks for medical personnel who need them the most. That's how Fauci finishes — to save masks for "people who really need them."[14]

Unfortunately, these words create a mustard seed of mistrust that will grow and grow — we are destined to keep hearing about Fauci's "no mask" statement.

CDC soon recommends just the opposite — and then tries to collect evidence that masking the population works — they will not prove the point.

Despite the HHS mask equivocation, there is a rush for masks and sanitizer. The *Ravalli Republic* runs two articles about a disease that has yet to arrive and has already led to shortages and worries about closing down. "Bitterroot sees mask, sanitizer shortage." The owner of Bitterroot Drug says, "Suppliers can't give... a date" for a restock of face masks that had been sold out a week before. The druggist says he is going to be making his own hand sanitizer.[15]

Several people take mask matters into their own hands, sewing for them and their friends. The *Ravalli Republic* shows three women with cloth cutouts, ready to make face masks, noting that a Hamilton Christian Academy[16] instructor has put a pattern on Facebook. My wife, Mary, is quite the seamstress and researches how to make homemade masks that perform almost as well as N-95 masks "if they didn't leak around the edges." She sews a cloth mask with a pocket so you can slide a vacuum cleaner HEPA filter into it.

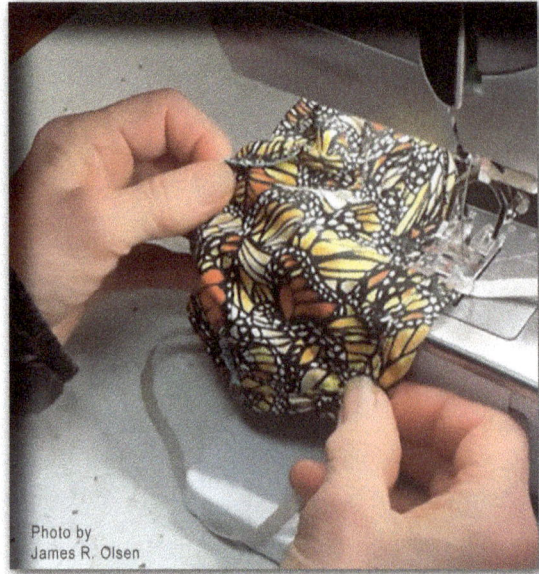

Photo by James R. Olsen

*Mary Byers Sewing Mask*

From the news pouring in from Italy and other places, people know government institutions will do something but don't know what. They ask. The County Health Nurse is working with schools to help develop plans when the coronavirus arrives. The Health Nurse says they are getting calls from parents. Are they going to close schools? "We are going to just have to wait and see what happens." The Stevensville School Superintendent says they have "already begun to disinfect common-use areas to dampen the spread of influenza."[17]

A spokesperson says that the hospital is always prepared for an influx. They see other respiratory diseases that impact the community as much as they expect from the coronavirus — I don't buy it.

Doctor Marshall Bloom, Associate Director of Rocky Mountain Laboratories, adds,

> RML doesn't have a role in helping prepare for an outbreak. Our role is research.

Doctor Bloom is a Bitterroot Valley institution. While local newspapers and the rest of us use a shortcut title for Doctor Bloom, "Lab Director" or "Lab Chief,"[18] his official title is "Associate Director for Science Management, Rocky Mountain Laboratories and Chief of Biology of Vector-Borne Viruses Section." Where this position fits is hard to find in the Byzantine organization of NIAID. Anthony Fauci holds the reins close; he is officially the Director of RML.

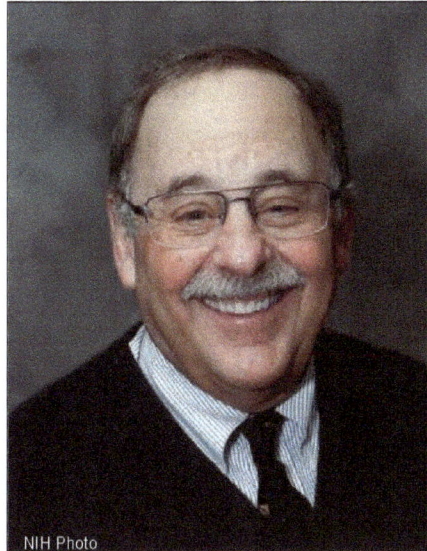
NIH Photo
*Doctor Marshall Bloom*

Doctor Bloom is the face of RML.

So, with RML more or less out of the picture as far as a local response is concerned, state and local governments start preparations with the resources at hand, inadequate though they are. The newspaper announces, "Governor Bullock forms coronavirus task force." Adjutant General Matthew Quinn, Director of Military Affairs, will lead the task force. The task force includes the Montana Department of Health and Human Services (DPHHS). The article says it is modeled after a previous response to Ebola — which never arrived in Montana. However, everyone is sure that new coronavirus will come.

He is known as Doctor Trout, an avid fisherman and involved in Trout Unlimited, serving as President of the local chapter in the '80s — which means weighing in on public land use occasionally — sometimes we're in agreement, sometimes not.

Marshall moved to the Bitterroots in the 1970s. His NIAID Biography gives a summary of his professional career:

> Dr. Bloom received his M.D. in 1971 from Washington University School of Medicine in St. Louis, Mo., and then joined the Rocky Mountain Laboratories (RML) of NIAID in 1972 as a research associate...
>
> He is a world expert in the molecular biology and pathogenesis of parvoviruses and is considered an authority in biocontainment...
>
> In 2002, Dr. Bloom was appointed associate director for RML in NIAID's Division of Intramural Research and supervised the permitting, construction, and staffing of NIAID's first biosafety level-4 facility.
>
> In 2008, Dr. Bloom was named associate director for science management for RML in NIAID's Division of Intramural Research.[19]

By mid-month, headlines announce, "County preparing for coronavirus," as the Health Nurse reports that only one test had been requested, saying we were "surrounded by states with cases, even though Montana has no cases yet." It isn't clear from the article if the person requesting the test had been accommodated. Montana does not have any test kits, so samples will still need to be sent to the CDC in Atlanta.

It's as if Lily Stomper is at the rodeo, a bucking bronc of some renown. Many a cowboy wants to ride that horse.[20] If they can stay on Lily's back for 8 seconds, they have a good shot at the prize money. It seems as if every researcher also wants to mount up to get their name on a COVID paper. Lily throws all but the best. One wishes there was a Lily on every peer review panel — the quality of many COVID studies has suffered in the rush to publish. That's not the worst — sometimes, a paper affects government policy and people's lives.

The texture of life has already changed, the future is unknown. A nascent black market emerges followed quickly by Internet irony. I get a video in an email from a friend.[22]

---

**CORONAVIRUS — EPISODE 1 - BLACK MARKET SUPPLIES**

1 **EXT. RESIDENTIAL STREET - DAY.**                              1

Camera looking out the side window from the drivers POV, residential district. Car slows to an intersection. a bare lot.

DEALER Standing at the corner, scruffy looking in a hoody, a black leather jacket over it, jeans. holds a lit cigarillo.

Car rolls to a stop. rolls down window.

> **DESPERATE BUYER**
>
> John?

DEALER looks at him, not responding.

> **DESPERATE BUYER** (CONT'D)
>
> John? Are you John?"
>
> **DEALER**

Looking furtively. ambles toward car window.

> Who sent you?

THEY EXCHANGE A SECRET PASSWORD

> **DEALER**
>
> Did he tell you how much?
>
> **DESPERATE BUYER**
>
> Yeh. (Breathless)

Hand to the glove compartment.
pulls out an inch thick stack of $100 bills.
hands cash through window.

> **DEALER**

Looks sideways down the street. looks at money. unzips jacket.
pulls out a roll of toilet paper. hands through the window.
pulls out a bottle of sanitizer. hands through the window.
slips cash into his pocket. takes a drag on the cigarillo.

> Now get the hell out of here. You didn't see me.

Car starts down the road.

> FADE TO BLACK

---

A report from the Imperial College in London predicting 2.2 million deaths in the United States make headline news; it is based on a model. The model is driving policy in the US and UK as it predicts sick people will overwhelm medical capacity as in Wuhan and Italy, causing policymakers to consider lockdowns.

The model provides some insight into how various interventions might work. The parameters fed into the model are reasonable. Though the study modeled interventions short of a lockdown, it seems to have assumed thorough contact tracing, and ideal widespread testing — factors that may or may not have affected the outcome significantly. [22]

As I learned in several modeling courses and some models of my own, betting on specific prediction values is almost always a bad bet, especially when news people and bloggers carefully note the predictions and gleefully announce when the predictions fail to match reality. The model's utility in revealing the relationships between interventions and outcomes quickly gets lost in the hoopla.

A year later, the modelers, also realizing the cautionary presentation that would have helped in the original paper, write,

> It is important that uncertainty in the parameters governing the transmission dynamics, and its influence on predicted outcomes, be robustly conveyed. Without it, decision-makers will be missing meaningful information and may assume a false sense of precision. [23]

Suppose the authors had included this caveat in bold type in the original paper. Decision makers may have looked more closely at how decisions affect outcomes.

The word *Hydroxychloroquine* (HCQ) breaks from its bindings from its resident home, the medical dictionary. Like a Brahma bull breaking out of the rodeo pen, HCQ bursts forth as a political meme in the Twitter wars, the source of much

anxiety, conflict, and destruction when let loose. [24] It starts as doctors struggle to treat patients with no case-studied approved medicine. In this event, a doctor will judiciously try some treatment for a patient with COVID-19 heading toward the ICU. An FDA-approved drug can be prescribed for COVID-19 "off label."

A lab experiment using hydroxychloroquine in a test tube shows promise. [25] This experiment offers much hope for a quick end to the pandemic. Politicians take it up, including President Donald Trump. Grabbing onto hydroxychloroquine as the nation's salvation is as in character as a red ball cap.

But, the unanswered question from an in vitro study is, "Will the body actually deliver the presumed dose to a cell?" French researchers, led by Didier Raoult, [26] tried out HCQ on 20 patients. The paper claims that people test negative sooner when using HCQ.

Unfortunately, the methodology used has a flaw. Testing on the HCQ group is done daily; tests on control groups are done every other day — assuming the positive test results remain positive on the off days. But, if one considers the test became negative on the off days, the results would change — the method of counting results is artificially biased in favor of HCQ. [27]

This and other small trials and test-tube experiments create a whirlwind of events, with the President trumpeting the results and scientists saying that the data is too sparse. Lost in all this is that the French paper itself does not claim a definitive answer:

> Our study has some limitations, including a small sample size, limited long-term outcome follow-up, and a dropout of six patients from the study, however in the current context, we believe that our results should be shared with the scientific community. [28]

Meanwhile, Raoult presses the issue in a video presentation on the success of HCQ in French,

# Twitter - Hydroxychloroquine

**Promoting (n=140, 36.4%)**

- Explicit promo (n=69, 49.3%)
- **Total countering critique (n=107, 76.4 %)**
- Supporting Trump and promoting (22/22)(100%)
- 79.5% of all bot accounts
- 75.8% of all QAnon-linked accounts

**385 most retweeted tweets with "hydroxychloroquine" from April 21st – 28th, 2020**

**Critiquing (n=163, 42.3%)**

- Explicit (n=87, 53.4%)
- **Total countering promo (n= 136, 83.4%)**
- Critiquing Trump and critiquing (110/116) (94.8%)
- 5.1% of bot accounts
- 0% of QAnon-linked accounts

**Neutral/Unclear (n=82, 21.3%)**

- 15.4% of bot accounts
- 24.2 % of QAnon-linked accounts

Alessandro R. Marcon,Timothy Caulfield  UIC Journal, First Monday, The Hydroxychloroquine Twitter War...

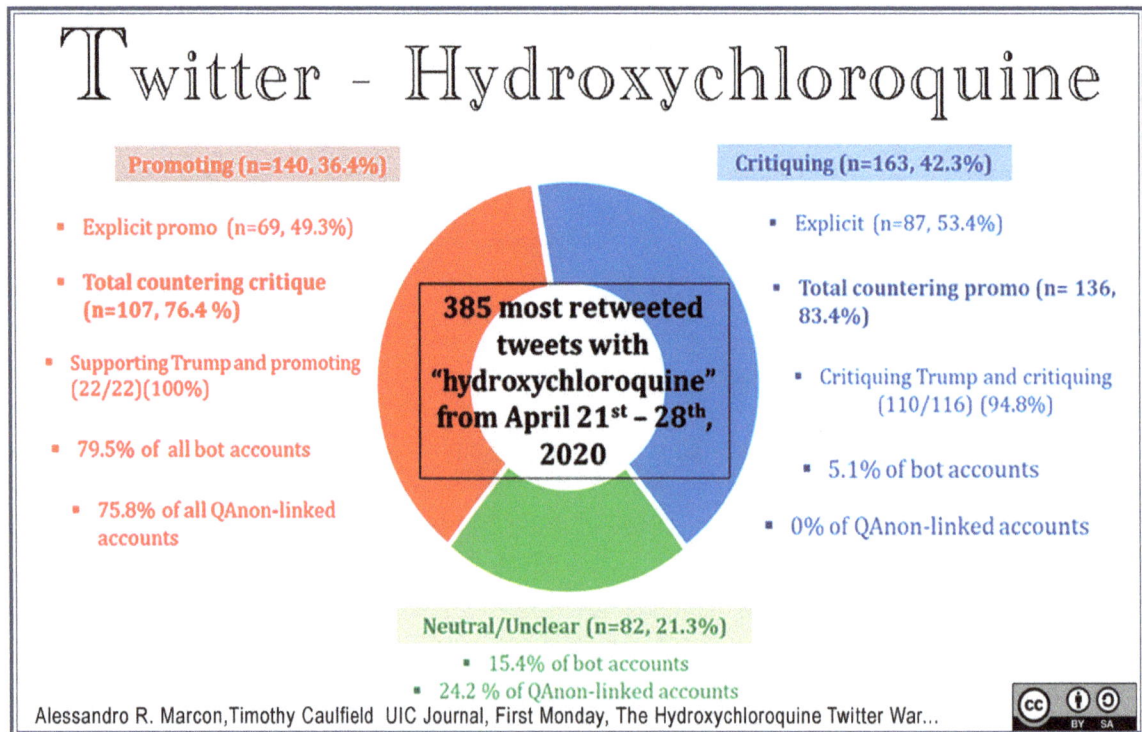

*Hydroxychloroquine on Twitter*

skipping over any caveat. His talk is quickly picked up by bloggers providing headlines in English — HCQ certainly works. He cycles through the alt-narrative interview circuit.[29]

A medicine used by millions of people for things like malaria becomes politicized, so promotion or criticism is now a partisan event.[30] Papers that hint that it works or say that it doesn't are looked at as political statements, including scientific papers. On Twitter, most promoters and critiquers line up along partisan lines. Bots are let loose in Twitter-land to deepen the partisan divide further.[31]

The critics complain about flaws in the design of the trial. The results of Raoult's paper, such as they are, show promise and should be followed up — they will be, in part, but not before the whirlwind becomes a category 5 hurricane.

Science becomes a driving force in HCQ arguments, political science, that is. Peter Navarro takes it all the way to the situation room, with Vice President Pence at the head of the table, as

he dumped a folder of papers on the conference table.

> I hear you're saying that hydroxychloroquine doesn't work. I've got a bunch of papers here that say it works. I have all the evidence in the world that hydroxychloroquine works. And by preventing people from getting it, you have blood on your hands!

Fauci responds,

> Those papers are not properly reviewed, they're anecdotal, and they really don't prove anything.

Navarro shouts,

> There you go again. You don't know what you're talking about. You guys have blood on your hands.[32]

The news amplifies the divide in the following months with stories like "Yet Another Study Shows Hydroxychloroquine Doesn't Work against Covid-19."[33] However, the study cited in the story is a retrospective look at patient records — yes, it is a larger sample, but not randomized as in a clinical trial.[34]

In the end, Raoult's group would do a somewhat more extensive study that, by 2021, shows the drug to be effective against COVID-19, shortening the time for a patient to test negative. However, the number of participants is too small to examine the risk of hospitalization or death.[35]

A hospital in Belgium, with no approved treatment, adopts HCQ along with azithromycin as their standard. They will later publish a paper that shows their patients fared better than other hospitals in the country in terms of mortality, dropping it from 26% to 16%. The "standard of care" for a patient diagnosed with COVID-19 is the institution's rules for medical care. However, the standard of care for other hospitals is unstated.

On the other hand, a large-scale UK clinical trial will find no benefit, but again, the standard of care for the control group is unstated. The standard of care treatments and HCQ may both have some benefits.

A double-blind trial with the control group getting a for outpatients will finally be published in 2024, which shows a decrease in the risk of getting symptoms. Even with over 4,000 participants, this trial was not of sufficient size to measure the risk of hospitalization.[36]

HCQ has been known to have antiviral effects, including for SARS. Yielding to political pressure, the FDA gives an emergency approval of HCQ for the treatment of COVID-19 — a move not really necessary since it is already being prescribed off-label by leading Medical Doctors and other physicians — because there are no approved treatments.[37]

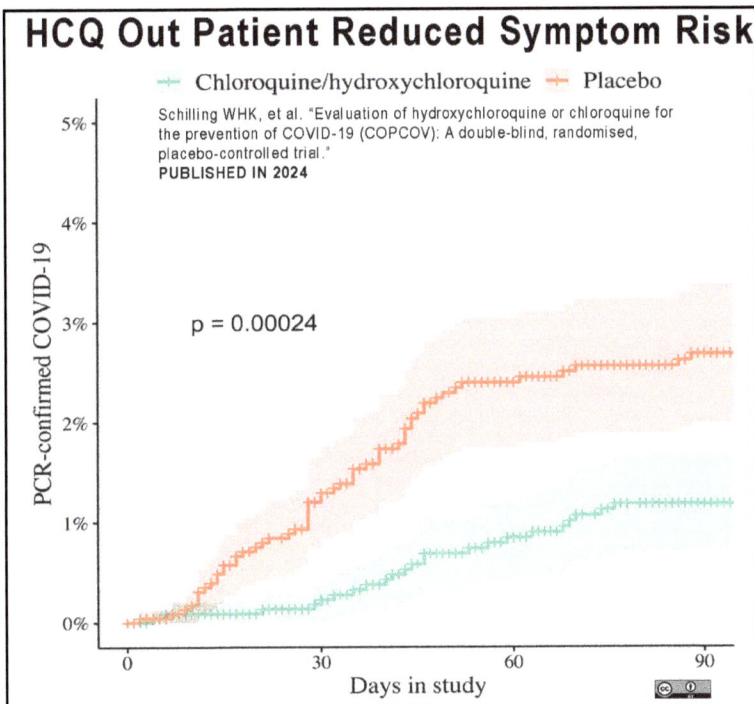

**HCQ Out Patient Reduced Symptom Risk**

Chloroquine/hydroxychloroquine — Placebo

Schilling WHK, et al. "Evaluation of hydroxychloroquine or chloroquine for the prevention of COVID-19 (COPCOV): A double-blind, randomised, placebo-controlled trial."
**PUBLISHED IN 2024**

$p = 0.00024$

*HCQ Out Patient Reduced Symptom Risk*

Self-medication turns fatal. When someone dies from taking chloroquine phosphate, a treatment available for aquarium fish, the agency finds itself issuing statements that one should not take the drug unless it is prescribed.[38]

It doesn't take long for Congress to get in the act. You can see the American people beginning to separate into two camps, with two very different COVID narratives on the floor of the Capitol. Representative Carolyn Melloney, Democrat, opens with a partisan swipe at the Trump administration,

> ...strategic errors and a failure of leadership... testing plagued by misstep, and resulting in substantial deficiency in who may be infected...CDC had tested about 4,900 people.

> By comparison, South Korea tested 66,000 people in just one week.[39]

Representative Jim Jordan, Republican, opens with an appeal to support the Trump administration,

I hope our experts will explain today that

the risk to the American people of contracting the coronavirus remains low ...

The risk remains low because of the actions of the Trump administration, in particular imposing a travel ban on travel from China."[40]

Unfortunately, the risk is not still low. We will soon find out. Jordan goes on,

An increase in test kits will inevitably show an increase in positive cases ... we should not play politics with the coronavirus.[41]

At one point, a Congressman asks Doctor Fauci where the best place is to get the correct data, noting the fake news on social media. Fauci's answer is the CDC website, saying that the CDC is a data-driven organization.[42] Unfortunately, the CDC website is not always data-driven; it will be noticed and grow into an immense problem in public trust by many citizens.

When Doctor Fauci is asked, "Where should America appropriately set their gauge?"

"Their gauge should be set to: *we have a serious problem...*"[43]

That is true. Some predictions are "grave." Professor James Lawler from the University of Nebraska Medical Center predicts that US deaths in the hundreds of thousands. Another, Tom Frieden, who oversaw America's response to the H1N1 epidemic, gives a worst-case scenario where half the population is infected and one million people die.[44] No one wants to believe it.

Fear and anxiety are starting to drive the rhetoric, personal decisions, and sometimes even the science. Fear is a good thing amid imminent danger. But, fear of what-has-not-happened-yet is soul-crushing — we call it anxiety. At The Table, the consensus on dealing with COVID fear, both the disease and government response, is to replace fear with caution. We do our best.

People at The Table are all abuzz about Event 201, citing an @JasonBermas video. Jason gives a counter-narrative interpretation of Event 201, which occurred in October 2019, making some good points on the flaws of globalism that such an event will uncover.

Jason's central theme, though, is that this exercise is a rehearsal of how government and industry are controlling our lives right now and planned ahead of time: The 2019 Event 201 was a rehearsal of a plan, not a speculative scenario.[45]

Event 201, which Jason is commenting upon, is a tabletop exercise of a simulated global reaction to a simulated coronavirus outbreak. It came up at The Table, so I watch the video. Having participated in military simulations and exercises, my response at The Table is that you have to prepare for your mission, and the best preparation is an exercise. A prediction that the next pandemic would be a coronavirus seems like an obvious choice.

The exercise tended to let the simulated disease quickly have its way — making it more of a "sales pitch" than a decent tabletop exercise designed to examine decisions and their expected consequence.[46]

It must have been in the air. As we've seen, the federal government and some states ran their own tabletop exercise, *Crimson Contagion,* in 2019, simulating a novel flu epidemic from China, estimated to cause over a half million deaths. *Crimson* pointed out the lack of readiness and the fractured authority at the federal level, which makes a quick, effective response difficult. With its long list of to-dos, because we are not ready, the report was released internally in January 2020 with a note to not distribute it further.

It was already too late to "prepare." The exercise has become real. Contagion has already arrived from China.[47]

The conspiracy theories keep coming. Another narrative, as the New York City outbreak gets reported, is that the crisis is fake. The hashtag #FilmYourHospital is exploding, spawning amateur sleuths.[48] I get forwarded several videos and pictures of empty hospitals and empty parking lots, which have been pulled from YouTube and Facebook.

It doesn't seem plausible — too easy to fake or misconstrue if the citizen reporter doesn't actually go into the hospital. Why would New York City hospitals be empty when hospitals are full in Wuhan, with hospital tents sprouting up like mushrooms to add beds? To be fair, as March is coming to a close, I email someone with contacts in New York City who grew up there and would give me a straight answer — I wait for a response.

My friends at The Table forward an email link that reads like a rant, "Coronavirus: what real science would look like, if it existed" from *NexusNews*,

> When you look at the justification for all the lunatic measures being taken to "stem the tide"..., you come to ...CASES. How many cases are there? ... are "infected?"[49]

I agree to look at it at The Table while not buying into the word "lunatic" to cover all of the measures, as some seem good and others seem misguided. *NexusNews* claims there is no evidence that PCR tests work, comparing these tests to taking your car into the shop and having the mechanic hook up a diagnostic machine. Should you trust a machine if the device itself has not been tested?

When I look at the medical literature, it is a valid question. There are different approaches to verifying the accuracy of the PCR test. One is if a patient has a specified set of symptoms. The problem is that many symptoms could be something other than SARS-CoV-2. A second verification method is to verify a negative test by re-sampling and doing another PCR test. But that assumes the test works in the first place.

A third method is to compare a PCR test against another approach, such as antibody tests, which look for antibodies the body makes in response to the disease instead of a strand of RNA unique to the SARS-CoV-2. But, antibody tests are not the gold standard. In fact, PCR tests are held up as the gold standard for testing in many trial and observation studies — but, as other scientists have noted, there is no gold standard for testing the COVID-19 PCR test.

The correlation of COVID diagnosis may be the best independent method of verification of the PCR test. Results from other countries are mixed. I have not found a paper on the subject from the United States.[50] Early reports from China of hospitalized patients comparing medical observation using Chest CT-scans[51] with PCR testing suggest that both are needed, but neither a magic bullet:

> Of the 1014 patients, 601 of 1014 (59%) had positive RT-PCR.
>
> Of the 413 patients (60%) with negative RT-PCR results had CT-Scans that were highly or probably COVID-19.[52]

In other words, a patient can test negative for COVID-19 but can be in the hospital with the disease — leading to the need to retest often. In the end, the belief in the validity of the PCR test rests at least partly on believing the chemistry of the PCR test.[53]

So, *NexusNews* wants us to toss the PCR test out. But throwing the baby out with the bathwater does not seem wise based on a theoretical argument without a close look at the real world.

When you get to the point of questioning every assumption, like when Road and Track writes an article that says, "Don't trust your speedometer," *NexusNews* would have us chasing the "golden standard" — to drive a measured mile at a constant velocity. But, if you did, how accurate would you be at looking at your watch at precisely the right time, keeping your foot on the accelerator just right? If you get a ticket because a police officer followed you while looking at the

# Reverse transcription polymerase chain reaction (RT-PCR)

In RT-PCR, The RNA population is converted to cDNA by reverse transcription (RT), and then the cDNA is amplified by the polymerase chain reaction .The cDNA amplification step provides opportunities to further study the original RNA species, even when they are limited in amount or expressed in low abundance. Common applications of RT-PCR include detection of expressed genes, examination of transcript variants, and generation of cDNA templates for cloning and sequencing.

**Sample from Patient — Matches Intended Pattern**

**RNA**
RNA consist of Start codon AUG and ends with poly A tail

**Nucleotides in a unique order**

**Oligo dT Primer**
Oligo dT Primer is binding to RNA poly A tail

**Reverse Transcriptase and dNTPs**

**Nucleotides roving around in solution**

dNTPs

**Reverse Transcriptase** is an enzymes binds to oligo dT primer and synthesises the cDNA by adding dNTPs

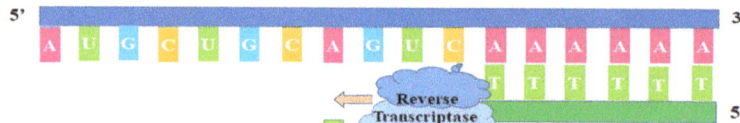

**Enzymes match intended pattern by binding roving nucleotide**

**RNA hybrid formation :**
First - strand cDNA synthesis

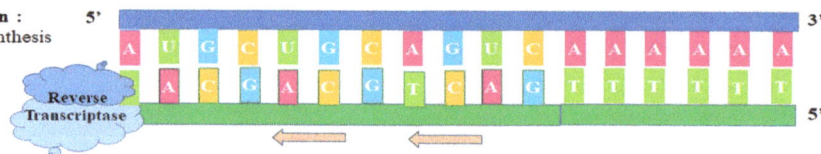

**Heat — Split strands**

**complimentary DNA**

**Make a bunch**

**Amplification of cDNA** with Specific Primers and Taq Polymerase

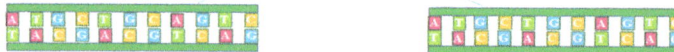

**Hyrbidize with agent to make it floresent**

*RT-PCR Chemical Process Illustration*

speedometer, I don't think this will fly in court.

PCR tests are designed to detect the presence of a specific segment of the SARS-CoV-2 virus RNA. The accuracy of PCR testing is questioned in the literature and counter-narratives. Though PCR test technology may not have been tested well against an independent standard, that doesn't mean PCR tests don't work — they have been helpful to many people.[54]

A PCR test requires the presence of a virus marker where it is swabbed in the upper nose. You may have acquired the disease, but it has not yet spread to your upper nose where the swab is taken. In the first week after exposure, false negatives are likely; if you have COVID-19 and recover, you can still test positive because the detritus of the virus may still be there. Hospitalized patients may test negative more than once before getting a positive result.[55]

There are three types of COVID-19 tests. Each detects a chemical in the body; two are detritus from the virus itself, and one is a marker made by the body's immune system. Figure 5.1 illustrates what is detected by each type of test.[56]

The RT-PCR test is designed to amplify a tiny bit of RNA from the virus, allowing it to detect small quantities in the sample. It discriminates only the unique RNA code for SARS-CoV-2 — a very sensitive test but still has false positives in the real world, with most studies reporting around 5%.[57] Antigen tests look for a virus protein without amplification, leading to a low false-positive rate.[58] However, it can result in a false negative when a person should test positive, so an at-home test should be taken at least two or three times over several days.[59] An antibody test indicates your body's immune system responded to the virus but does not mean the virus is still present.

Montana finally gets a few COVID-19 PCR test kits on March 9th. Four positive tests in Montana are confirmed on March 11th, which, not coincidently is the same date of the first COVID-19 case in Montana. Because of the delays in developing test kits, the date of the first COVID-19 case says more about the arrival of test kits than the arrival of the virus.[60]

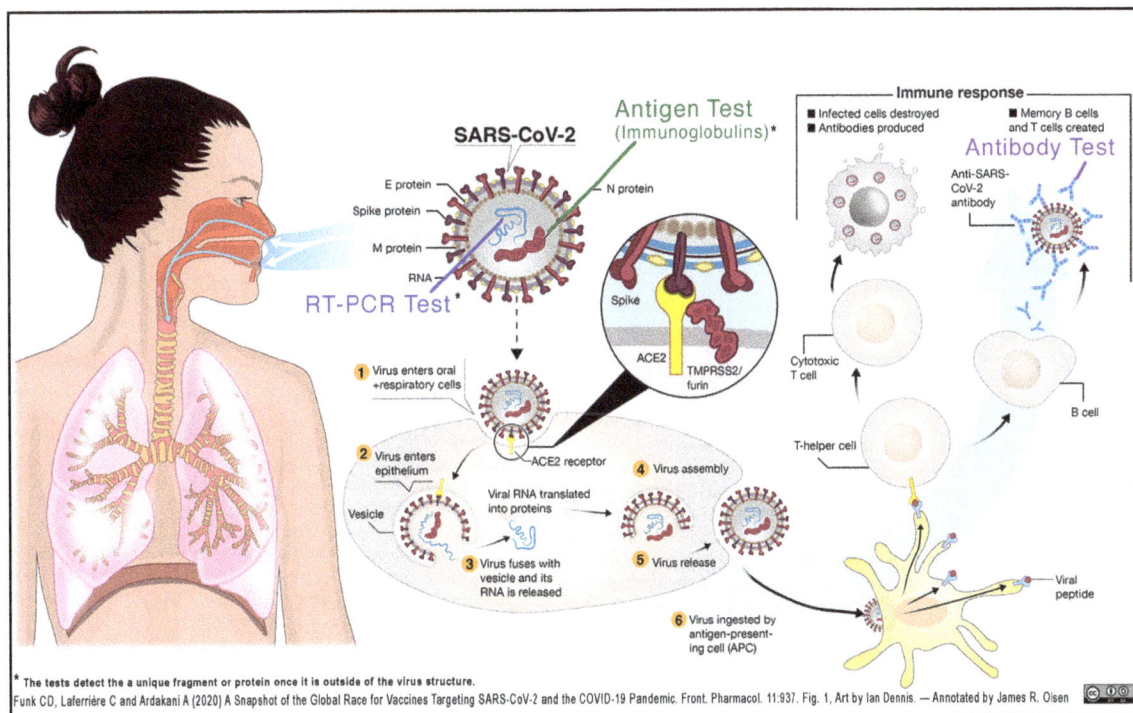

*Figure 5.1 — COVID Tests – What They Find, Where in the Immune Cycle*

In a press interview, Doctor Fauci responds to a question about how few test kits are available to states. It is almost an apology when he says the initial CDC test is meant to be used in a doctor's office rather than "geared more to a public health situation," moving his hands almost to make quotes. "Traditionally, that kind of approach for infectious diseases works." Spreading his hands palm down, "What is clear now is that it doesn't work when you essentially need to blanket the country. The only way that works is if you get the private sector involved." Adding that they were going to get many, many more tests available in a week or so.[61]

Case counts are reported daily. Government officials make policy decisions based on the case counts. Without widespread testing, case counts are a poor metric for how many people have been exposed to the coronavirus, been infected, or are now breathing it into the air. This is the case, pun intended, when one small Colorado town, using private funds, offers tests to the entire town.[62]

When case counts rise, decision makers, with no better metric to grab onto, react as if the disease is suddenly on the rise. Yet case counts are driven by the testing regime, the rules for getting tested, the availability of tests, how many tests were performed; these driving case counts as much the percent of people who have the disease.

"Coronavirus Tally May Be The Tip of the Iceberg as Sick Go Untested," reads a headline in *Bloomberg*. "Testing only the most severe cases is a disastrous public-health decision," says Ralph Baric.[63]

Let's say that 10% of the people in a county of 100,000 have COVID — that's 1,000. On average, only 300 of them will report having symptoms. Let's say that all the people with symptoms are tested. The case count will be 300, or 3% of the population. Now, say you test the 300 people with symptoms and 3,000 more. The case count will likely double to 600 if the test is perfect. Not because more people have COVID, but because you tested more.

"Flatten the curve" enters the vocabulary.[64] However, if public health officials want to know what is happening, it is critical to test asymptomatic people for a disease like COVID-19. In our example, if you did 1,000 tests randomly and 10% are positive, you have a decent guess that 10% of the population has the disease. Then, public health can use the risk of hospitalization to get in a position to be proactive instead of simply reacting, hoping the hospitals won't fill up. The bottom line is that the case counts are the tip of the iceberg. The more testing, the more of the iceberg you will see.[65]

In contrast to the United States, South Korea never had a total lockdown. Its comprehensive response to the pandemic, massive testing, transparent and truthful public information, thorough contact tracing, and isolation, saves lives. But it does come with a price. In South Korea, people who are ordered into self-quarantine are given a $300 subsidy. There is also a severe penalty for violating in the form of an $8,000 fine and up to a year in prison. There is a high level of compliance and few violations.

Of course, forced quarantine for a disease like COVID-19 won't fly in most counties in the United States. Even if we did everything else from South Korea's program, we would have to think about a less draconian quarantine program — but the United States never gets that far. Without effective and broad testing and a comprehensive contact tracing program, the quarantine question is moot because too many undetected infected people are walking around.

Figure 5.2 shows the delay between infection and as confirmed case; South Korea is making public health decisions quickly because of thorough contract tracing while the United States waits for a confirmed case report. In Figure 5.3, South Korean cases reflect the infection rate more accurately because more people are tested, and

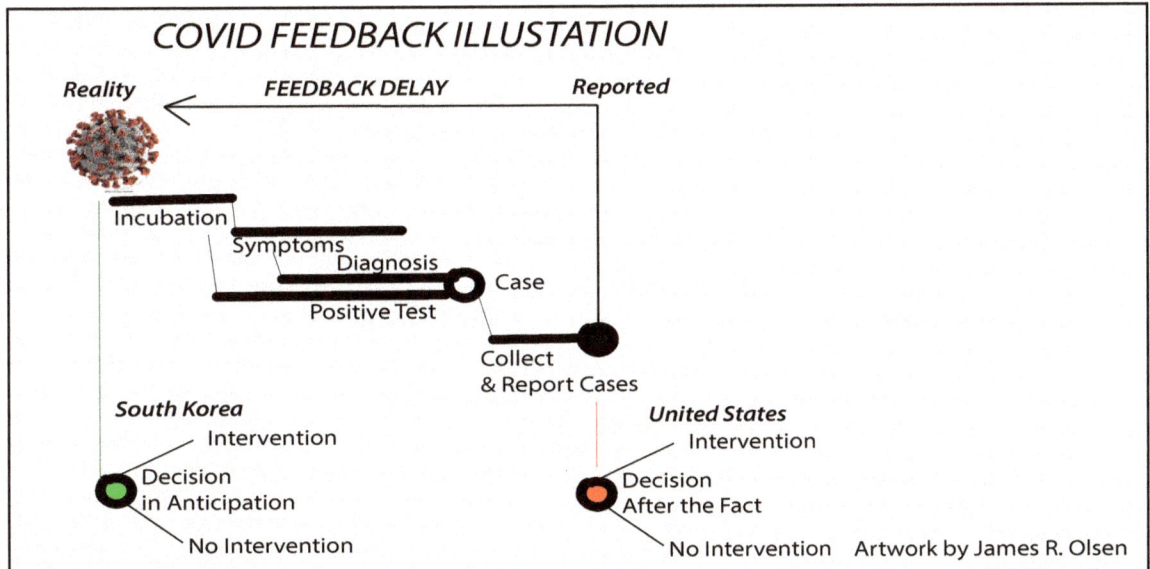

*Figure 5.2 — COVID Feedback Delay*

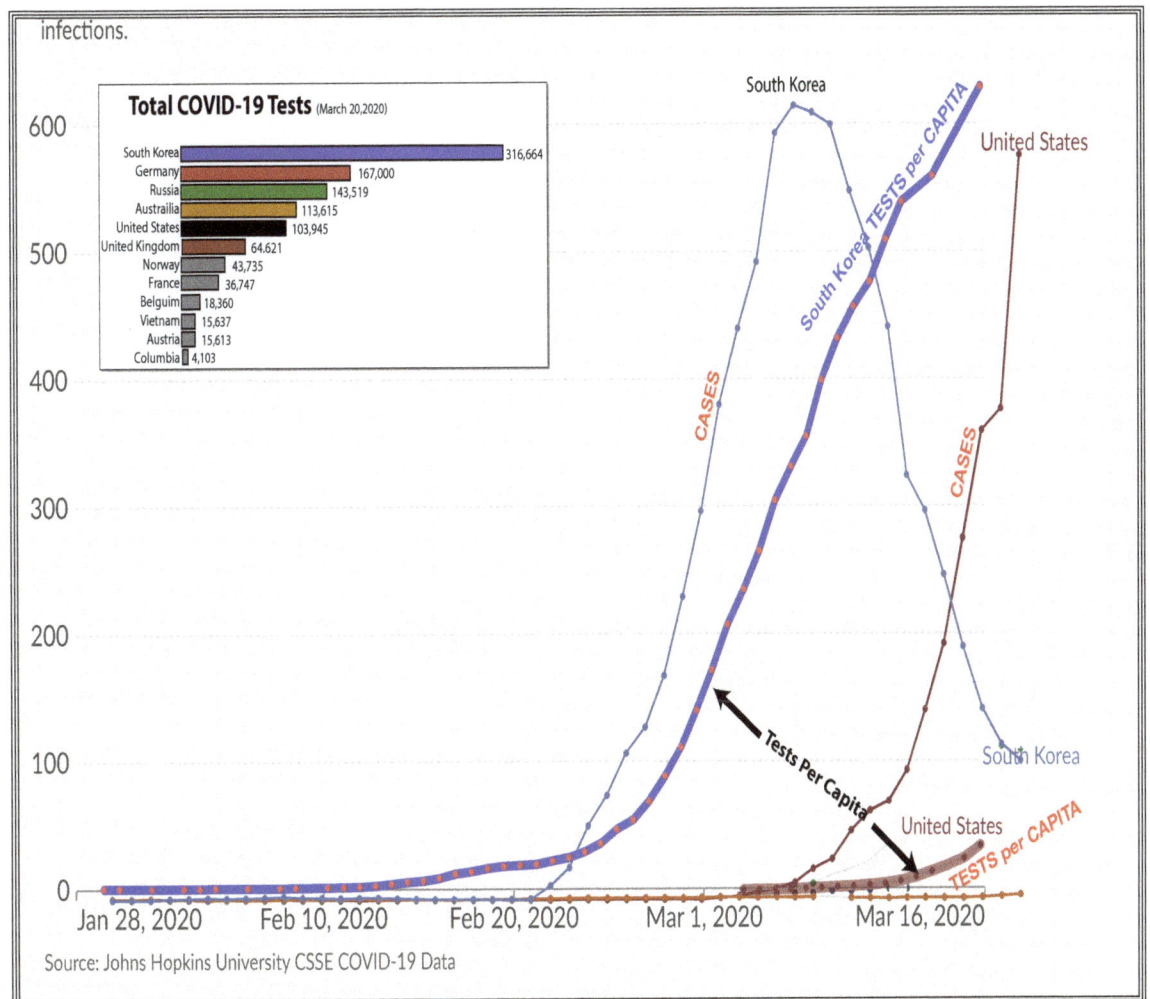

*Figure 5.3— United States,South Korea, Cases and Tests Per Capita*

the case counts drive contact tracing and isolation. You can see the South Korean cases rising as more testing is done. Case counts come down quickly while continuing to test at a rate independent of case counts.

On the other hand, given the much lower testing rate in the US, it is a good guess that we have way more infected people than shown in Figure 5.4 — we just see them later. While South Korea has more "cases," it does not necessarily mean more infected people. It means they have a better idea of what is happening. All of this comes together for South Korea because they are identifying, tracing, and quarantining and have put the trained staff in place to do it nationwide.

By mid-month, deaths from Confirmed COVID-19 has only visited 79 souls in the United States, while the Worldwide number is 7,155, over half of them reported by China.[66]

The lockdown in Wuhan ends after 14 days with no cases. Doctors struggled to treat a disease with no proven medicine. A team in one Wuhan hospital did a small observational trial, turning patients on their bellies — prone position — for people struggling to breathe. There were improved outcomes.[67]

The Government announces that people can pick up the ashes of their loved ones at the funeral homes. The long lines belie the official Chinese death toll. Government-sponsored celebrations are held with much fanfare, a victory in controlling the disease. Is there merit in this — even if the numbers are not to be believed?[68] If the death toll is, say, 200,000 — a world record if divided by 11 million people of Wuhan. But, less than half the death rate of the United States if divided by 1,400 million people of China.

While China did limit the outbreak to Wuhan, the Chinese Government's missteps, falsehoods, and draconian suppression of free speech are blamed for exacerbating health outcomes during the crisis.

Now it is the United State's turn. We'll use our free speech to make trustworthy information harder to find than the proverbial needle in a haystack, then try to censor with little success.

It is fair to say the dire coronavirus news helped me sell some books. My first book reading of

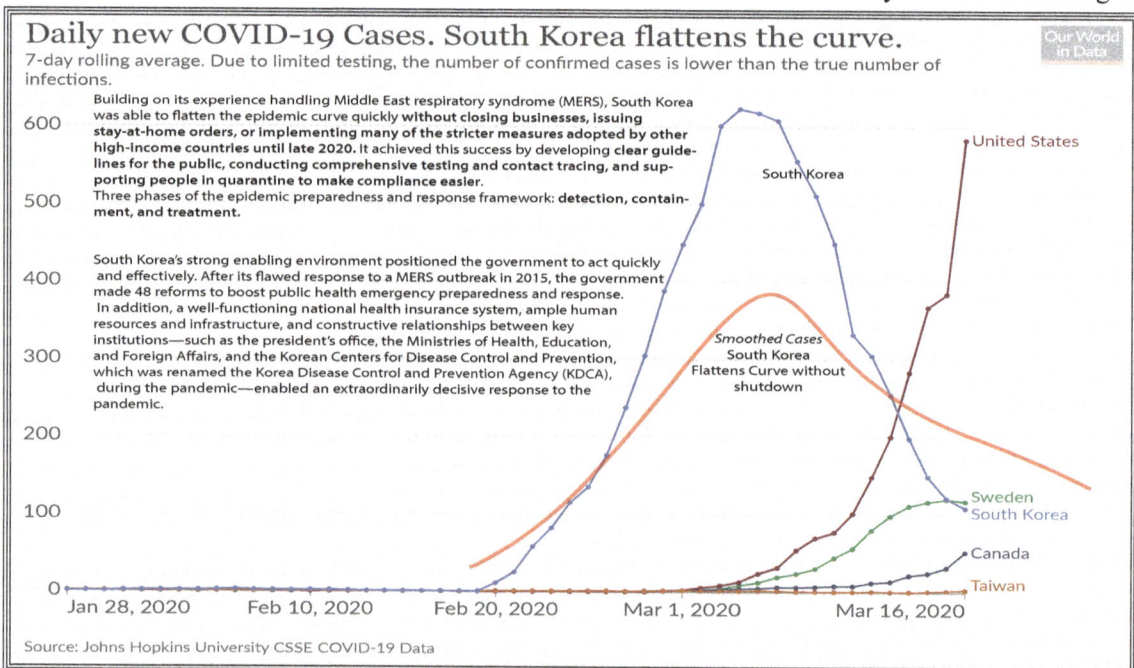

Daily new COVID-19 Cases. South Korea flattens the curve.

7-day rolling average. Due to limited testing, the number of confirmed cases is lower than the true number of infections.

Building on its experience handling Middle East respiratory syndrome (MERS), South Korea was able to flatten the epidemic curve quickly without closing businesses, issuing stay-at-home orders, or implementing many of the stricter measures adopted by other high-income countries until late 2020. It achieved this success by developing clear guidelines for the public, conducting comprehensive testing and contact tracing, and supporting people in quarantine to make compliance easier.

Three phases of the epidemic preparedness and response framework: detection, containment, and treatment.

South Korea's strong enabling environment positioned the government to act quickly and effectively. After its flawed response to a MERS outbreak in 2015, the government made 48 reforms to boost public health emergency preparedness and response. In addition, a well-functioning national health insurance system, ample human resources and infrastructure, and constructive relationships between key institutions—such as the president's office, the Ministries of Health, Education, and Foreign Affairs, and the Korean Centers for Disease Control and Prevention, which was renamed the Korea Disease Control and Prevention Agency (KDCA), during the pandemic—enabled an extraordinarily decisive response to the pandemic.

Source: Johns Hopkins University CSSE COVID-19 Data

Figure 5.4 — Confirmed COVID-19 Cases through March 16, 2020

*I Ching of a Thousand Doors* is on the 11th, and it will be the last one for quite a while. I set up at Chapter One Bookstore,[69] drawing a decent crowd. The *I Ching* is both a book for divination, similar to Tarot Cards in the West and a book of wisdom. For a book about divination, a demonstration is inevitable.

After reading some passages, I ask the audience what question to ask. The response is universal: "What will happen with the coronavirus?"

Having volunteered to help, Hollie writes the question on the flip chart. We cast "Turning Point."

> Travel goes away; the kings of antiquity closed the passes; merchants and strangers did not travel.[70]

I grab my pen to start signing some books.

On March 12th, the Governor of Montana, Steve Bullock, Democrat, calls a COVID-19 Emergency,[71] probably aware that the President is preparing a national emergency declaration, which the President signs on the 13th. On the 15th, Deborah Birx, with Doctor Fauci's support briefed President Trump on a plan to recommend a "layered" approach to shutting down the country for 15 days to flatten the curve. The President concurred.[72]

Governor Bullock announces two COVID-19 cases in Missoula, the city just north of us, and orders schools K-12 to close.[73] In a Letter to Subscribers, the *Ravalli Republic*, on page 2, suggests washing hands, maintaining social distancing, avoiding touching your face, and covering your mouth with tissue or sleeve when sneezing while also announcing free coverage of COVID-19 on the newspaper's website.[74]

On March 17th, the Ravalli County COVID-19 Incident Management Team (IMT) is formed. The Health Officer, Doctor Carol Calderwood, and the Director of Public Health meet with County Commissioner Jeff Burrows, who serves as Incident Commander.[75] Burrows also serves

as the chair of the Ravalli County Health Board; they soon add others to the IMT.

We are used to "Incident Teams" coming together during fire season most years, which used to begin in late July or August but has been creeping forward into June. Sometimes, the incident team works with an international group fighting wildfires, like in the Fires of 2000, the largest wildfire in the country that year. It will be a wild fire all right, the coronavirus, jumping every pulaskied[76] and dozed fireline.[77]

Doctor Carol Calderwood, MD, has been the Ravalli County Health Officer for over a decade, a part-time job that seems to call her up during occasional outbreaks such as H1N1 in 2009[78] or someone bringing a dead rabies-infected bat to school[79] or, in the face of a fair bit of controversy, prohibiting children who were not vaccinated for Whooping Cough from going to school during an outbreak it 2012. Montana Public Health awarded her the Elaine Fordyce Award for her efforts during the Whooping Cough outbreak.[80]

Carol is expressive when she talks, always in professional terms, using technical terms sparingly, often using her hands to give shape to a complicated explanation, always seeming to be willing to smile if given half a chance. COVID-19 will be a real pressure cooker for Doctor Calderwood.

Doctor Calderwood received her medical degree in 1989 and served in the US Air Force during her residency at Keesler Air Force Base in Biloxi, Mississippi. She practices at the walk-in clinic at Marcus Daly Hospital.

Jeff Burrows, sporting short-cropped red hair with a close-cropped mustache and beard, presents a forthright demeanor when running a meeting. A product of Hamilton High School, about 40, he earned an engineering degree. Jeff married a young woman who was two years behind him in school. He went into business with his father-in-law.

But, anti-government sentiment simmers in Ravalli County. Jeff got his start as a commissioner when the Republican Central Committee nominated him as the candidate to fill a vacancy on the Commission in 2012. When Jeff was interviewed for the position by the sitting commissioners, the then Sheriff, Chris Hoffman, and the undersheriff showed up to quiz him about his father-in-law's Bitterroot Survival Outfitting Systems Co-op blog. One of the commissioners had been alerted by an anonymous source to one of the blog posts. The quote in question was about a post from Jason Hommel titled, "American Cowards! (Like me!)."[81]

What caught the Sheriff's eye was his father-in-law's response to a statement by Hommel. Hommel blogged that the US Supreme Court has ruled that citizens have the right to resist unlawful arrests "using guns, if necessary!" asking,

> How many are willing to arrest an officer when they are breaking laws 'like aiding and abetting the irs (doesn't deserve caps!)'?

Of course, this isn't really true; the 1900 Supreme Court case in question simply reduced a murder charge to manslaughter[82] — and later decisions noted there were more peaceable ways to resolve the issue. So, if you try it here, you might get an all-expenses paid vacation to Deer Lodge, the home of the state prison.[83] What really bothered the Sheriff was the blog entry:

> Anyone who can honestly reply that they are not a coward at this time, please reply so we can arrange and opportunity for you to prove you are not.[84]

The reason is that the Sheriff had to deal with just that with our homegrown militia movement in the mid-90s, when a judge had to spirit her children to a secret location because of threats over a traffic ticket.[85]

Rather than push it, like Ruby Ridge, where law enforcement killed an unarmed civilian during a firefight,[86] Ravalli County took it slower — after some behind-the-scenes negotiations —

and a public show of support for the justice system in front of the Court House. The local leader turned himself in.[87]

The undersheriff recalled that they "dealt with that kind of thing in 1995," saying the officers live and die in this community to protect people. "This is a horrible thing to bring into this county."

Jeff Burrows replied, "I don't agree with this, I think it's crap," saying that even though he provided "technical" help for the website, he did not read everything on it and had not seen the blog post.

That's what the undersheriff needed to hear. Everyone was satisfied with his answer, and he was confirmed as Commissioner.[88]

It was a good pick. Jeff, along with the other two commissioners, my opinion, not everyone agrees, to be a breath of fresh air in dealing with the public — though there seemed to be a learning curve. In Montana, everyone has a right to speak before any decision is made and talk about any subject they choose before any public body — welcome or not.[89]

In this town, citizens insist on this right and the right to examine public documents — and will sue to enforce it.[90] Of course, it is not unusual for public bodies to limit time for even moderate crowds to one to three minutes. But even in a gym full of angry people in 2016, anyone who wants to have their say on the favorite wedge issue of the day, immigration, will have their say. The commissioners will sit through every one of them — 500 people, some holding up the Quran, citing sword verses in a raucous meeting, which I heard was pushed by a get-out-a-crowd campaign by someone in Bozeman. The crowd often tried to shout down people they disagreed with — threats were made, one instance being a group of grown men threatening a teenager who had calmly said his peace. Since then, it seems that the Commissioners have made more of a

**County Health Officer: 2020**
**Montana Code Annotated (MCA)**
**Powers And Duties Of Local Health Officers**
50-2-118. Powers and duties of local health officers. Except as provided in subsection (3), in order to carry out the purpose of the public health system, in collaboration with federal, state, and local partners, local health officers or their authorized representatives shall:
(1) make inspections for conditions of public health importance and issue written orders for compliance or for correction, destruction, or removal of the condition;
(2) take steps to limit contact between people in order to protect the public health from imminent threats,including but not limited to ordering the closure of buildings or facilities where people congregate and canceling events;
(3) report communicable diseases to the department as required by rule;
(4) establish and maintain quarantine and isolation measures as adopted by the local board of health; and
(5) pursue action with the appropriate court if this chapter or rules adopted by the local board or department under this chapter are violated.

MONTANA DEPT. PUBLIC HEATH AND HUMANT SERVICES REGULATION
37.114.307 QUARANTINE OF CONTACTS: NOTICE AND OBSERVATION
(1) If a communicable disease requires quarantine of contacts, a local health officer or the department shall institute whatever quarantine measures are necessary to prevent transmission, specifying in writing the person or animal to be quarantined, the place of quarantine, the frequency with which possible or known contacts must be medically observed to determine if physiological signs of the disease are occurring, and the duration of the quarantine.
(2) A local health officer or the department must ensure such contacts are medically observed as frequently as necessary during the quarantine period.

*Health Officer Authority 2020*

point of tamping down the shouting and clapping — in admittedly smaller crowds — we'll see if it sticks.[91]

The Health Officer brings her authority to the Incident Team — an authority that the Sheriff does not have, nor does the Governor — the right to order someone to be quarantined.[92]

The COVID-19 Incident Team meets every morning, with membership quickly growing to include emergency services, the Mayor of Hamilton, law enforcement, the County Attorney's Office, and others. They issue daily press releases. Today, there are still no cases and no business closures. "We are taking this one day at a time."

On the 18th, states begin closing doors to restaurants, gyms, schools, movie theaters, and concert halls — to keep people out. They then start closing doors to houses, apartments, condos, and resident homes, ordering a "shutdown" or "lockdown" — to keep people in.

It begins with California, first an emergency declaration, then a state-wide shutdown — state after state falls in line, but not Montana — yet.

On March 20th, the three Ravalli County Board of Commissioners, Greg Chilcott, Chris

Hoffman, and Jeff Burrows, all Republicans, declare an emergency, which they have to do if they want the emergency funds flowing down from Washington and Helena — but no shutdown like some other Montana counties.

The Commissioners clearly worry about a visceral reaction to business closures when they announce a special meeting. Board Chair Chris Hoffman says, "The purpose of the meeting is not to discuss or propose mandatory closures of Ravalli County businesses. Rather, we are discussing the costs associated with managing this emergency."[93]

Hope still reigns with business owners talking to the *Ravalli Republic*, picturing the owner of Nap's Bar and Grill,[94] known for great burgers and beer, saying, "We will adapt."[95]

But, some businesses didn't wait. I get an email saying, "Our beloved coffee shop is closed," the venue for The Table, Big Creek Coffee.[96] Big Creek announces they will return to window service in a "week to 10 days," taking the time to do some planned renovations.

Since I buy their coffee beans, I announce, "Big Creek Coffee Field Site Open for Business" in an email inviting people over to my back porch for coffee and cigars, figuring outdoors is safer, with much less risk of transmission for small groups with a breeze blowing.[97]

The virus can be transmitted outside, but the risk is much lower. Sometimes, a headline will pop up claiming otherwise. Still, I found no case studies or measurements of infection rates experienced in the real world. For example, one, "Landmark Study Finds Coronavirus Easily

Transmitted Outdoors," references a model of particle movement, not the interaction of a viral particle with a human organism; I discount that study to zero.[98] Outside in small groups seems a safe bet.

This gives me an idea: the Back Porch BBQ, usually on Sunday evening, is announced with a text to family and a friend or two.

From: James Olsen to...
Back porch BBQ, Sunday, 6:30 or so. Come if you can.

I crank up the smoker. The feast, mostly locally grown — typical spread: chips, avocado dip, pulled pork, smoked corn on the cob, butter, green beans, homemade potato salad, homemade BBQ sauce, water, red wine, white wine, plates, silverware, cloth napkins, water glasses, wine glasses, small whiskey glasses, ice bucket, the propane porch heaters if it's cold. And some good bourbon and premier cigars for those who want to indulge. Sometimes, it was just a couple of people, occasionally a dozen.

Good company keeps sanity. While there is a slight correlation with mandates, it is the community that counts — the medical literature suggests getting out and about during the pandemic reduces depression. Montana is faring well by one metric — families. In all but six states, divorce is going up and marriage going down in 2020. Montana does the opposite by a good percentage, fewer divorces, and more marriages.[99]

Even during the shutdown, we BBQ; the food seldom all gets eaten:

From: James Olsen to...
Left-over day. Tuesday. 7:00.

It doesn't take long. On Sunday, March 22nd, Marcus Daly Hospital sees the "first case of virus" in Ravalli County.[100] On the 23rd, I send a letter on "the nature of the data" to the County Commissioners.[101]

I think it is critical for decision-makers to know the limitations in case counts and why they don't reflect reality very well. The letter summarizes South Korea's response as a model to strive

for since they have had one of the best outcomes: massive testing, thorough contact tracing, and no lockdown needed. Some highlights:

*Perception is a delayed reflection of reality* – this is inherent in the problem. What is reported in the news is what is perceived — and what people respond to and act on. But what we are seeing is old information.

*Perception is an imperfect reflection of reality* — this is inherent in how you measure. Not everyone is tested; tests can have faulty results. To make matters worse, the information on COVID-19 cases has been driven by the availability of test kits and the rules for deciding to test a person.

*Errors in the information are inherent in the problem.* It is easy to measure deaths; it is currently impossible to measure how many people actually have COVID-19 since a very large number have mild or no symptoms and do not get tested. In other words, when t saw fatality rates, we were seeing an accurate number divided by a guess, which equals a guess.

The cluster effect... MIGHT BE TRUE,

- The virus is transmitted faster and more readily among clustered social groups...

- The virus is transmitted more slowly between groups...[102]

One will have to look all the way to Alabama to get a good briefing from Doctor Jeanne Marrazzo of the University of Alabama discussing cases, knowing where they are, and how many there are:

If you come to the ... emergency room, we're going to test you. If you are feeling a little sniffly, you're probably going to stay home and not be tested. Until we have widespread testing, it is and will remain a challenge.[103]

Foreclosed — house sold on the courthouse steps. The new owner does not want a lease and wants them out, so he hires Datsopoulos, MacDonald & Lind, a well-known premier law firm in Missoula,[104] who sends a letter ordering Peter Moore to get off the property in ten days.

They send a letter to Laurel Ferriter, giving her until April because she is a tenant and the house was funded with a HUD[105] loan. A tenant of a HUD loan property is covered by a Federal Law.

Laurel Ferriter and Peter don't have money for a lawyer. They have no place to go, as the Bitterroots have become a destination for COVID flight. There are no rentals for what they can afford, $800. There are only a couple of rentals in the county for any price.

Laurel is disabled and cannot get out of a chair without help. Peter is her caretaker. They are poor, especially after Peter was let go last year for missing work as a pesticide applicator for having one too many migraines. In due course, the house payments lapsed on the HUD-backed Deed of Trust.

The court hearing is on the 16th.[106]

It began when COVID began. The text just before Christmas wasn't a good start.

> Laurel Ferriter text:
> Happy Sunday! So, I found out the Monday before Thanksgiving our house was going up for auction on that Wednesday. Someone came by the house to ask if we wanted to sell b4 the auction... We thought the property hadn't sold but we finally got a registered letter... saying we should be out by the 12th. ... We have no resources. Counting on prayers for the moment.

>> Jim Olsen text:
>> Wow! I didn't know you were in arrears. If you want to make an appointment with Attorney _____, I will raise the money for his assessment and options. ... Are you in arrears because of a mortgage or taxes?

> Laurel Ferriter text:
> It's too much to text. Mortgage. Peter knew but thought HUD was buying...o talk to a trustworthy lawyer asap...

>> Jim Olsen text:
>> Happy day after Christmas. We went and watched the teenage granddaughters open presents. _____ and I are ready to meet. Peter indicated 2:30 at Big Creek today and _____ will be available as well as me. He should bring all of the paperwork including the mortgage papers.
>> ....

> Laurel Ferriter text:
> Sounds good! We're getting ready. Peter says we will be there as close to on time as we can! I haven't left the house since this started ~ afraid to leave the house in case someone is watching & they change locks on us!
> ....Do you think it's safe for today for me to come also??? ... What do YOU THINK???

>> Jim Olsen text:
>> Not sure. I am wading through the laws.[107] I assumed you were coming but maybe it would be good to have someone present on the property at all times.

> Laurel Ferriter text:
> Peter wants me along to fill in gaps...I guess I'll decide closer to leaving...he will check the mail & see if there is anything new...[108]

Laurel decides to come. Laurel was about to get in the car in the parking space next to their house when a car drove by. Afraid someone was watching them, fearful that they would lock them out, remembering when someone burnt a trailer, killing the cats she cared for in flames, she ducks down; she falls. Peter has trouble getting Laural back on her feet; a passing car stops; a kind person gets out and helps.

When they arrive, Peter walks in alone and tells us that Laurel's arm hurts. Dr. Binder was there, so we walk out to the truck. After a brief discussion and examination, Tim says something might be broken. Daly Hospital is a few blocks away, so off to the emergency room.

I follow a bit later. Several hours later, we emerge, Laurel in a sling to immobilize her broken shoulder.

I remember Laurel sitting on my living room floor, long flowing hair, making her pitch — a Super Two Highway with a bike path instead of the undivided 4-lane the Montana Department of Transportation (MDOT) had planned to upgrade the current two-lane road for our section of Highway 93. I noted that the first thing their group would have to deal with is the perception that Highway 93 was the most dangerous road in America, with many cars sporting bumper stickers that said, "Pray for Me, I Drive Highway 93."

I made a little speech I had used before that had scared more than one person off:

> There is no such thing as a campaign that doesn't last a year. Are you willing to put in the work?

Laurel was undeterred. She, Peter, and others were willing to do the work, so I helped them, agreeing to help them through the National Environmental Policy Act (NEPA) process.

It took three years, not one. Our lawsuit was lost, but the campaign was a success. Successful because MDOT added the bike path and engaged with citizen groups that came together to make the 4-lane safer. Citizens read highway design manuals, pressing MDOT engineers to move toward their design ideas in what were called "focus group" meetings.

When you drive from Lolo to Hamilton on Highway 93 today, you will see roadside landscaping, welcome-to-our-town signs, turn bays in Lolo, Florence, the Stevensville turnoff, and Victor, which citizen-driven improvements of the original plans. When you see the bike path along the highway, you see a citizen vision. As a plus, the Coalition raised money to bring an expert to town. He introduced a very reluctant MDOT to the safety and efficiency of roundabouts.

Laurel and Peter put themselves on the front lines, joining a group that put sufficient pressure on MDOT to engage the public instead of dictate.

Laurel is smart and learns quick. Growing up, starting in Sidney, Montana — my summary — it was a rough, impoverished childhood. As she was pulling herself up, she'd had a car accident and a brain injury — a setback that never left. Over the years, her health slowly degraded, often in pain. Peter is a mechanic and has a real feel for it to this day — though his health, lungs, and crushing migraines have gotten worse with time.

Peter, Laurel, and I attend Peter's hearing at the Ravalli County Justice Court, Department 1, Justice of the Peace Jennifer Ray, presiding. The judge has the paperwork ready. Peter and Laurel had both sent the court responses describing Laurel's situation, that Peter is her caregiver and must be on the property every day.

The judge did not have much to say about the information presented, making a statement to the effect that she had the unfortunate duty to give the property over to the new owner.

She signs the order:

> If the Defendant fails to vacate the premises within 72 hours..., then it is ordered that the Sheriff ... or any other peace officer... shall enter the... premises, by force if necessary, and physically remove each and every person detaining the premises and restore same to Plaintiff by *April 7, 2020 @ 11:59 pm* (handwritten).

So, we looked and looked for an affordable place to rent in Ravalli County, then Missoula, then contacted an agent in Butte who Bill Good knows, Bill having helped Laurel and Peter out before, to look for places around Butte and Anaconda.

Peter files an appeal to the District Court, providing $300 for the "sufficiency" stated as the minimum — creating conditions for an automatic stay of the eviction order. Judge Ray signs the automatic stay called for by Montana law.

The next day, Governor Bullock issues a directive suspending evictions.[109] This is a temporary relief — the directive includes Laurel because of her medical conditions. The case moves to District Court, which means we wait.

The Governor also seems to be trying to ease into a lockdown by first ordering restaurants, health clubs, and theaters to close.[110]

On the 28th, The hammer drops. As anticipated, as many predicted, Governor Steve Bullock issues a Stay at Home Directive, closing all "non-essential" businesses:

- Stay at home or place of residence.
- Non-essential business and operations to cease.
- Gatherings outside of home prohibited.

- Travel prohibited except essential activities.

Permitted: Health and Safety — supplies and services — outdoor activities — work to provide minimum basic operations —care of others — health care — human services — essential infrastructure — government operations — a list of business operations.[111]

I am bothered when I read it. I notice immediately when I read the shutdown order that religious gatherings are not listed as an "essential business," nor are political gatherings. All of the churches I know of had suspended services on their own, but that isn't really the point. These are protected by the First Amendment of the United States Constitution and the Montana Constitution, with even more robust language.[112] Indeed, an accommodation, or some rules about meeting size and distancing, could be made similar to other businesses that were "essential," which include, with some irony, liquor stores and real estate brokers.

Attempts to call and email Governor Bullock's staff get no response. With only a million people in the whole state, it doesn't take much in Montana to become known to politicians, so the Governor knows who I am. However, if he has an inner circle, I am so far away I may as well be *Shane*[113] riding for the far away hills. I have been to a couple of state conventions. I suspect they don't like to see me coming.

Mentioning that we are acquainted on a voice mail, I finally get a call back from the Governor's Assistant Chief of Staff, Adam Schafer. I tell him the order is not constitutional the way it was written and suggested a fix. He says he will pass it on. I don't hold my breath because I suspect they were trying to drink from a fire hose. I never hear back.

I send a message to the Governor:

Dear Steve, You need to amend your order to include the right to travel and gather for religious and political purposes - as the Governor of New Jersey did.

1) It is a 1st amendment right.

2) I believe in the right.

3) Call me regarding the interpretation by some of your latest order in Ravalli County.[114]

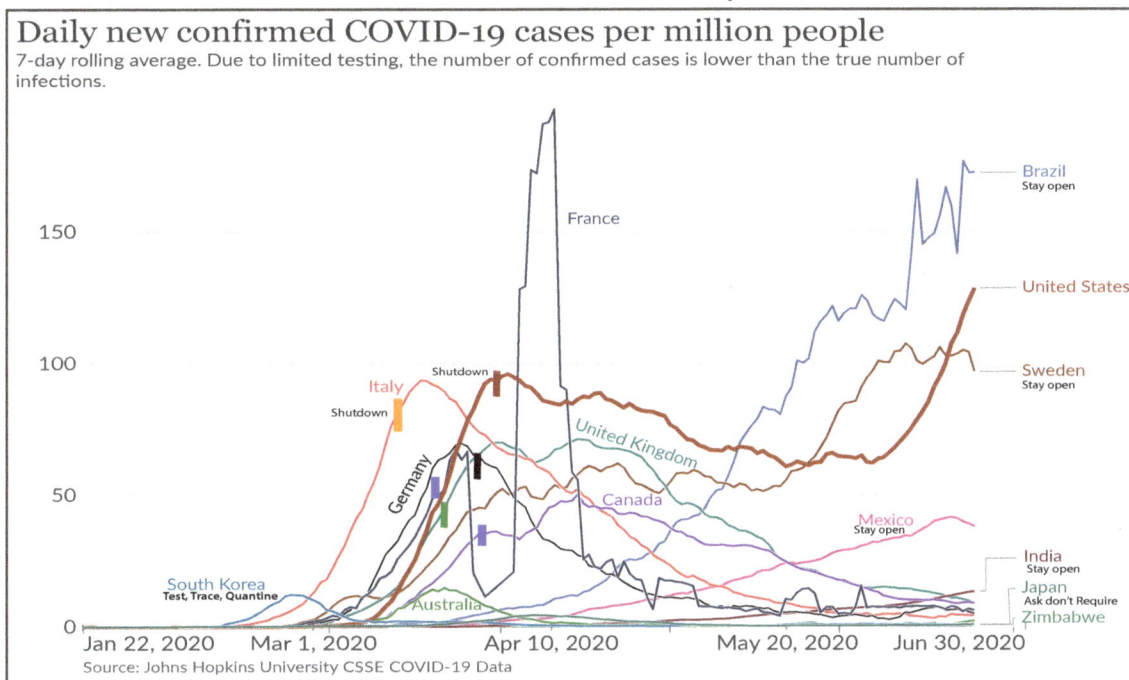

## Daily new confirmed COVID-19 cases per million people
7-day rolling average. Due to limited testing, the number of confirmed cases is lower than the true number of infections.

Source: Johns Hopkins University CSSE COVID-19 Data

*The World Experiments With Lockdowns*

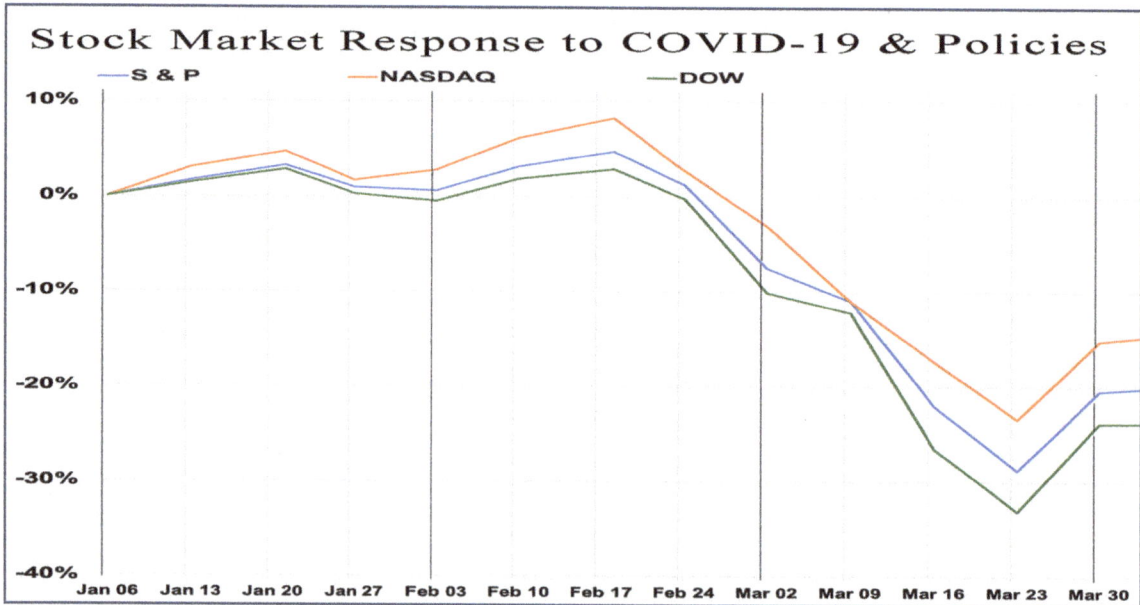

## Stock Market Response to COVID-19 & Policies

*Stock Indexes Jan. to March 2020*

Radio silence from Helena. The time will come when people in other states will rope that runaway calf and prevail in the Supreme Court.[115]

The Coronavirus is non-partisan. It continues its march across the country. More and more, the news and the literature are saying this is more than a lung disease, affecting the body in many ways, including loss of the sense of taste and smell.[116]

Amid the all-to-familiar partisan bickering in Washington, Congress passes relief bills, including the Coronavirus Aid, Relief, and Economic Security Act (CARES Act). The Act extends unemployment and orders directed payments of $1,200 for each adult and $500 for each child for everyone making less than an adjusted gross income of $75,000 a year.[117]

The stock market makes a prediction, as is its nature. By the end of the first week of March, most stocks are down 5-10%, then accelerate like a skier sliding out of control right down the middle on a black diamond slope at the Lost Trail Ski Area:[118] Will it recover?

By the end of the month, the market votes again. This time, voting with money to en-dorse the CARES Act, unemployment subsidies, and various other subsidies — voting that they are likely to save the economy. Stocks start to bounce back, and I managed to get in a little bit at the bottom.

Some pieces in the medical literature suggest that Chinese medicine and nutrition could be important.[119] Fang reports a message from a doctor friend:

> Right now there are many hospitals in which the entire treatment ward is managed by doctors who practice Chinese medicine, and ... achieved positive results.
>
> Of course, those traditional Chinese medicine doctors also employ some Western medicine and Western medical treatment practices.
>
> This mixture of Chinese and Western medicine... has also won... approval with... agencies.[120]

This comes on the heels of an email from Hillery Daily, who, much to my sorrow, had retired from her practice as a Naturopathic Doctor and Chinese medicine. I had gone to her for years, as well as to medical doctors. She is very good at diagnosing mystery diseases. She is not afraid to pick up the phone and call a Medical Doctor to collaborate.

The email says she made tinctures for immune support, a Chinese medicine formulation from an NIH website.

*Montana Botanicals*

Since we were in lockdown and she lived up the West Fork,[121] she arranged to meet on the street in front of the friend's house.

---

**CORONAVIRUS — EPISODE 2 - ELIXIR**

1 EXT: RESIDENTIAL STREET - NIGHT 1

   HILLERY in winter jacket.

   dimly lit beside a street light.

   car pulls up to curb.

   **DESPERATE BUYER**

   Hillery?

   HILLERY nods.

   holds up the tincture bottles.

   DESPERATE BUYER hands over cash.

   HILLERY puts tinctures in plain brown paper bag.

   hands bag to DESPERATE BUYER.

(The only thing missing are the final words)

      **HILLERY**

         Now get the hell out of
         here. You didn't see me.

            FADE TO BLACK

---

The first case in New York City, a city of over 8 million people, is recorded on the 1st. Mayor Bill de Blasio says people should go about their business. He is confident that the largest public health operation of any city in the US will contain the disease — he does complain to the news media that they have limited tests — by the 10th, they have enough testing capacity to satisfy the demand — hundreds of tests a day. He is frustrated that the FDA has not approved automated testing so they could do thousands per day.

Cases multiply. Within a week, the Mayor's hopes for a Saint Patrick's Day parade are dashed when he signs an emergency order closing down large gatherings and forcing 50% capacity for small venues. In response to a question at a press conference on the 16th, Governor Cuomo does the math for hospital beds needed versus the 50,000 they have, saying,

> You'll break into a sweat. You'll feel great anxiety. Panic attack. And you'll be right.[122]

By the end of the month, he is proven right.

As the news of the outbreak in New York City starts to look like a trailer for *Outbreak*,[123] no one seems to notice that amid the rush to get ventilators, a report from the Seattle area that 12 of 18 ventilated COVID-19 patients died.[124]

31 Mar. 2020: 5,359 COVID-19 Deaths Reported

| 6 | 7 | 8 | 12 | 13 | 15 | April 2020 | 24 | 26 27 | 28 | 29 | 30 |

Doctors, Civil Rights Groups urge CDC to release race and ethniciy COVID data.

Chicago Tribune: 68% COVID deaths in African American Communties

American Heart Assoc. of possible heart issues with HCQ & others caution

EASTER

US will stop sending funds to WHO

First $1,200 stimulus check deposited

New York: 1st Mask Mandate

Georgia, Alaska, Oklahoma partially reopen in spite of concerns it is too early

UK doctors report children with severe inflammatory syndrome

MT Lockdown over, gyms stay closed

Young, poor with symtpoms avoid medical care

Remdesvir shows promise

Project Warp Speed

## COVID-19 Treatment Guidelines[1] (NIAID COVID-19 Treatment Guidelines Panel)
~Not all recommendations shown. Specific protocols for childrean, co-occurring disease such as HIV and Kidney, testing and diagnostics not shown ~

• Does not recommend pre or post-exposure prophylaxis • Insufficient data to recommend for or against antiviral drugs

Ms Corona  by Mary Byers
Photo by James R. Olsen

# 7

# TRYING TIMES IN NEW YORK CITY

*Like art, political action gives shape and expression to the things we fear
as well as to those we desire.[2]
— Madeleine M. Kunin*

**APRIL 2020**

Her moniker is Melody, a seamstress's model who is sometimes fitted with a wedding dress of a bride-to-be or a witch's hat as October nears its end. She is now a Goddess of Caution made by someone who really can not afford to catch COVID-19. I walked out to the shop, where she stood, Ms. Corona 2020. On Earth Day, Melody is a hit at the 22nd annual Recycled Art Show at the Art City Artist Coop.[3] Mary is a long-standing "Art Citizen" and always comes through with a new piece of art every year. This year would be different: customers would have to look through the store-front windows.[4]

On the 1st, I get an answer about empty parking lots in New York City: the outbreak is "sadly true." It came with a link to *Slate Magazine*, which published first-person reports by two doctors who work in New York Emergency Rooms. My correspondent finishes the email by saying, "Tell them to stay the f *** home." The parking lots aren't empty; a host of white refrigerated trailers serve as morgues. Here is what New York doctors have to say:

> She has a meager PPE[5] allowance: one N-95 mask for the week and one surgical mask for the day.

> She sees less "normal" pathology. Patients come in with a broken bone have X-rays consistent with COVID. "Everything is COVID now."

"There's a gross lack of scientific guidance," so she does her own risk analysis.

"One after another, the presentations[6] are similar." A slight fever and body aches. The body aches get worse. Then, more coughing, shortness of breath, and more "severe respiratory issues."

"We're scared, too."[7]

They are short of ventilators. New York approves putting two patients on each ventilator.[8] They are short of ventilators. New York approves putting two patients on each ventilator. She arranges for patients to share the ventilator pump.

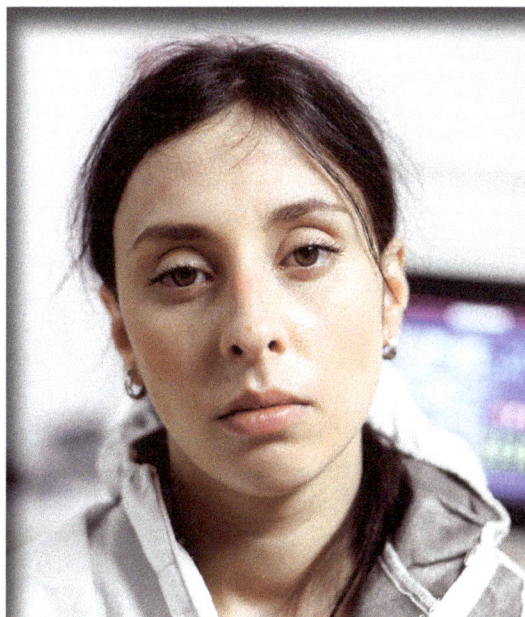

*Exhausted*

The white refrigerated trucks show up now so frequently she is no longer taken aback. "It feels inevitable, now." She notes that some now have shelves so the bodies won't have to be piled onto each other.

She worries about her patient's isolation. Not seeing their family, "the lack of touch." Patients see faces obscured by face shields and goggles. They hear voices muffled by masks. She laments, "I worry about my patients feeling as though they are alone."[9]

At the same time as the Empty Parking Lot meme, the opposite theme is overwhelmed hospitals "murdering" patients "because they don't care."[10] A video by a nurse goes viral and is pulled from social media, only to bounce over several counter-narrative sites, headlined as "these-are-the-true-facts."

Pick your poison, empty or full, but some practical critiques, a few of which are similarly pulled by social media, actually lead to better outcomes. In Wuhan, Italy, and New York, in particular, there are people, Doctors, and Nurses facing a new disease in a very high-pressure situation.

Some valuable nuggets are inside this confusing, often false social media. It rang a bell, what Doctor Cameron Kyle-Sidell says in his video about his experience in the Emergency Room (ER).[11] It disappears from YouTube; it is quickly picked up by other social media sites, including a post on "Take Back Your Power."[12] YouTube reposts it.

The Doctor argues that the Standard of Care in his hospital is killing patients. "I believe we are treating the wrong disease," he says. It is not pneumonia; it is more like high-altitude sickness. He is reading reports from Italy and attempting to figure out the best treatment for what he is seeing. The critical question for him becomes, "when to intubate."[13]

Another conspiracy theory? It turns out that Cameron is right. There is a problem.[14] Kyle-Sidell realizes that COVID-19 isn't like pneumonia, leading to a lack of strength to breathe. The patients can breathe but are still starved of oxygen. He isn't sure what to do beyond high-flow oxygen, but he knows the standard tube-down-the-throat intubation of someone lying on

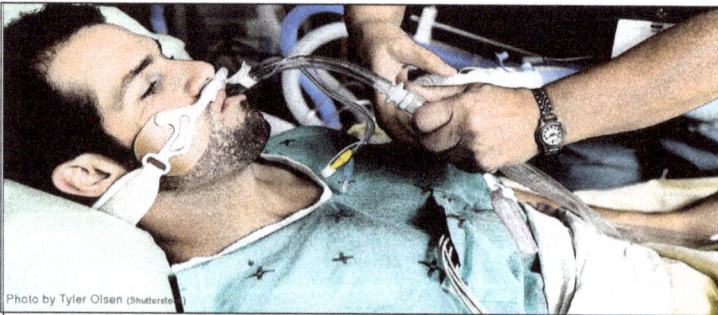

Photo by Tyler Olsen (Shutterst...)

*Intubation*

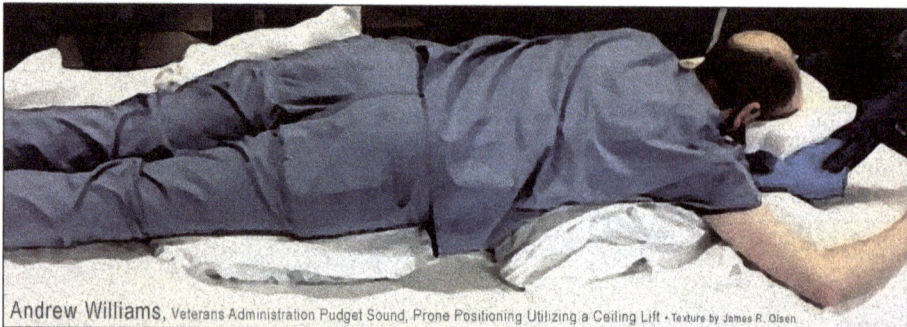

Andrew Williams, Veterans Administration Puget Sound, Prone Positioning Utilizing a Ceiling Lift • Texture by James R. Olsen

*Awake Prone Position*

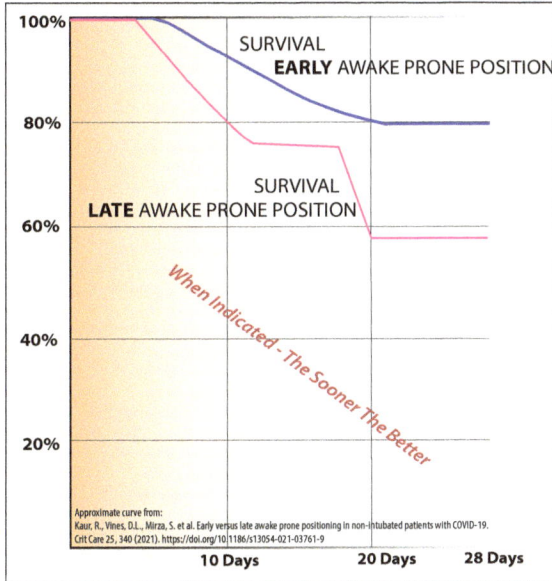

*Survival Rate Awake Prone Position*

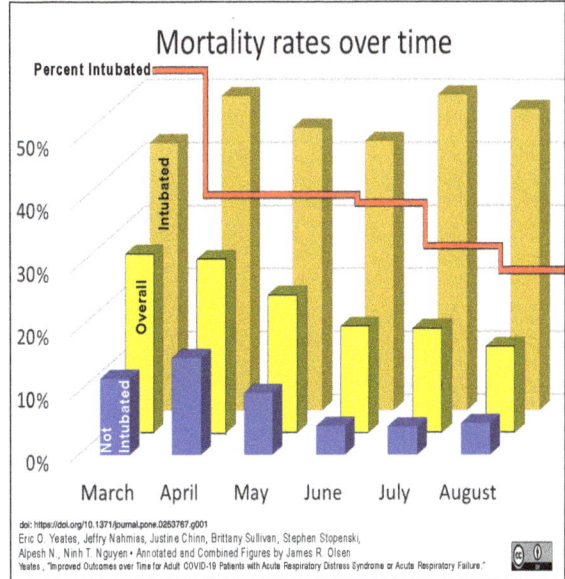

*Improved ARDS Treatment Over Time*

their back isn't working. Within a week, he is on Web-MD.[15]

The symptoms Kyle-Sidell is seeing are more like Acute Respiratory Distress Syndrome (ARDS). A syndrome means a combination of symptoms — there may be more than one cause. The benefits of the awake prone position for an ARDS induced by COVID-19 become evident. A 2013 trial published in the *New England Journal of Medicine* also showed prone position improving survival significantly for ARDS.[16]

Two conclusions will emerge:

1. There is little difference in mortality between intubating early or late.[17]

2. There *is* a significant difference in mortality for someone who can breathe on their own but is oxygen deficient *when* placed in the awake-prone position.[18]

It takes too long for this information to be sent to every clinic and hospital in the country. What is missing is a well-coordinated relationship between medical providers and researchers and between NIAID and the institutions determining the Standard of Care. Kyle-Sidell points out that reading the works of the "people who have come before us," the Chinese and Italians who experi-

enced the outbreak before New York, yields excellent insight

NIAID funds and conducts a lot of research. When NIAID makes recommendations, there is no well-defined, fast-reaction connection with clinicians. Doctor Fauci's remedy is to organize a COVID-19 Treatment Guidelines Panel, a panel of several dozen experts from the research and medical community who will issue Coronavirus Disease 2019 (COVID-19) Treatment Guidelines — a good idea, but still lacking the clinician connection.[19]

An unknown number of patients die who would have otherwise survived if the awake prone position had been used earlier — which becomes part of the standard care for COVID patients with low oxygen levels in most places as the year progresses.[20] This is backed up by evidence that mortality fell as improved treatment methods for COVID-induced ARDS are adopted. Researchers at the University of California at Irvine concludes,

> Additionally, we suspect that prone positioning, high-flow nasal cannula,[21] and a change in intubation triggers have also contributed to this noticeable reduction in mechanical ventilation rates.[22]

## Behind the Curtain in New York City——

New York City is a story of Wuhan revisited — but without the protective gear. Streets are empty, and hospitals are full. The doctors use pseudonyms "to write freely" in the *Slate* article because most hospitals pressure, reprimand, discipline, or fire an employee who publicly makes the institution look bad. This is another twist on the "code of silence" surrounding malpractice, an issue that goes back to the precode 1931 movie *Night Nurse*.[23]

Some American hospitals' behavior is eerily similar to the Chinese Communist Party in Wuhan; hospitals in the New York City outbreak area and around the country order their staff to keep quiet about conditions, administration missteps, and misdeeds. 30,000 employees at the NYU Langone Medical Center get this email:

> Speaking to the media without permission will be subject to disciplinary action.

> Please share positive and uplifting messages to support your colleagues and our organization.

While we decry a Chinese hospital turning their employees into propaganda ministers, the facts fit in America as well. To make matters worse, trustworthy information is missing inside the hospital walls. Adam Witt, "former nurse," notes that there is never an honest communication about supply status and how to deal with it. Instead, the CDC rules change almost daily with no scientific rationale — "Obviously, the rules are driven by supply, but they never say that." He comes to work every day, and there is a new standard.

"The moment you open your mouth you become a target." A Union Rep., Nurse Witt takes a day, as is the custom, to support another nurse facing disciplinary action for speaking out. That becomes the excuse given for firing him.

Nurse Diana Torres, who works at the Mount Sinai West Hospital in Manhattan, says,

> We've all been exposed... We don't know who has it. We can't get tested. I can't get my family tested, even though I have confirmed exposure. ... the people in Chi-

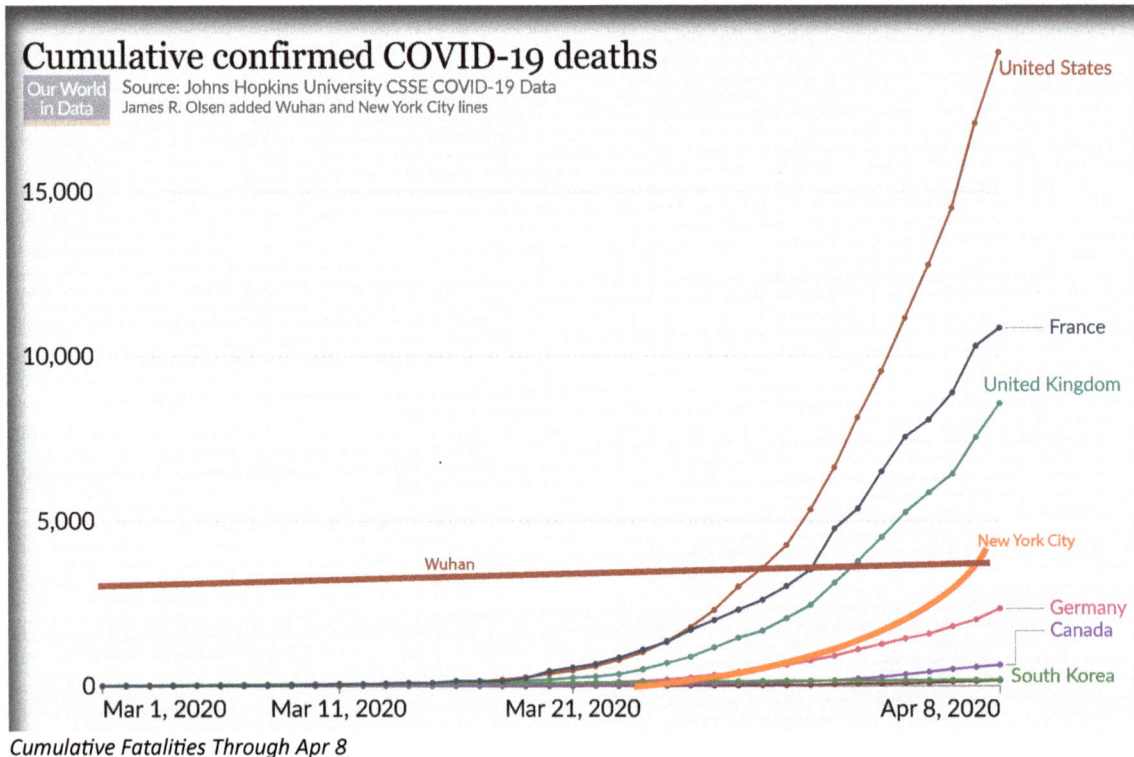

*Cumulative Fatalities Through Apr 8*

na... Italy... have better PPE than we do by far, and they still got infected.

If I make it through, I can do something else. If I am dead, there is no going back... we already lost one [nurse].

When they shut down Wuhan, Nurse Torres had asked what the hospital was doing to get ready. She asks, "What are the protocols?"

Her management responds, "Yeh, Yeh. Yeh, we're preparing."[24]

They weren't. It is only when the first patient arrives that they train the nurses how to don the PPE that they do have. She isolates herself from her family, coming home to live in the attic of her house for three months. She is accused of causing trouble and spreading rumors. Sound familiar?

Deciding to speak up, she says to herself, "I will probably be fired." It may not be a crime in the United States, but it can affect your livelihood — though Diana kept her job.

She stands on the street, the news cameras rolling,

> The CDC went as far as saying that if you didn't have a mask, just wear a scarf. What kind of agency are you? How can you have any credibility? How can you expect we will believe you or trust you?

Diana is not alone. Dawn Kulack, "former nurse," says the same about the CDC. Dawn holds up a supposed medical mask they gave her that looks like it is made from tissue paper. She brings in her own N-95 mask and gloves, having promised her son she would protect herself and her family. She is ordered to take them off. She speaks up; she is fired.

Tears run down their cheeks when these nurses talk about their patients, as they watch them die,

> We didn't know what to do. We tried our best, and it didn't make a difference for a lot of them.[25]

While in New York City, COVID is going exponential, the stay-at-home order in Wuhan is over, with around 3,800 reported deaths, though more likely tens of thousands. Regardless of the cost, the draconian Wuhan shutdown appears to have run the curve to near zero — but it has taken armed guards at the entrance to the city, taken volunteers bringing food to houses, masked and gloved, taken a surge of outside help, with tented medical clinics sprouting like white mushrooms in the parks and parking lots. China has implemented a Zero-Covid policy that will continue for two years.

Fang Fang finishes her blog about the city where she grew up, a city of 11 million souls, with a hope and a lament. The hope is that the professional doctors practicing allopathic medicine, which she calls "Western" and traditional medicine, will work together to solve the problem.[26] The lament is that Fang has collected trolls:

> Vile and lowly as they may be, for the past several years now, these ultra-leftists have been gradually spreading throughout our society, just like the coronavirus.[27]

The Wuhan shutdown was hard to sustain even in China, a country ruled by an autocratic Government. Evidence shows that shutdowns reduce fatalities but come with a price, as they encourage political unrest, challenge mental well-being, and interrupt economic activity.[28] Now, it is America's turn.

Like most communities, people are working through COVID-19, often as a community response. The headline in the first issue of the *Ravalli Republic* this month reads, "Bitterroot Valley residents work together to face pandemic crisis."[29]

The ever-present Cornerstone Church[30] provides a meal. Those who are on the edge of sleeping outside navigate the intricate network of churches, non-profits, and private citizens. It is well known to people staying at the Big Sky Christian Center who show up at The Table —

there is no other place for a single male in the Hamilton area who needs a long-term bed and doesn't have a rental or place to park a trailer — both of which are getting scarce and expensive for working people. Triple-Creek,[31] a resort up the West Fork, which has won awards for one of the best resorts in the world, steps up with their cooks to provide sandwiches to Cornerstone.

Volunteers multiply. A woman from Florence organizes mask-making and volunteers at the church feed. I know one of the cooks and am tempted to go myself because he is one of the best chefs in town — a real feel for food that would make him a champion on *Chopped*.[32] I don't want to encourage more COVID-flight, but we have more than our fair share of excellent chefs.

The coffee shops are closed, like most businesses. Some types of businesses are allowed to stay open. Other businesses are not "essential" according to the Governor's order.

*Ford's Department Store Sign.* Photo by James R. Olsen

Alan Ford owns Ford's Department Store on the corner of Main and 2ⁿᵈ in Hamilton. Allan has an unassuming manner and a voice that resonates with a friendly reserve. He describes it as a family department store, saying it's "a little bit like JCPenney® from way back," clothes, work-wear, western-wear, kids, women's, "and we try to do something of domestics, pillows, sheets, towels."[33]

A retailer at heart, Alan exhibits a trait that cannot be duplicated by a big box store or online retailer — he knows his customers, and his customers know him. They closed for four weeks. "It was good to find out what it would be like to be retired," not having to show up in the morning, though still "doing stuff" about four hours a day, getting deliveries and things like that.

A product of Hamilton High School, Alan says, "there was a lot to do here" — hunting and fishing when he was growing up. While his brothers moved away, he always wanted to live in Western Montana. Now, in his late '60s, he recently married for the first time — who he describes as "a younger woman" — "she works Saturdays."

A couple of weeks after the shutdown started, Alan leaves the door unlocked while not turning off all the lights. "I did a little before" the shutdown was over. "All you have to do is sell dog food (and you can open) to avoid the whole thing," noting Ace Hardware and Murdoch's are open — the Governor's order allows sellers of farm goods and building supplies, though "they have more customers." Alan is glad he didn't go through what his brother did in Portland, being shut down for a long time.

His CPA gets him to take advantage of the pay-day savings. As soon as Ford's opens up, there is a pent-up demand. None of the help gets sick. Ford's has a good year in 2020.[34]

Mara Lynn Luther co-owns Chapter One Book Store. She knows her customers, and her customers know her. Chapter One Bookstore is indie-bound.[35] Chapter One is a Texas Longhorn, a rare breed, but still hardy even in lean times — keeping the rustlers at bay, first in the face of the '90s *You've Got Mail*[36] big box chains, then Amazon.

Chapter One does well in 2020, better than since 2008. Mara Lynn speculates that people have more money and are not spending it on other things, though she is unsure whether they read more. "Our downtown association had the money for us to pay our employees when we were closed. People saw that their favorite businesses were struggling. If you care about something you make a real effort to come to it."

The legacy of the bookstore began as an enterprise in an 11 x 90-foot nook at the Roxy Theater on 2ⁿᵈ Street — founded by Russ Lawrence and Jean Matthews. You can see the Roxy sign in the opening shots of *Yellowstone*,[37] Season 4, Episode 1, with the siren-wailing police cars responding to a bomb explosion at the end of Season 3. The Roxy opened as a movie theater in the 1930s and was still playing Hollywood favorites when I moved here, but has since folded up the silver screen.

The ownership of the bookstore has changed, but not the legacy — first, the person works there; you meet them behind the counter; then, one day, you realize they are an owner. Shawn Wathen worked at the store in the '90s and later bought it. In 2010, Mara Lynn, having worked there for five years, partnered up with Shawn. They will sometimes buy a book and put it on their shelves because they know one of a customer who is interested in

*Photo James R. Olsen*

*Roxy*

it — I am sure I have been the beneficiary of that idea. "They'll either buy it or not, but we will do that. And we have the ability to do that as an independent bookstore," said Shawn to a reporter.[38]

Mara Lynn notes that they are not afraid to put what they think on their window—banned books have been a standard; a rainbow appears, slowly doing a shape-shifting glissade across the window; slogans pop up as issues get hashed out in the public square.

During the pandemic, Mara Lynn is, in the words of a fancy management consultant, "agile." They do curbside like a lot of outfits but also deliver. They even hand out books as people point at one through the window, doing everything they can to make their customers comfortable and safe. So, maybe it's not that much of a surprise when Mara Lynn says, "People came in and handed us their $1,200 check, saying we 'want you to be open on the other end.'"[39]

*Chapter One Bookstore*

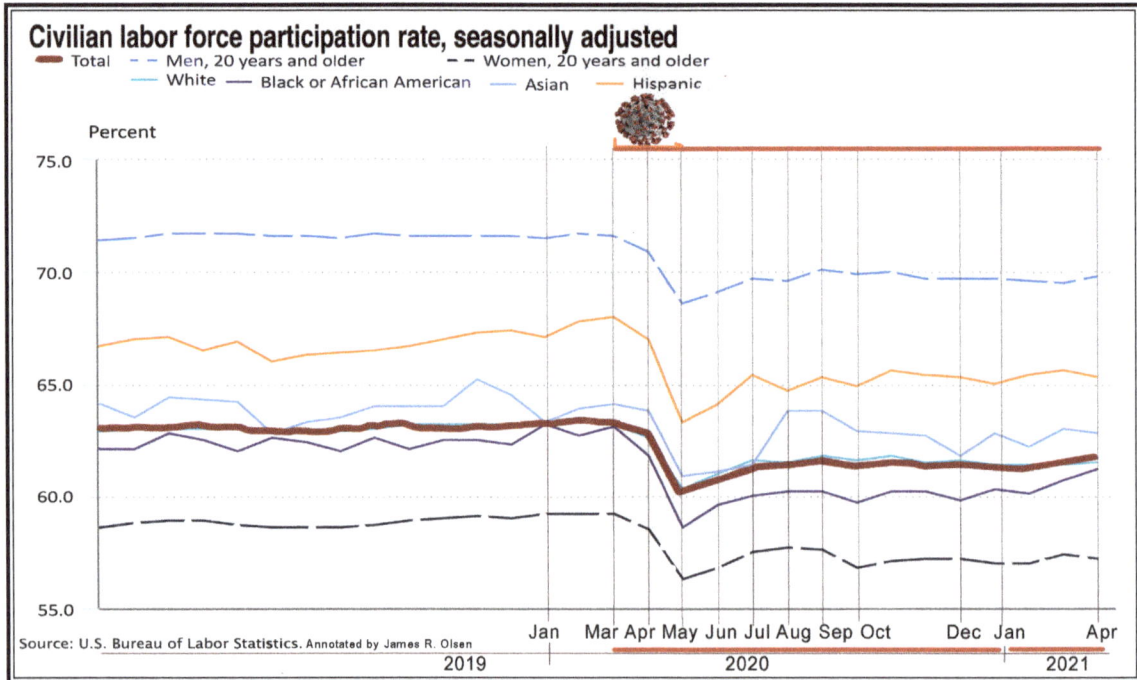

**Civilian labor force participation rate, seasonally adjusted**

*Labor Dip 2020*

I draft a letter to the Health Board ahead of the meeting scheduled for April 8th. They must be tired of hearing from me by now, but I send it anyway, as its importance outweighs the risk of being considered a gadfly. The letter includes a section on food supplements and some treatments being written about in the medical literature:

> Get Healthy... – both the NIH and Naturopathic Doctors agree. Supplements are recognized by the NIH are: Vitamin C, Selenium, Vitamin E, Zinc, avoiding B-6 and folic acid deficiency, D-3 (esp. respiratory disease).[40]

Regarding vax mandates, I know the kettle is already heating up. So, in for a penny in for a pound, I provide another input to the Board:

> The right to consent to treatment and preventative medicine... the risk/reward choice is up to the individual, not the provider. Most people accept the side effects risk — but it is their choice, not the provider or the Board of Health.[41]

I don't expect to get any response from my letters, as that is how it often works when sending input to government agencies. It's like putting

super glue on unwanted mail. They may want to throw it out, but it sticks; the letter "is on the record." My previous experience is that, over time, a series of inputs often affects decisions made. I can also reference them when pressing a point at public meetings.

We are on lockdown, so the County has closed their building to the public. The Health Board can include the public because they meet in the Commissioner's Meeting Room, in the Ravalli County Administration Building, the old hospital building, on 4th Street in Hamilton,[42] where several years ago they installed a system of cameras, mics, video screens that feed a real-time Internet feed. The system is called Granicus.[43]

The public faces the five-member Board, the members sitting behind a semicircular table usually occupied by the Board of County Commissioners.[44] The room is on the second floor of the County Administration Building, a historic brick building built as a hospital in the 1930s, named Marcus Daly Memorial Hospital.[45] The hospital and Marcus Daly's name moved to a new, modern hospital in 1975, which serves the entire County.

I plan to watch at home. April 8th arrives; it is time. Commissioner Jeff Borrows, the board chair, calls the April 8th meeting of the Ravalli County Board of Health to order.

The three County Commissioners appoint the Ravalli County Health Board members. While state law gives some flexibility regarding how it is organized, a County Commissioner chairs the Board in our County. Commissioner Jeff Burrows is the chair.

The Health Board spends most of its time with wastewater permits and waivers. They worked for a couple of years on an update to the local regulations, accompanied by some controversial public input. The Board has dealt with local outbreaks, seasonal flu and whooping cough, with most of the actual work being done by the Health Nurse and staff, who often work with the schools.

Then came COVID-19.

The members are County Commissioner Jeff Burrows; Nurse Katie Scholl, RN; Doctor Michael Turner, MD; Doctor Wayne Chilcote, MD; and Roger DeHaan, PE.

Nurse Katie (Katherine) Scholl moved to Florence almost two decades ago from Florida; her husband is a retired firefighter. Katie has worked in emergency rooms and testified as a forensic nurse in court. She expressed a high degree of motivation to be involved in the community when she and her husband donated defibrillators and training to the Sheriff's Department. The newspaper article shows Sheriff Chris Hoffman (now a County Commissioner), Undersheriff Steve Holton (now the Sheriff), and a patrol lieutenant each holding a new defibrillator.[46]

Doctor Michael Turner is around my age, in his 70s, and listed as an orthopedic surgeon practicing in Hamilton. I heard that he is a veteran.

Doctor Wayne Chilcote is a radiologist practicing for over 50 years.

Roger DeHaan lived here when I got here — my wife has been backcountry skiing with Roger and his wife more than once. He is the Professional Engineer on the team, a civil engineer who designs septic systems — the expert for what the Health Board does most of the time. A liberal outlook, he has run for local office as a Democrat in the past.

Doctor Calderwood is proud as she begins her prepared brief on the Ravalli County COVID-19 IMT (Incident Management Team) to the Ravalli County Health Board on April 8th. The Incident Team had been working daily for three weeks, setting up after the first case in Missoula, "the big city" 45 miles north of Hamilton.

With the announcement of the first case in the County in March, the question is, "Can it be contained?" After a brief introduction about when the team formed, Doctor Calderwood goes on to say,

> I am proud ... the team. And I'm not exaggerating when I say, "I believe I could hold them up against teams of large communities..."...
>
> I am also very heart-warmed by the sacrifices of the public ... especially regarding preservation of our healthcare capacity.
>
> I would like to read our intent:
>
> To reduce morbidity and mortality for COVID-19 in Ravalli County by implementing current best practices, to communicate, educate, reduce viral transmission, isolate patients with suspected or confirmed COVID-19, quarantine or monitor people based on likelihood of prior exposure, and to support the delivery of the health care infrastructure...[47]

What a great start for a part-time Health Officer. Now, it seems the Health Board will jump in and help marshal the resources to round out what has started. Presumably, they are suitably impressed. I am disappointed.

After a pause, in a raised-pitch voice that sounded like a question, Carol says, "Ready for questions." This will probably be one of the most thoughtful meetings of the Health Board during

2020 in that the members truly explored proactive ideas they could do to protect the community. But it got off to a bad start.

In response to Doctor Calderwood's statement that she is ready for questions, Katie asks her what she wants from the Board. Calderwood says, "That's really up to the Board," noting an overlap between the Board's duties and hers — which is true. Katie notes that the communication with the Board "was slim."

Doctor Calderwood responds with a narrative that has the team going at breakneck speed.

The Doctor Turner responds,

> We are responsible for the health of the community. That's what we are by state law. It's not the County Attorney, the Sheriff, or anybody else.

That's not entirely the case — the duties and authority of the Health Board and Health Officer overlap.

The tension rises. What is probably bothering Doctor Turner is an "example" of something that was to be brought to the Board, an email that said "it has been decided." That is decided without leave from the Health Board. Pointing around the room, "None of you are on the board."

Doctor Calderwood notes that they needed some tools (some enforcement mechanisms). She could do it, but she wanted to bring the Board in. But then she realized it had to be on the agenda for two days, so she just decided to do it — she indeed has the legal authority to impose civil penalties using the courts.[48]

Jeff Burrows says he was the one who put it on the agenda and then took it off. The reason he pulled it

was that it had gotten attention and could well become very contentious.

I had already gotten an email on the 4th from someone upset about the agenda item, urging me to show up:

> Discussion with possible decision on adopting Governor Bullock's orders to allow criminal citations on non-compliance; Appointment of back up Health Officer; Adjourn

Burrows undoubtedly got some harsh feedback and took it off the table. Criminal citations in Ravalli County are simply asking for trouble you don't need.

It becomes a poker game, the stakes being who's in charge.

Katie says, "We need all of the information," demanding minutes be sent after every meeting. The ante is in, and the opening bet is made.

Doctor Calderwood struggles to figure out how to do daily minutes for the Board without the staff and with more important things to do. She's about to fold.

---

## Local Health Boards: 2020

**Montana Code Annotated (MCA)**
**Powers and Duties of Local Boards of Health**
50-2-116 Powers and duties of local boards of health.
1) In order to carry out... in collaboration with federal, state, and local partners, each local board of health shall:

...

(f) identify, assess, prevent, and ammellorate conditions of public health importance through:
   (i) epidemiological tracing and investigation;
   (ii) screening and testing;
   (iii) isolation and quantine measures;
   (iv) diagnosis, treatment, and case management;
....
   (vii) collecting and maintaining health information;
   (viii) education and training of health professionals; or
   (ix) other measures allowed by law.

...

(i) bring and pursue actions and issue orders necessary to abate, restrain, and prosecute the violation of public health laws, rules, and local regulations.

...

2) Local boards of health may:

...

(c) adopt regulations the do not conflict with 50-50-126 or rules adopted by the deparment:
   (i) for the control of communicable diseases

*Health Board Duties*

Burrows raises. It is as much of a challenge as an invitation; Burrows asks if the board members want to attend the 8:00 AM meetings.

No one takes him up on the invitation — although I think Katie may have attended some of the meetings. Though not responding to Borrows' invitation, the challenge dampens their ardor. They fold.

The meeting gets more productive. The discussion gets into testing: 177 tests, three positive.

Then the health nurse discusses contact tracing and how time-consuming it is for each case — interview the person, then interview their contacts, go back almost two weeks, and record any place they stopped or got groceries. After all of that, "He's in quarantine and his close contacts are in isolation."

Doctor Calderwood is happy to report that we only had two more cases, and the contact tracing is complete. She calls this "containment." She hopes that a community spread, not knowing how the person got the disease, can be held at bay, even though Missoula already had a few community spreads. People in Ravalli County work, commute, and shop in Missoula, with stores such as Walmart and Costco that we don't have up the Bitterroot.[49] Since testing is so limited in the County, there is no way of knowing if there is community spread until someone with symptoms goes to the clinic and gets tested — COVID-19 might have been spreading for weeks before that happens.

Concerned that the staffing will become overwhelmed, Doctor Turner asks about the health nurse's plans if they have another five cases tomorrow. She has 8 or 9 people on call. But what about the 10 cases or 20 cases?

Doctor Calderwood talks about Marcus Daly Hospital preparing for a surge — she thinks they are doing great. Still, she is not privy to the details. They have their own incident team.

The COVID money from the state is on a reimbursement basis. What a way to run a railroad. The County doesn't wait, though. The County Commissioners fund a PCR test machine for the hospital, which is slow in coming because they are delivered on a priority basis.

Then, the Board discussion devolves again, going on for an hour, to how to advertise, message, and what type of media to use. The Board seems to be avoiding their primary job. My frustration is rising. Enough — I call in, saying,

> I wonder why the Board didn't take control of this three weeks ago. They should roll up their sleeves and get the job done... I don't think the team has enough resources... What I see happening is you're thinking of *what you have* instead of *what you need...*
>
> You have the authority to integrate the hospital into your team... You are in charge of that. *You said it.*[50]

I suggest that the Board integrate the medical community into a coordinated response to COVID-19, noting that the community had done this before for mental health crisis intervention.[51]

I was pleasantly surprised when Doctor Turner and Katie agree that integrating the hospital and county teams was worthwhile. Turner notes that if they don't, the Board has the authority to order them to do it. It ends up with a commitment to make a phone call.

I never learned why nothing would come of it. The forceful language is probably met with a hospital administration with a tough hide. It is destined to be *All Hat No Cattle*.[52]

Hollie calls in, shifting the conversation from disease to health,

> With all of the advertising you're talking about, why not put something out there about building up your immune system, especially for those who are immunocompromised. What about actual health?[53]

After discussing the water-solubilty of vitamin C and fat-soluble D, the Board moves on to money.

But, Hollie lit a small flame under the steam kettle of healthy living. The kettle will slowly build steam in the coming Health Board meetings until it threatens to boil over.

Letters and emails begin to flutter onto the field from the grandstand, a notable one from Iiona and Kristina Bessenyey, descendants of Marcus Daly, with a gracious offer:

> These are certainly risky and difficult times. This email is to let you know that if there is something our family or the Bitter Root Stock Farm can do to help the County during the pandemic, please do not hesitate to reach out to us. Thank you for your leadership during this unprecedented emergency.[54]

On the 22nd, a Phased Opening plan is announced for the State of Montana. The phases are much like the rest of the country. Phase 1 will commence on the 26th with a partial list of types of businesses and churches, conditioned on imposing a six-foot separation. Outdoor activities are to maintain separation and sanitize frequently.[55]

The directive to reopen comes as a surprise. The Health Board holds a special meeting to determine what to do with the Governor's directive. The Health Nurse reports that only five people have positive tests in Ravalli County; she's getting one or two daily test results. The hospital recently opened a respiratory clinic where people can get tested if they have respiratory symptoms.

Doctor Carol Calderwood gives a summary of the new Phase 1 opening directive. They had been working on a reopening plan, but the Governor's plan is pretty complete. The directive says that the Board can only do something more strict.

On a remote feed, Roger interjects to ask if the County is meeting the Governor's stated requirement for "Core Preparedness Responsibilities." They flip to the slide from the Governor's directive. It includes:

- Ability to test symptomatic people for COVID-19 and trace contacts for COVID+ results.
- Ensure sentinel sites are screening for asymptomatic cases and contacts for COVID+ results are traced...

While Katie says that Public Heath has been "taking care of that," it is unclear if "that" includes the second bullet.

The two MDs on the Board begin thinking about the possibility of increased infections. They agree there would be a "3 or 4 week" delay, giving the disease time to spread. No one on the Board knows if the spread will stop, even if everything stays closed: "we just don't know enough about the new disease."

Burrows is thinking about the impact on people's pocketbooks, observing that it is arbitrary that some businesses were allowed to open and some not, the "nots," including hairdressers and gyms. He says if some science is behind it, that would be okay. Of course, there is no science. Burrows implies that who gets to open has some politics behind it — he is probably right.

The next day, the Board meets again, having had time to digest the surprise directive from the day before. Holding his hands up, Burrows finishes with a challenge,

> Did the IMT discuss the legitimacy of the order as far as a rationale for why this business versus this business? If we adopt this, we better be willing to go to the public and say, "There is no rationale to it. We just did it."

Doctor Calderwood responds that the IMT team did discuss it. She would like to hear the science behind the different treatments of different types of businesses. Health officers from around the state are trying to call the Governor to answer these questions.

There is no science in deciding which businesses to close. Short-term shutdowns seem to correlate with slowing the disease and tamping

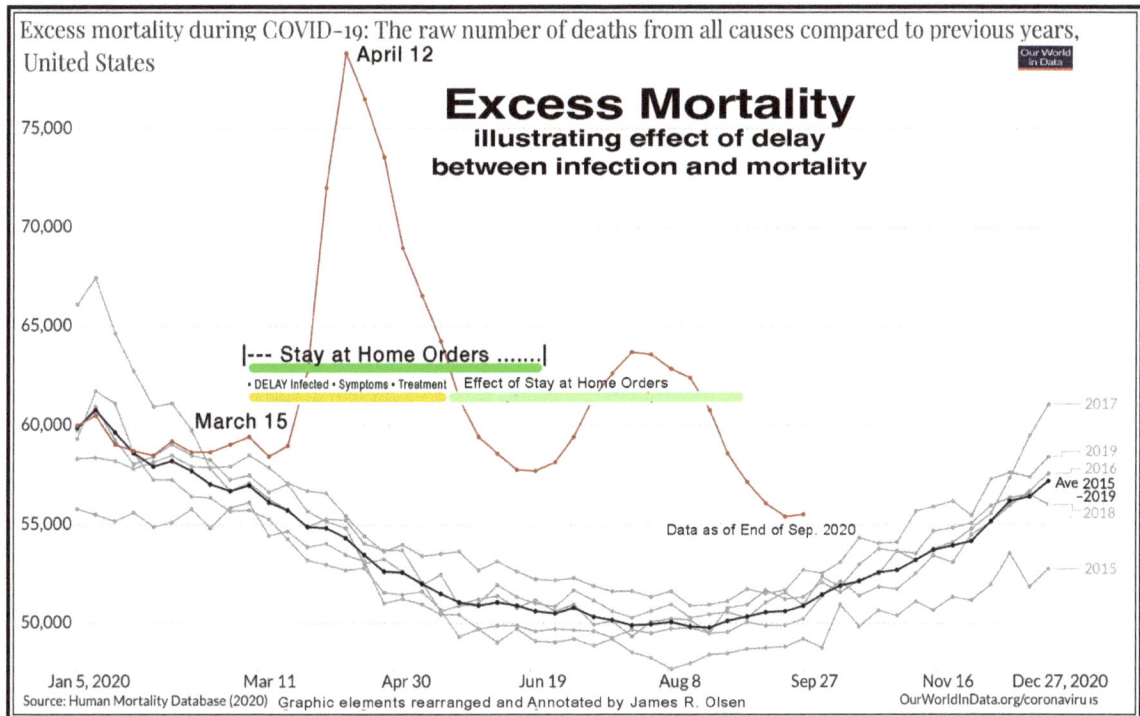

Excess mortality during COVID-19: The raw number of deaths from all causes compared to previous years, United States

# Excess Mortality
## illustrating effect of delay
## between infection and mortality

*US Excess Mortality vs Stay at Home Orders*

down fatalities, though some scientific papers say yes and some no. However, excess mortality data provides strong evidence that the answer is yes. The economy can probably handle a week or two, but some states keep the lid on longer — the economic and social consequences will linger.

Commissioner Hoffman comes up to the mic and reports that Governor Bullock returned his call. The Governor's bottom line is he got more flak about reopening than not reopening. There is no science; he is simply "trying to space things out."

Bowing to Governor's rationale of pressure politics, the Health Board does not adopt the order. Instead, they "acknowledge the order."[56]

On April 27th, the lockdown is over by the authority vested in Montana Governor Steve Bullock.[57] It appears that it did flatten the curve of excess deaths.

I get a call from the Market Master of the Hamilton Farmers Market. We already planned to open the market on the first Saturday in May in anticipation of the end of the lockdown.

This is my fourth time on the Farmers Market Board. It is a great market, with free music from local musicians, selling food and crafts, in downtown Hamilton, three blocks south of Main. Over the years, it has become the place to meet old friends, an echo of the promenades around the public square of old.

The market has grown, now delivering around a half million dollars in total sales every year in a city with a population of 4,800. The market opens on the first Saturday in May. Of course, we must get a street permit from the City of Hamilton. The Mayor wants the Market Master to enforce six-foot distancing. I say to tell the Mayor it is too dangerous to ask our employees to do it; if he wants someone to enforce it, he needs to send someone.

We have dealt with intimidating, angry people on a few occasions. I spent most of a season a couple of years ago dealing with the controversy of an AR-15 auction at the Republican booth. Though not unusual at the County fair, the weapon was set up on a tripod in close quarters. It ended up in an open hard case. Representatives from the Republican Central Committee met with the Farmers Market Board and were very helpful. No real problems, but a lot of work.[58]

No, there was no way we're going to get our staff involved in enforcing a Government mandate that makes a good number of people livid. With no way to manage it, we posted signs around the market asking customers to "maintain their distance so we won't be shut down" — it worked well enough.

The Mayor is a nice guy when you meet him. He is probably simply dealing with some pressure since, compared to the County, the City Council is more liberal. The market does have a trump card — no one is going to wipe out $500,000 worth of business without paying the political piper. The Board decides to open cautiously with food vendors only; we will have our own "phased opening."[59]

It is poor policy, turning to working people to police the mandates coming out of Helena,[60] demanding employees, the people behind the counter, to be their police force. This policy is already driving the political divide and setting neighbor against neighbor.

Herd immunity enters the common vocabulary. Not surprisingly, "herd immunity" started with a herd, a herd of cattle, and a disease now called brucellosis.[61] Brucellosis causes cattle to abort fetuses. A brucellosis-free herd is a big deal for beef producers worldwide.

In 1918, herd immunity was shown to be one of the tools to control the disease in livestock when there was no cure. If a cow gets the disease, you let them live but refrain from bringing in cows from other herds. As the infection spreads through the isolated herd, the disease will find itself without any viable hosts, as enough cows become immune for long enough — herd immunity.[62] Later, veterinarians observed that herd immunity was achievable if a vaccine is administered.[63]

Brucellosis is a familiar word in Montana. Years ago, cattle gave brucellosis to wild elk and bison. The elk have returned the favor. The Yellowstone Bison Herd has never done so.[64] Of course, elk are exposing cattle to brucellosis, so a fundamental principle of herd immunity is violated unless you shew away the elk or kill them off — not going to happen to our favorite game animal. So, the control of the disease depends on vaccination, testing, contact tracing, and isolation of the sick.

That would seem to be the end of it. But politics steps in. There is an ongoing, decades-old battle in the Yellowstone Park region. Yellowstone Park hosts the only continuously wild bison herd. The bison have not been taught to read a map, so they wander off the park into their traditional local feeding grounds.

Using brucellosis as a rationale, the size of the bison herd has been dictated in the Park Service Bison Management Plan. This may sound nice on paper until it gets implemented. In the *Yellowstone* series, the Department of Livestock agents are an extension of the fictional Yellowstone Dutton Ranch. In real life, the department agents got the assignment to cull, capture, slaughter, hunt, and haze bison that leave the park. The hazing is done with motor vehicles and on horseback. The conflict is as much about which critter gets to eat the grass as it is about disease control.[65]

When Doctor Fauci wrote about herd immunity in 2019,[66] he used a graphic that has been rolling around for several years at the NIAID.

Despite flu vaccines, the flu comes for a visit every year. Doctor Fauci's hope to get nearly everyone from 6 months and older to get a flu shot is not going to happen.

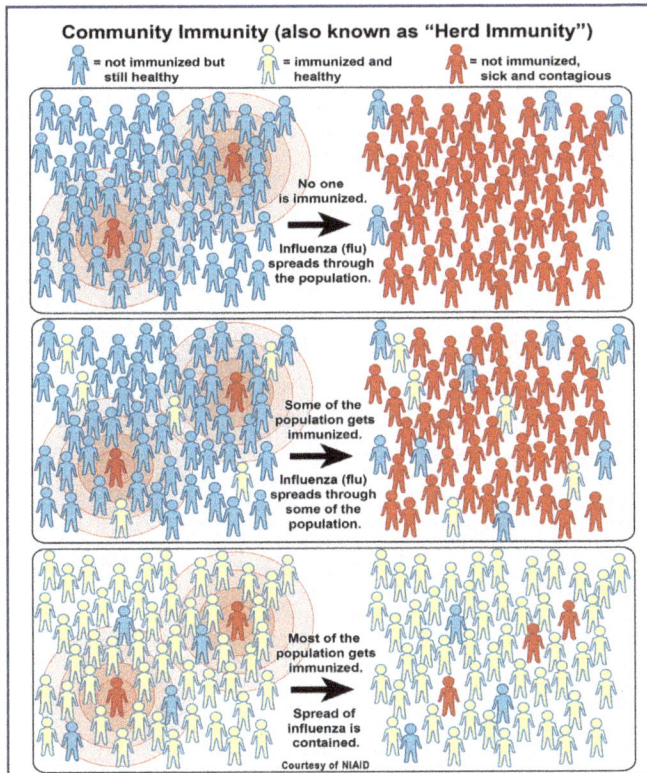

*Herd Immunity*

As the politicization and polarization take hold, COVID-19 vaccine acceptance gets less likely as the pandemic unfolds. Currently, 27% of adults tell pollsters they don't intend to get vaccinated for COVID-19.[67] This puts vaccination rates for COVID-19 at the low end of the 70% to 90% estimate to get herd immunity — and the estimate itself is guesswork because the effectiveness of the vaccines and the rate of mutation of the virus are still unknowns. If the virus mutates, vaccines must be upgraded like vaccines for seasonal flu; if the vaccines wear off, they'll have to be renewed.[68]

Washington being Washington, everyone with a connection has an idea. If he hadn't been busy enough, Doctor Fauci reacts to events sprouting like plants in a desert after a thunderstorm. Among these were alternative treatments passing through Doctor Fauci's email inbox.

Many he simply acknowledged several, such as one from Teresa McPeak:

There is a wealth of medical research papers about the association between Proton Pump Inhibitors (PPI) and increased risk of community acquired pneumonia. Long term use of PPIs reduce the pH of the stomach from 1 to 5. This reduction in stomach acid reduces the body's ability to fight viruses...[69]

Then there is "one old Marine to another Marine," in an email to Matt Pottinger, Deputy National Security Advisor, suggesting that grounded aircraft with pressurized cabins be used for Hyperbaric Oxygen Therapy (HBOT), which has been in use for over 100 years, and was the only effective treatment for the 1918 flu pandemic." The statement has some merit.[70] But the claim that it is being used for treating severe COVID-19 with a "success rate close to 100%" is overstated.[71] This is from someone Fauci couldn't ignore. He responds, "Let's discuss this."[72]

It takes a disciplined mind, even with 3 to 4 hours of sleep,[73] to spend time planning for the long haul amid the chaos.

Doctor Fauci is invited to a "special late breaking session" on April 26th as the second of the six featured speakers of the Annual Meeting of the Academy of Science[74] on Zoom.

The content of his speech is as precise as his enunciation; the rasp in his voice, usually barely noticeable, is more predominant than usual today. After mentioning that we are in a very difficult situation, he moves to the situation in the United States. Putting up his slide show, he reports that the US has nearly a million cases and 50,000 deaths, noting it probably began in Washington State, coming from China.

The ongoing outbreak in New York City, "we now know" with sequencing data, came from Europe. Showing daily case and death counts, he emphasizes that even though cases are stabilizing in New York City, the deaths are still rising because of the lag between getting the disease and becoming a "case."[75]

His hands are clasped before him, almost as if in prayer. The case and fatality rate chart by race is a surprise; he does not know why. There will be papers suggesting genetics, but they are speculative — noting differences in ACE-2.[76] There will be no good answers in the literature to come; it is as likely that the difference is due to differences in location, living conditions, stress, and access to health care. These can also lead to an increase in COVID-19 comorbidities, which, in turn, leads to higher mortality.[77]

Doctor Fauci notes that test kits got a late start in the United States, "as everyone knows," because of the publicity. We are now better off than a week ago and will be better off even more in a week because we now have the private sector involved. "We are learning more every week," says Fauci as he introduces an NIAID Strategic Plan. Ultimately, "we need vaccines" because "that is how we are going to end it."

Fauci speaks of his time on the White House Coronavirus Task Force, led by Vice President Pence. "We meet every day, seven days a week, sometimes into the evening."

But the honeymoon is already unraveling between Fauci and Trump, Trump contradicting Fauci and Fauci "a real time fact checker."[78] Few Presidents will tolerate that, certainly not President Trump.

Before he wraps up, Doctor Fauci delves into treatments and vaccines. He presents a new promising medicine, remdesivir. Remdesivir has been successful for monkeys injected with SARS-CoV-2, but it's "not a knock out drug."

This isn't news in Montana because the RML had done press around it a week earlier: monkeys were infected with SARS-CoV-2 right here in Hamilton.[79]

The animal facility at RML is decades old. A fluff piece in 1942 reported a newborn monkey, "6-Day-Old Monkey Going Strong."[80] Going from pens and zoo-like enclosures to

Fauci National Academy of Science Presentation

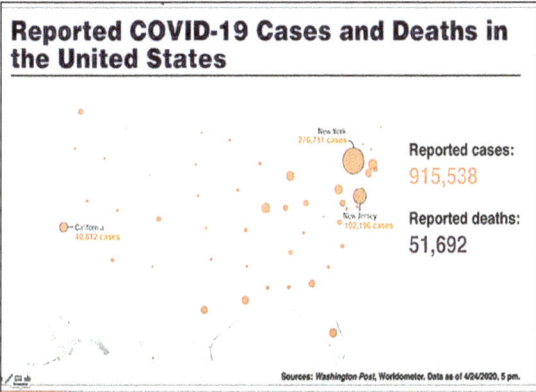

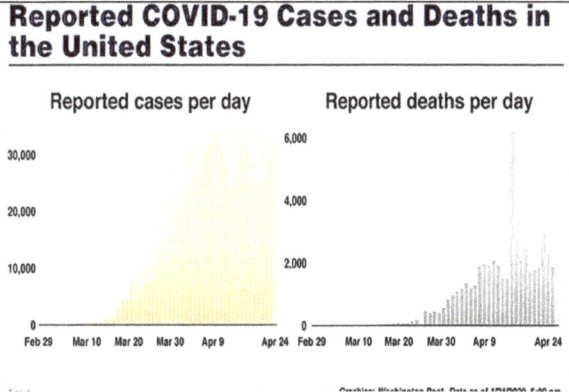

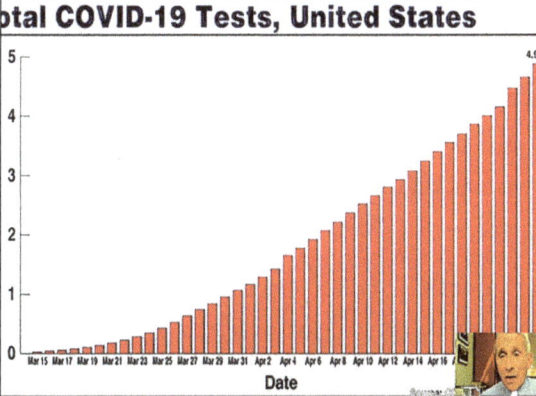

## Reported COVID-19 Cases and Deaths in the United States

Reported cases:
915,538

Reported deaths:
51,692

Sources: *Washington Post, Worldometer.* Data as of 4/24/2020, 5 pm.

## Reported COVID-19 Cases and Deaths in the United States

Reported cases per day

Reported deaths per day

## Total COVID-19 Tests, United States

Date

### NIAID STRATEGIC PLAN FOR COVID-19 RESEARCH

APRIL 10, 2020

- Improve fundamental knowledge of SARS-CoV-2 and COVID-19 disease
- Develop diagnostics and assays
- Characterize and test therapeutics
- Develop safe and effective vaccines

**NIH** National Institute of Allergy and Infectious Diseases

---

March 24, 2020
Volume 117
Number 12

**PNAS**
Proceedings of the National Academy of Sciences of the United States of America

**Prophylactic and Therapeutic Remdesivir (GS-5734) Treatment in the Rhesus Macaque Model of MERS-CoV Infection**

E de Wit, H Feldmann et al.

---

CSH Cold Spring Harbor Laboratory

**bioRxiv**
THE PREPRINT SERVER FOR BIOLOGY

April 15, 2020

**Clinical Benefit of Remdesivir in Rhesus Macaques Infected with SARS-CoV-2**

BN Williamson, E De Wit et al.

---

**NIH** National Institute of Allergy and Infectious Diseases

February 25, 2020

**News Release**

# NIH Clinical Trial of Remdesivir to Treat COVID-19 Begins

- First clinical trial in US to evaluate experimental treatment for COVID-19
- Initial trial participants are Americans repatriated after being quarantined on the Diamond Princess cruise ship
- Study can be adapted to evaluate additional investigative treatments and to enroll participants at other sites

---

**NIH** National Institute of Allergy and Infectious Diseases

March 16, 2020

**News Release**

# NIH Clinical Trial of Investigational Vaccine for COVID-19 Begins

*Study Enrolling Seattle-Based Healthy Adult Volunteers*

Trial of vaccine candidate mRNA-1273 will enroll 45 healthy adult volunteers ages 18 to 55 years over approximately 6 weeks

Photo: AP

Apr. 25, 2020, YouTube (Frueh, Sara. "NAS Annual Meeting: Experts Discuss COVID-19 Pandemic and Science's Response"

---

# White House Coronavirus Task Force – Announced Jan. 29, 2020

**Chair: VP Mike Pence**
**Response Coordinator: Deborah Birx**

| | |
|---|---|
| Jerome Adams | Stephen Hahn |
| Alex Azar | Derek Kan |
| Stephen Biegun | Larry Kudlow |
| Robert Blair | Chris Liddell |
| Ben Carson | Steven Mnuchin |
| Ken Cuccinelli | Robert O'Brien |
| Kelvin Droegemeier | Matthew Pottinger |
| Anthony Fauci | Robert Redfield |
| Joe Grogan | Joel Szabat |
| | Seema Verma |
| | Robert Wilkie |

a modern 10,000 square-foot facility, the lab and its monkeys contributed to AIDS and Ebola research,[81] though a heater-gone-wild roasted some monkeys and hamsters — a news piece picked up nationwide.[82]

The remdesivir trial on monkeys, *Rhesus Macaque*, the species called a "model" in the literature because, in many cases, the biological mechanism is similar enough to a human to serve as a good predictor for how a human will respond to an infectious disease. *Rhesus Macaque* has been used to study about 100 human diseases.[83]

What that means in practice is that eight Macaque are given SARS-CoV-2. The progress of the disease is tracked with samples and tests, the euthanized and dissected. The results show the disease's progress and effect on the lungs and other tissues.

The authors conclude that treatment should be "initiated as early as possible." Researchers administer the drug 12 hours after giving

the monkeys the disease, which the paper noted was "close to the peak of virus replication in the lungs." Obviously, this time between infection and treatment will not be replicated in the real world, where people will have had the illness for days or weeks before receiving the drug: exposure plus incubation time plus the time for onset of symptoms plus the time to get to a hospital. All biology being equal, it is unsurprising that the Macaques have better outcomes for remdesivir than humans will experience. The authors find it noteworthy that despite the improvements, there is no difference in viral shedding or contiguous virions.[84]

Fauci gets to the crux of the matter as he talks about the large clinical trial of remdesivir,

> The only way you are not going to get perpetual ambiguity is by doing a randomized, controlled trial.

The Chinese, on that same day, the 29th, publish a study that meets Fauci's golden standard, a dou-

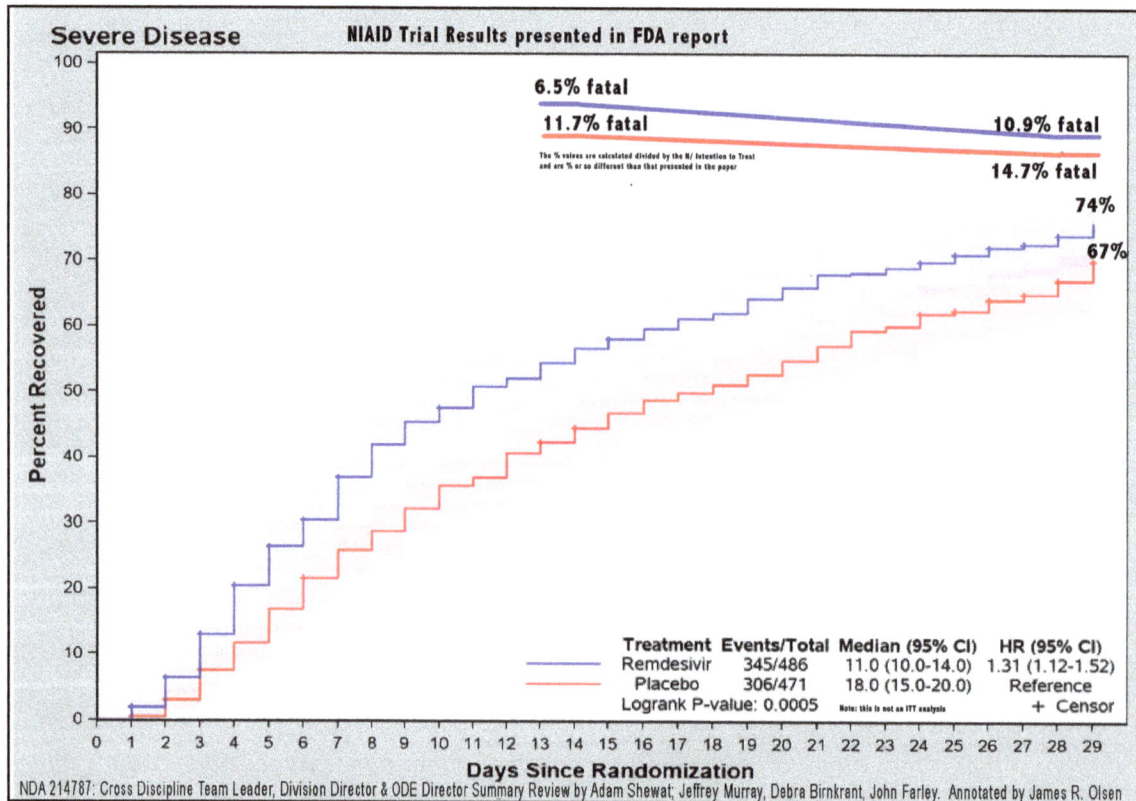

*Remdesivir Severe COVID Recovery and Mortality*

ble-blind, randomized, placebo-controlled trial of hospitalized, laboratory-confirmed COVID-19 patients selected for respiratory symptoms and radiologically confirmed pneumonia. For severe COVID cases, at least as well as 237 cases can tell — remdesivir and the placebo group are nearly identical: mortality, 15% vs. 13%. The isn't much difference between the Chinese trial and the evidence that will be presented to the FDA.[85]

The result that will be presented to the FDA is that recovery is faster for some, and the fatality rate favor treatment by a few percentage points. Still, it is not statistically significant to say for sure. The drug made no difference in the recovery rate for intubated patients. The results of the NIAID trial are reported in the *New England Journal of Medicine*.[86]

The recovery rate at the end of 29 days is 74% vs. 67%, though we don't know the final outcome. The mortality rate of 10.9% for remdesivir treatment seems to be in the range of the actual experience in the United States for fatalities for hospitalized patients, which ended up at 10% by the end of the year — but is almost double that for people over 75 years old.[87]

If we play Monday Morning Quarterback, history will show that the lack of centralized coordination meant that many trials for COVID antivirals were doomed from the start. According to Dr. Janet Woodcock, the Food and Drug Administration acting commissioner, trials were often too small and poorly designed to provide valuable data.[88] Testing other drugs, some of which would prove more effective for severely ill patients, is left to other countries during 2020 and beyond.[89]

The NIH admits that no well-designed trial has been done for ivermectin, which is becoming the counter-narrative's favorite sibling. The failure to have a trial, even if to prove the negative, is driving confrontational politics, driven by rumors, claims, and observational studies and trials of varying usefulness. All the NIH can do is ad-

mit that the effect of ivermectin on COVID-19 is unknown.[90]

One study shows that early treatment with hydroxychloroquine (HCQ) reduces the chances of hospitalization and mitigates symptoms. Others show no effect.[91]

There is some evidence that ivermectin is as effective as a prophylactic or for mild disease when given early. However, two studies showing reduced fatalities are suspect, one the authors' conflict of interest and another that was pulled by the journal. Despite the uncertainty, there is a tremendous amount of off-label use — people are taking these drugs.[92]

Of course, a large-scale, controlled trial for both drugs is in order. The fact that NIAID does not take up a randomized trial of ivermectin leaves the issue to fester. It's as if the law of physics applies, the Conservation of Energy. Instead of focusing on science, we expend energy in at least equal amounts on confrontational politics.

Another angle of attack by the critics is that NIAID put a lot of effort and money into remdesivir. This leads to questions:

- ❑ Did the taxpayer get treated fairly?
- ❑ Did Fauci make money unfairly?
- ❑ Was the selection of what drugs to spend money on biased?

The first question is taken up by Congress and assigned to the Government Accountability Office (GAO), which was set up by Congress as the last bulwark of nonpartisan analysis. Surprisingly, the Department of Defense (DoD) spent much of the Government's $162 million for trials of a drug patented by a private company, Gilead, during the Ebola scare.[93]

If the taxpayer helps fund an invention, the law says, in most contexts, that the taxpayer holds some of the rights.[94]

The GAO found, for remdesivir, that the drug had been invented before any Government help and that no patentable inventions were filed for

**Timeline of Select Events and Federal Contributions to the Development of Remdesivir**

| Basic Research | Drug Discovery | Preclinical Research | Clinical Trials | Drug Review and Approval |
|---|---|---|---|---|
| Scientists identify biological mechanisms of a disease | Chemical compounds are synthesized and screened to identify a drug targeting the disease | Drug is assessed for safety and efficacy in laboratory and animal testing | Drug is tested for safety and efficacy in human volunteers | FDA reviews the drug application and approves the drug |

| Before 2009 | 2009-2013 | 2013-2020 | 2016-2020 | 2020 |
|---|---|---|---|---|
| • Academic, federal, and industry researchers investigate molecular building blocks of RNA and viral replication. | • Gilead scientists synthesize parent compounds and remdesivir. <br><br> • Gilead begins patenting the parent compounds and remdesivir. | • CDC and DOD scientists, in collaboration with Gilead, perform antiviral testing of remdesivir for Ebola and other viruses. <br><br> • NIH scientists and NIH-funded university scientists, in collaboration with Gilead, conduct preclinical studies of remdesivir's antiviral activity against coronaviruses. | • NIH funds two Ebola clinical trials involving remdesivir in 2016-2019. <br><br> • NIH funds a clinical trial of remdesivir for COVID-19 in 2020. | • FDA grants emergency use authorization, based in part on NIH-funded COVID-19 trial results, for using remdesivir to treat COVID-19 in May 2020. <br><br> • FDA grants approval to use remdesivir to treat COVID-19 in October 2020. |

- Not reviewed by GAO for this report.
- Reviewed by GAO for this report.
- Following invention, Gilead conducted its own remdesivir research and clinical trials, initiated collaborations with federal agencies, and made substantial contributions to the research performed in those collaborations.

Source: GAO analysis of information from the Centers for Disease Control and Prevention (CDC), Department of Defense (DOD), Food and Drug Administration (FDA), Gilead Sciences (Gilead), and the National Institutes of Health (NIH). | GAO-21-272     Credit: Government Accounting Office. Report GAO-21-272

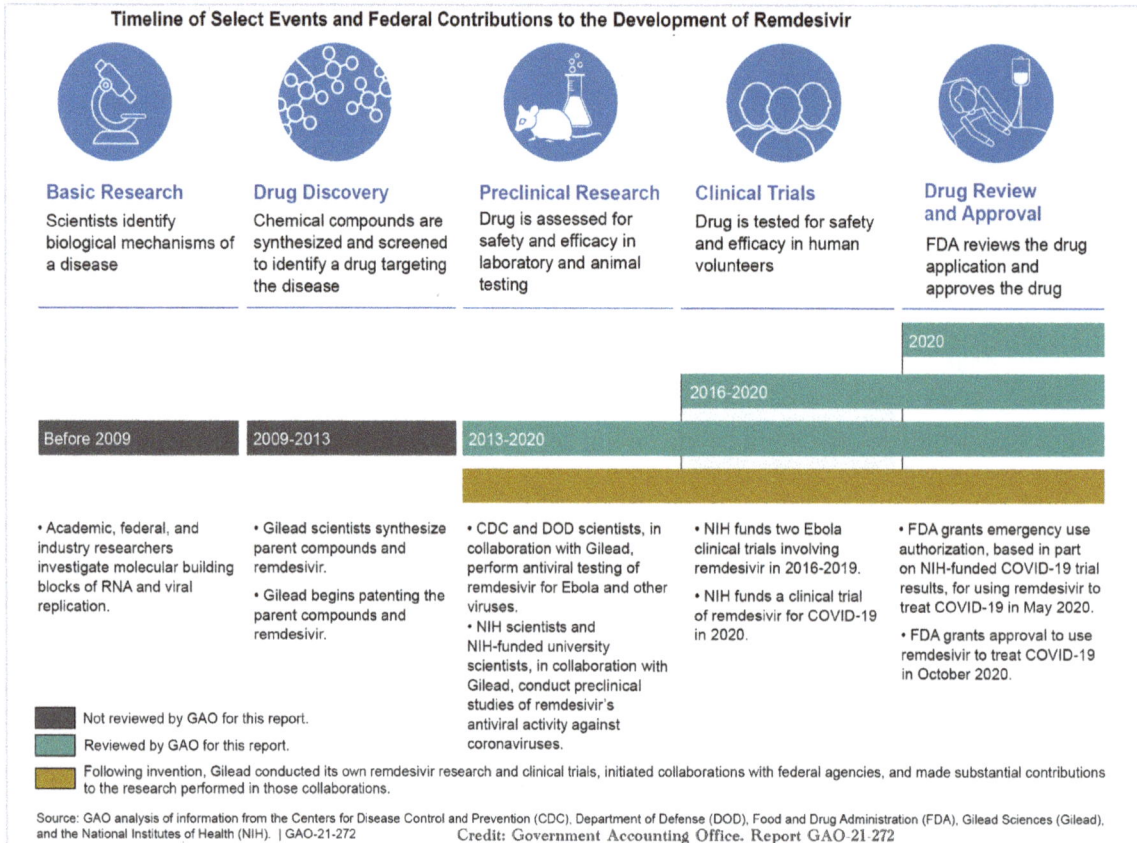

*Remdesivir Development*

other possible inventions, such as delivery methods.

Did the taxpayer get treated fairly? It seems so.

Did Fauci make money using his position unfairly? His financial reports answer the question. No. It does not contain any targeted investments in pharmaceutical companies or other companies with which the NIAID does business. Instead, the investments are in mutual funds and the like.[95]

Was the selection of what drugs to spend money on biased? Probably. There is no direct evidence of bias. Like the Military-Industrial Complex that Eisenhower talked about, preferences can be found in relationships created by colleague networks, the revolving door,[96] trade shows, and conferences. These can become a self-perpetuating money-making machine while perverting the free-market theories in the *Wealth of Nations*.[97]

The inner workings are often innocent enough — but not always. More lawyers than doctors have been appointed as Secretary of HHS. The current head, Alex Azar, is a lawyer who became a businessman, having done noteworthy work while in HHS during the Bush Jr. Administration. In 2007, Azar went to Eli Lilly, becoming president of the US division. During his watch, the price of insulin tripled.[98] Rich Bright filed a whistle-blower complaint, which names Azar's assistant secretary, Doctor Kadlec. He would win his case with the decision noting that HHS,

> allowed industry consultants to play in securing contracts that Dr. Bright and other scientists and subject matter experts determined were not meritorious.[99]

We have made little progress in regulating chemical concoctions since *Silent Spring*.[100] The parallels between the approval process for assessing pesticide risks a half-century ago, and the FDA's Emergency drug approval process is downright spooky. COVID-19 has once again highlighted a long-standing flaw in the American regulatory process of many industries, not just food and drugs.

This has been going on for decades for pesticide approvals. For instance, a product such as DDT, so effective at controlling mosquitoes diseases that the inventor won a Nobel Prize,[101] was approved and used widely before the impact became apparent to scientists beginning in 1947. But it takes decades to get enough political pressure to finally get members of Congress up in arms.[102] It took another decade before being pulled from the market in 1972.[103] The flaws in this process have been apparent for generations.

The process for the regulation of pesticides is similar to that of the FDA, including advisory panels of experts and the same controversy about the application of science in assessing risk.[104] The FDA approval process is governed by a maze of regulations.[105] It seems clear that:

- Good science proven by a clinical trial sometimes gets lost in the FDA approval process. Instead, the regulations are procedural, with the expert panel and regulators often deferring to the drug maker's judgment.
- The approval is supposed to be based on risk versus benefit, but the dangers of unforeseen issues due to small trials are often ignored.
- For an EUA, there is supposed to be no other effective treatment — this is sometimes ignored.[106]
- The trial data is provided by the applicant, who is motivated to design a study and provide an analysis that supports approval.
- There is no rigor, or even a requirement, to respond to substantive comments from the public.

This self-testing differs from the DoD's approach, where the contractor does most of the testing, a separate military organization designs and runs the final operational acceptance.[107] Even so, there are headline-making fiascoes. The

FDA's process does not apply consistent scientific rigor and must be improved.

Did any other country have a better plan? One country did, taking advantage of a medical system that doesn't exist in the United States.

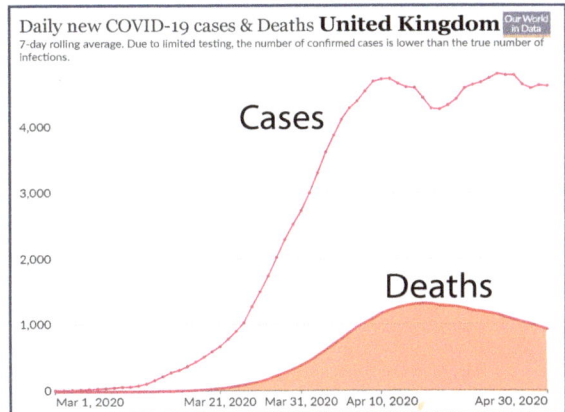

*United Kingdom Cases and Deaths Apr. 2020*

The British respond to rising fatalities in their country with an innovative nationwide drug trial system. It's too bad the British didn't present at the National Science Academy annual meeting because they had devised a better way to run drug trials quickly and efficiently.

In March, Martin Landray and Peter Horby from Oxford University dreamed up a massive clinical trial that would come close to meeting Fauci's criteria while trying out multiple possible treatments. The idea had its genesis in the 1980s when "a group of Oxford scholars was dissatisfied with the lack of treatments for heart attacks."[108] They contemplated a trial using 10,000 to 15,000 patients instead of the usual 100 to 1,000 you see in most trials, "but a large clinical trial in the midst of a pandemic is novel."

Sir Peter Horby is Chief Investigator of the RECOVERY Trial and a Professor of Emerging Infectious Diseases and Global Health. Referring to the SARS and the 2009 flu epidemic, he said, "Hardly any trials started. It was appalling. It just was a failed model..."[109] They did better with Ebola, getting trials underway within a few months, but the outbreak was under control before we got "any good answers."[110]

When the COVID-19 outbreak began in Wuhan, Horby was working with Chinese scientists to get randomized trials going in Wuhan using protocols he had helped develop for MERS.[111] They began enrollment in Wuhan on January 19th to try out two HIV drugs and remdesivir — but the "very aggressive public health measures in China" caused new cases to "plummet." So, he got more funding and thought of extending the Wuhan trials. But they ran out of patients. The funders said to come on back to the United Kingdom — there were plenty of cases there.[112]

At this point, the Director of the Welcome Trust,[113] Jeremy Farrar, brings Horby and Landry together. Horby had "never done a clinical trial over about 200 patients," Landray said, "I'd never, ever worked on infectious diseases."[114] Together, Horby and Landry developed a massive multi-drug trial called RECOVERY — Randomised Evaluation of COVID-19 Therapy; the Chief Medical Officer of England, Sir Chris Whitty, signs on when he hears the pitch.[115]

The key to RECOVERY's success is simplicity. Upon checking into the hospital, a research team staff member approaches a patient. If they agree to participate, a one-page form puts them into the program, joining 35,000 participants in 2020. Horby walks us through it:

> We randomize patients; we toss a coin to decide whether patients will receive a new experimental drug or the usual standard of care. That gives us a very unambiguous and clear answer about whether the drug is beneficial... or even harmful. These kinds of trials are really the gold standard....[116]

What a difference. Doctor Fauci certainly knows the gold standard and implemented it for remdesivir, but that and monoclonal antibodies[117]

*Professors Martin Landray and Peter Horby at Oxford — Photo by Alicia Canter*

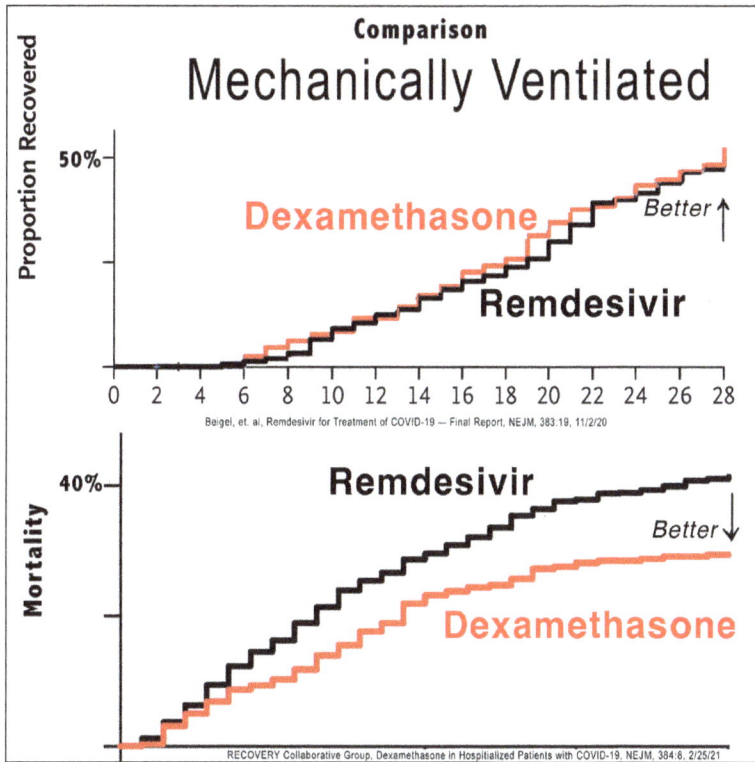

**Comparison**
# Mechanically Ventilated

Beigel, et. al, Remdesivir for Treatment of COVID-19 — Final Report, NEJM, 383:19, 11/2/20

RECOVERY Collaborative Group, Dexamethasone in Hospitalized Patients with COVID-19, NEJM, 384:6, 2/25/21

*Comparison Ventilated Patients remdesivir, dexamethasone*

will be it for COVID-19 treatments in the United States in 2020. Doctor Fauci did not have the same logistical starting point as the UK RECOVERY organizers, who took advantage of a national public health system with uniform national procedures, compared to the United States with a state-by-state, hospital-by-hospital patchwork, each with their own patient management process.

The RECOVERY initially tries drugs on patients arriving at a hospital. PRINCIPLE (Platform Randomised Trial in the Community for Epidemic and Pandemic Illness) is kicked off by Oxford University, a platform that will be used for outpatient trials.[118] It will take until 2021 in the United States for Duke University to organize a large-scale medical trial for outpatients.[119]

As we'll see when we get to summer, a treatment that works better than remdesivir came out of this massive trial for hospitalized patients. However, dexamethasone is a steroid with a long list of side effects, especially for prolonged use.[120]

We have already begun arguing, protesting, and suing over three drugs. Let's call them the three siblings: remdesivir, ivermectin, and hydroxychloroquine. But, there is a fourth that works — it will take another year for the FDA to approve the one that showed the best for severely ill COVID-19 patients, dexamethasone.[121]

Anyone my age who tries to gut it out with COVID-19 may be gambling with their life.

One issue that crept up in the literature is that the timing can be critical. Of the three siblings, we've seen how remdesivir works and its limitations. Several trials show that hydroxychloroquine does not work for severely ill patients any better than other standards of care. Hydroxychloroquine has failed to demonstrate improvement in several clinical trials, but one study, though it suffers from too small a sample to show 95% confidence, does show an advantage for mild to moderate symptoms when taken early.[122]

Nutrition, herbs, and supplements that might keep one out of the hospital have not been the focus of RECOVERY or NIAID. They have been studied and reported upon various papers of varying quality.

I hypothesize that making sure not to be vitamin deficient and getting treated early improves one's odds more than waiting to see if you get really sick. More effective treatments are in the pipeline.

*30 Apr. 2020: 66,154 COVID-19 Related Deaths Reported*

1  2  4  8  9        15      May 2020   25—25      28

- genome sequences for varients.
- Data Base of SARS-CoV-2
- CDC SPhERES Public
- Remdesivir Emergency Approval
- WHO Renews Emergency Declaration
- COSTCO Requires Masks
- Whitehouse rejects CDC plan for reopening economy
- 14.7% Unemployment, 20 million
- CDC alert: child multisystem inflamatory syndrome
- MT Gyms, Movie Theaters opened
- Memorial Day
- George Floyd Killed
- Minnisota National Guard deploy to quell voilence during protests
- US Deaths pass 100,000

**COVID-19 Treatment Guidelines** (NIH COVID-19 Treatment Guidelines Panel)

• Hospital: Anti-blood clot; Severe: Remdesivir; Only for Complications: Convalesent Plasma  • Insuffient information to recommend for or against Hydroxychloroquine • Cannot recommend for or against Tocilizumab, an antinflamitory, currently under trials

Photo by Picturesque Japan

*Social distancing*

# 8

# KEEP YOUR DISTANCE

*The only areas under control were where we used some promptness in removing diseased trees. Where we depended on spraying, the disease was out of control.[1]*

*— Joseph A. Sweeney, Superintendent of Forestry, Toledo, Ohio, 1953*

**MAY 2020**

I remember the day in Thule, Greenland, February 11th, 1987, when none of us had seen a single photon of sunlight for months, the orange-lit clouds on the horizon having teased us for the last two weeks as the latitude of where the sun kisses the earth crept up from below the horizon. We took a break from work that noon, stepping outside, looking south over snow-covered tundra, muted as if in a giant shadow, which it was. Suddenly, a red flash from the top of the sun blazed over the horizon like a laser beam, grazing low over the snow to light up the sky. It was gone in minutes. We knew tomorrow we would get 20 more minutes of sun. With a visceral understanding of why our ancient ancestors celebrated the events of heaven, we said little as we went back to work. It is like that when the market opened on May 2nd, expressed on everyone's face. We had been released from the darkness.

Spring woke up from its winter sleep on schedule this year. Already at work for a week or two, the snow melt has been raising the water in the Bitterroot River. Today, a perfect spring day arrives in mountain country, the rising sun heating the almost-plant-killing-frost-air of the night to jacket-wearing weather by the opening bell. As the Farmer's Market proceeds, the sun reaches a hands-width from its zenith to deliver near tee-shirt weather under fair skies by the 12:30 closing bell.

The various food coupon programs are known by their acronyms, SNAP and WIC.[2] They are essential to a lot of our customers — the market had taken the time to set up its books and send the market master to the requisite training to accept these coupons. Thankfully, the Governor relieves some of the burden, including ordering automatic renewal of these programs for three months.

By far, the most popular topic is COVID-19; no surprise there. But a common refrain is, "I think I had COVID in February," or January. But, like me, no one will know because Montana has no test kits.

For at least a few weeks, the undercurrent of anxiety seems to have abated as people socialize again. I skip the May Health Board meeting, as did the rest of the public. COVID goes on, the future unknown. The reopening from the shutdown is a release, a respite, even though the virus marches on.

The CARES Act makes employment matters weird. The $600 extra per week, in addition to an average unemployment insurance payment of $380, means a low-wage worker will make more on unemployment than they did at their job.[3] This upset apple cart will ripple through the economy as it reopens.

The Health Board meets. The Public Health Nurse reports that most of the tests are presurgery. The hospital finally has the PCR test machine that the County paid for but has yet to calibrate it.[4] The Health Nurse has help from one nurse and three others on-call — just in case of a spike, which she predicts will be in the fall. They have five positive cases that they are contact tracing.

The Fair Board has just met. It is planning to have a full-on fair leading up to Labor Day. Burrows says, "I don't think COVID can live on fair food." Roger retorts, "They live on us." With that, the energy shifts back to septics.

But, within a week, the Health Department is dealing with an outbreak at the Stock Farm,[5] our gated community for the rich and famous, originally built as part of the Marcus Daly estates, now sporting a first-class golf course and country club, developed in 2009 by an investment group that includes Charles Schwab.[6]

Of course, many of us have not set foot there, while others have worked there. The Stock Farm managers have been generous with accommodations — which has been something of a tradition with performers booked by the Bitterroot Performing Arts Council, which brings Grammy Award-winning performers to our small town. The word is out in the music booking world — "performer, book an act in Hamilton, world class accommodations with incredible mountain views."[7]

The palpable relief of the Farmer's Market's opening day is short-lived. Fear-inducing government warnings are multiplied by fear-inducing stories being passed around the Internet, forwarded with a click and a tap, and passed around peer to peer.

Hollie asks what I think about a video called *Out of the Shadows*. By the time I get to it, it's been pulled off YouTube. I find a DVD for sale online and buy it. While I am waiting, I look up the video on IMDb:[8]

---

Hollie,

While waiting for the video, I did a quick check of who made it and their self-described credentials or online bios. Using the IMDb Site, which is the go-to site you linked to and is the one usually used for referencing movie credentials.

Mike Smith has a career doing stunts in Hollywood movies.

Keven Shipp appears to be Mike Smith playing Kevin Shipp.

Brad Martin is a professional skier who turned to stunt coordinating.

Liz Crokin made a career as a gossip columnist.

ALL of the other cast listed were from archive footage - meaning they were not an active part of the cast and may be in the movie involuntarily.

Jim

---

The comment about Keven/Kevin Shipp is my mistake. Kevin Shipp has written several books and is on the Alt-news interview circuit. The earliest story is a lawsuit Shipp filed against the CIA that says that his family had been affected by toxins at a base where they lived.[9]

The *Out of the Shadows* video is reposted on YouTube by the time I get the DVD. I watch it. It is another connect-the-dots concoction with a sleight-of-hand.

It's like a sleight-of-hand with cards. You can do plenty of amazing card tricks if you get good at a double lift with a regular card deck. You look like you are turning over the top card of the deck, but you are actually turning over two cards, showing the second card in the deck. It's a slight, a trick, the audience thinking they are seeing one thing when the magician is showing them another — the magician has a plan. The plan is to control what the audience thinks they see.

*Out of the Shadows* begins with a great question we discussed at The Table. How do we know what we know? We know most of what we know because someone told us. The example I use is that we "know" the earth goes around the sun — yet we have never actually been in the middle of the solar system and observed it with our own eyes. OK, a thoughtful beginning. Then, suddenly, "Is the CIA involved in Hollywood?" comes as a question from off-screen. "Yes," says Shipp.

After a story of a Hollywood stunt-man's personal quest, clips flash across the screen, lifted from news footage, Congressional Hearings about the CIA planting stories. OK — I remember those. Then it gets into World War II propaganda, which I researched in support of *Pearl Harbors' Final Warning* by Valarie Anderson — my sister. No secret here. I won't cite them, but there are several books on the relationship between the military and Hollywood during the war. In fact, some clips from the series are used to train incoming servicemen and women; *Why We Fight* is paid for by the military and made by leading Hollywood producers and directors.[10] You can be the judge of the merits of this relationship. But what does that have to do with the price of tea in China?

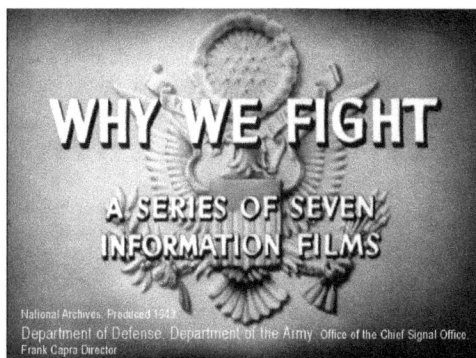

*Why We Fight Title Screen*

The movie slips into slights-of-hand. Its fast-paced transition moves into Pizza Gate, Epstein, child trafficking, satanism, and 666 symbols, visually connecting the images so that the visually linked images seem like connected facts — your first thought is that these images are logically related. On second thought, "How did these images get in the film?" This is the sleight of hand. They have nothing to do with each other except that the producer carefully selected them to build on the producer's preconceived theme. The video finishes with, "once we get people critically thinking...

now they are seeing the dots." What are the selection criteria for the facts and claims in the video?

1. Are the facts selected and arranged to support a preconceived conclusion or

2. Are the facts selected based on some criteria unrelated to a preconceived result and found to support the conclusion?

The first is propaganda. The second is science.

I suggest Hollie stop listening to whoever put her onto this — even if it is true. We agree that dwelling on it without acting on it will simply create anxiety — there is enough to be anxious about already.

The dangers of 5G is an ongoing discussion before the pandemic at The Table. There are potential dangers of exposure to electromagnetic radiation, from sunburn to cancer. The effects on the body are related to frequency, intensity, and duration. Radio waves, cell phones — which are themselves radios, TV signals, WiFi signals, 5G, X-rays, and light are all the same physical stuff at different frequencies. After reading a paper or two, the safety of the upper end of the frequency range of 5G does not seem to have been examined well. I say I will help organize a 5G campaign if a few people are willing to get into it and see where the data takes us.

With COVID-19 as a new dot, the claim is that 5G is causing COVID-19. I look at a video on the subject that I got in an email link. My email response:

> Does a pretty good job in the beginning of explaining electromagnetic energy (various frequency bands known as radio, light, x-ray, microwave, 5G band) although the video backgrounds conflate disconnected things - such as "phased-array" which is an electronic steering technique within this electromagnetic spectrum.
>
> ....he brings up some text that reiterates what I pointed out before - that 5G frequencies induce currents in a cell.[11] However, the voice narrative did not go into that, instead .... global conspiracy and Christian theology.

Of course we can have a discussion about our predisposed belief systems and apply it to the video... The key thing I always look for in this and "main stream" narratives that are "scientifically proven" is a citation.

There is no doubt that electromagnetic radiation can and does affect biological processes based primarily on intensity, frequency, duration, and dose.[12]

Johannes Hevelius, Prodromus Astronomia, volume III: Firmamentum Sobiescianum, sive Uranographia, table QQ: Orion, 1690
*Orion Constellation Connect the Dots*

Connect the dots is a term used to fit current events into a purposeful, nefarious effort to control our society. It comes in many forms. Some conspiracy theories get so familiar to those who follow them that they use shorthand. One day in May, I received a copy of a copy of a copy of someone's analysis [are my notes]:

Event 201 = Gates

[Event 201 is a desktop exercise of a new coronavirus outbreak, Bill and Malinda Gates being one of the sponsors, starting with the premise of an outbreak in South America that spreads around the world, the participants representing various agencies making recommendations as to how to control its spread — some of course were used in 2020 — so it was not a get-ready exercise, but actually a plan to create a pandemic][13]

ID2020 = Gates

[A plan for a world wide digital id program for refugees — that will become a Big Brother tracking program for all US Citizens][14]

Corona Virus Patent = Gates.

Covid vaccine = Gates

New economy based on Human Activity (Microsoft Patent 060606)[15]

Simulate a global outbreak

Make a very contagious, low mortality virus in Fort Detrick BSL-4 lab.

Weaponize it to make it much more fatal as a back-up plan

Transport it to a lab in Canada so that it can be smuggled to the BSL-4 lab in Wuhan (deniability)

Wargame the plan just before release (Event 201)

Fund the talking heads

Release it from Wuhan

Enact a series of quarantines and back-to-normal

Enact the new economy (Microsoft Patent 060606) using food/water/shelter to force submission.

It is as natural as humanity. Our ancestors connected the dots. They connected the stars in the sky, confirming their mythical god's presence in the heavens. Dot connection can start with what you already believe. This would make it to a Health Board commentary:

I worked on things like National Missile Defense, where we rehearsed World War III ... — that didn't mean we were planning to instigate a war, but it was our job to figure out how to respond to one.

The fact of a new disease outbreak is predictable — just not when or what. The probability that it will be a coronavirus was a good guess... When I looked at Event 201, it seemed that was their purpose, though poorly executed in terms of... realistic adaptations and responses.

To give it its due, the view of this narrative is that *if* ID2020 is implemented worldwide, it could require everyone to get a Digital ID to do business. Worse, this technology could be used to track and restrict your movements. While it seems a long way off, the distance is measurable and not as far-fetched as some think. It is real enough in the People's Republic of China, where the government uses a "health code, 健康码" with an embedded QR Code (Quick Response Code) and phone tracking technology. Available through several apps, the health code is required to enter most stores, use public services, and travel. Designed to control the spread of COVID-19, the code turns red if a person is infectious.[16] The Chinese government then uses QR Codes to inhibit travel by falsely flagging a dissident as "infected."[17]

Finally the Microsoft Patent appears to be a yetFinally, the Microsoft Patent appears to be a patent application that has yet to be granted. Having gone through the patent process myself,[18] the Microsoft Patent application seems too conceptual. he application does present a concept that has the potential for profound social manipulation, using body sensors and even brain wave sensing (not embedded, presumably externally sensed) to reward specified behaviors with cryptocurrency.[19]

*The Mask* joins the ranks of tribal icons, the *Raised Fist*, *Trump Flag*, *Rainbow Flag*, *Elephant*, and *Donkey*. A mask worn or not worn is becoming a political statement. You are now taking a stand whether you want to or not. The only thing both sides agree on is that it is harder to breathe with a mask on. I get this email:

Why are you wearing a mask?
You didn't when a million people died to tuberculosis last year, when research showed it spreads through the air from person to person within feet.
Because the media didn't tell you to.
Because the state didn't mandate you stay home.
Because you only worry about what the media tell you to worry about.
They pick what you should fear. They give you fear, you are afraid, you accept less freedom.

Everywhere you look, someone is instilling fear. For instance, Tucker Carlson, who, as I write this, pausing just long enough to take a look at his show on *Fox News* to have him tell me that "civilization is falling apart."[20] Caring for others is something we should all agree upon. But, too often, it morphs into self-righteous divisive anger when everyone else is not doing the same.[21]

Gilead Sciences gets their emergency approval on May 1st for remdesivir, a great press

Daniel O'Day
CEO Gilead

Dr. Stephen Hahn
FDA Commissioner

Official White House Photo by Joyce N. Boghosian          Annotated by James R. Olsen

*President Trump, CEO of Gilead, FDA Commissioner*

day for the administration, touting America's progress in fighting the pandemic; the theme: being aggressive is good. A few days earlier, President Trump gave notice that he wants the FDA to move as "quickly as they can." Now, the emergency authorization is being celebrated in the Oval Office.

Doctor Fauci says the median recovery time for patients taking the drug is 11 days, compared with 15 days in the placebo group. He adds that the mortality benefit of remdesivir "has not yet reached statistical significance... This will be the standard of care. What it has proven is, that it can block this virus."[22]

The vaccine race is on, quickly becoming a race for national pride, particularly in the United States, Great Britain, China, and India.[23] Although it will take a while for vaccines, it will get worse before it gets better. COVID-19 is making its rounds, with Brazil and Spain having newsworthy outbreaks as New York City and Italy begin to turn the corner.

As remdesivir is being approved for emergency use, the NIAID goes full-bore toward getting a vaccine out to the public. Much progress has already been made with the first three Priorities of the *NIAID Strategic Plan for COVID-19 Response* by the time it is published.[24]

While it is more of a list of tasks than a plan, the NIAID lays out the agency's priorities, objectives, and categories of tasks in good form.

Doctor Fauci is considering how best to organize the science part of vaccine development. It is Moderna's turn to get an infusion of help, $955 million, with HHS providing 100% of the cost to get through trials and FDA approval.[25] A decade ago, the company stole a march on an emerging vaccine technology using a feature in the body for all sorts of biological processes, the messenger RNA,[26] or mRNA.[27]

Messenger RNAs are always made in our cells, a natural phenomenon. Most of us have heard of DNA,[28] the string of genes that make us who we are. It replicates every time a cell splits so that there is one in each of our cells. DNA can transcribe the "instructions" to create a messenger RNA, which goes forth to make protein molecules.

SARS-CoV-2 has a spike protein that can penetrate a human cell membrane. The immune system will eventually learn to be on the lookout for these intruder proteins. So, why not make a bit of messenger RNA that makes the same intruder protein, stripped of all the other viral mechanisms?

NIH started funding mRNA research in the 1990s and has grown since.[29] Using mRNA for lab work, therapies, or vaccines has been tried

---

*NIAID Strategic Plan for COVID-19 Response* [bullet points under objectives omitted]

Priority 1: Improve fundamental knowledge of SARS CoV-2 and COVID-19
    Objective 1.1: Characterize fundamental SARS-CoV-2 virology and immunological host response to infection.
    Objective 1.2: Evaluate disease dynamics through natural history, transmission, and surveillance studies.
Priority 2: Support the development of diagnostics and assays
    Objective 2.1: Accelerate the development and evaluation of diagnostic platforms
    Objective 2.2: Develop assays to increase understanding of infection and disease incidence
Priority 3: Characterize and test therapeutics
    Objective 3.1: Identify promising candidates with activity against SARS-CoV-2
    Objective 3.2: Conduct treatment studies to advance high-priority therapeutic candidates
Priority 4: Develop safe and effective vaccines against SARS-CoV-2
    Objective 4.1: Advance promising vaccine candidates through clinical trial testing
    Objective 4.2: Advance vaccine development through assay and reagent development
    Objective 4.3: Advance vaccine development through adjuvant characterization and development.

and mostly abandoned as too hard, as mRNA is unstable. The mRNA needed a package to enclose it and be small enough to enter the cell. A lipid nanoparticle serves as the package that delivers the mRNA. As a few scientists pressed on, problems began falling by the wayside by 2010.[30]

Moderna was founded in 2010 and started down the trail of mRNA vaccines well before the COVID-19 pandemic.[31] The Moderna candidate,[32] already patented for other diseases, is already being tested on animals for COVID-19. Presumably, you just need to tweak the genome on the already developed "platform."[33]

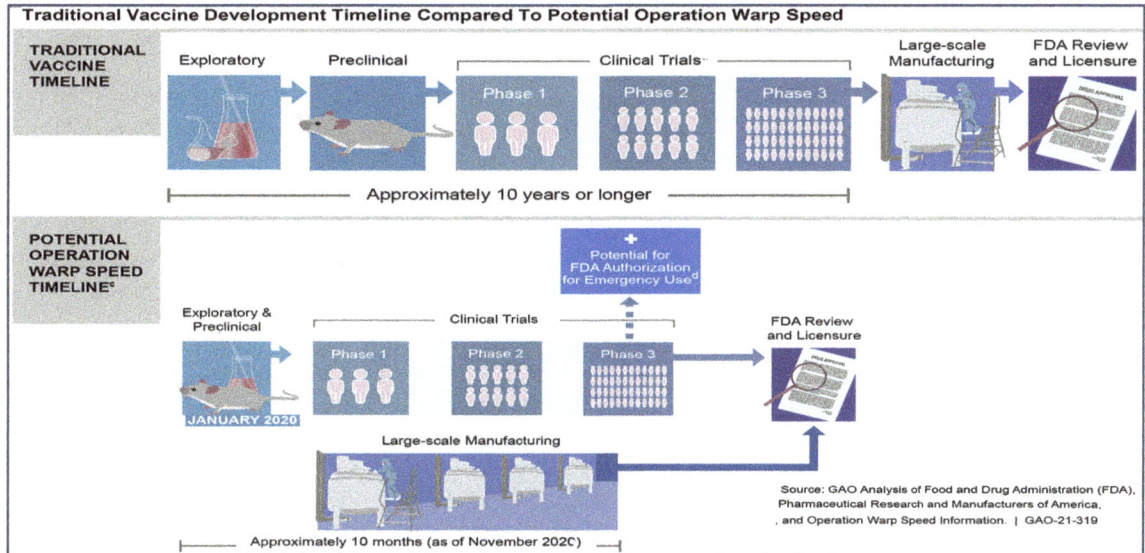

*Traditional Vaccine Timeline vs Warp Speed.*

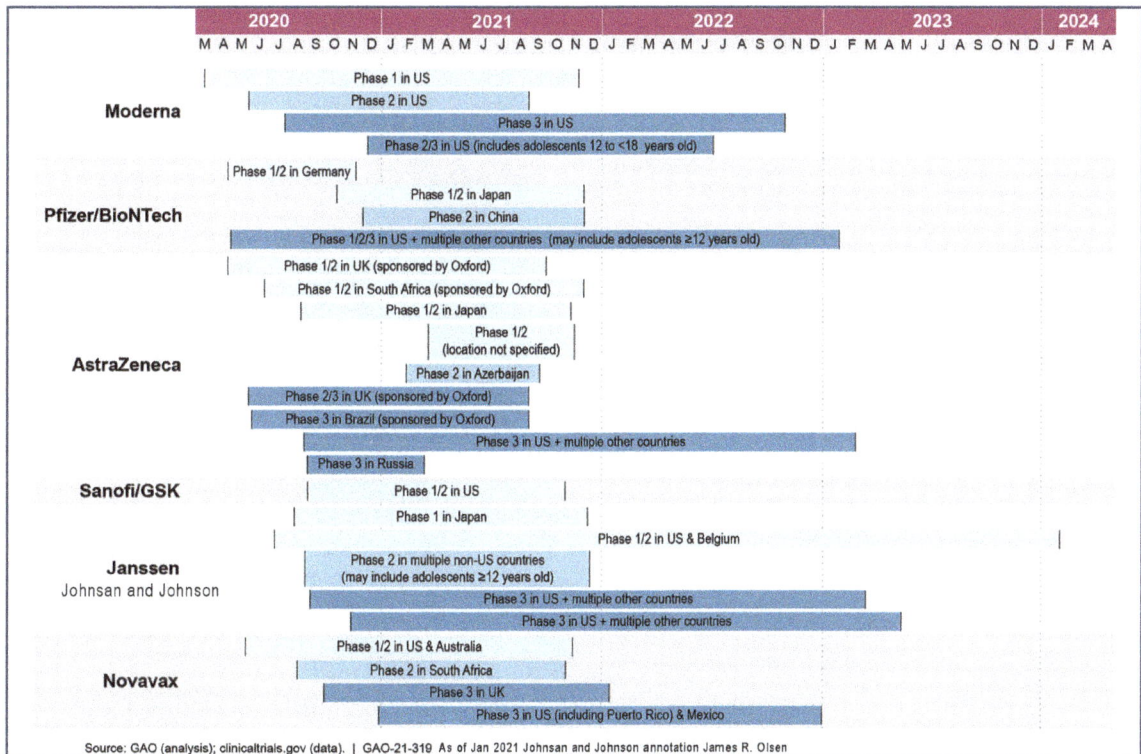

*Vaccine Trial Timelines*

In a C-Span interview, Marshall Bloom notes that "there is like a whole telephone book worth of potential vaccine candidates." Appearing to be speaking from notes, he says that any vaccine approved is "absolutely certain that no other infectious diseases are lurking," adding,

> The Latin Poet Horace, in a famous quote, says, "to make haste slowly." That means you move at deliberate speed, but you need to be very careful so that you don't create some side effect that was not anticipated. That's why in our country the FDA goes through three different phases.

Later in the interview, Doctor Bloom notes,

> Asymptomatic people who have antibodies are going to be important issue in the coming months.[34]

A question, which Bloom calls the $64 million question, is if you can get COVID-19 again after you recover.

Doctor Bloom responds that it is the subject of intense research, but they don't know yet.

Asked about when a vaccine would be available, Bloom replies that over the last 60 years, a number of vaccines haven't worked as well as expected, and getting one by the end of the year would be great. But, as the poet Horace said, "we have to make haste slowly and be careful."[35]

Messenger RNA vaccines are a relatively new technology being applied to a complex system, namely the human body, and to a worldwide population. After a convincing tour, the textbook *Development Betrayed* shows that the scientific method fails to be a good prediction method for complex systems. It suggests: try-a-little; see-how-it-works; adapt.[36]

The long-term unforeseen consequences for a particular mRNA vaccine may not have been detected yet — and may not be for a decade. Granted, there is a history of mRNA vaccine technology, but it is unclear how much effort and attention has been paid to unforeseen properties and behaviors. My take is that there is a slight but real chance of some unforeseen significant downside that could come forth for the novel vaccine technology.

Other approaches are being used as well. We have a history of treatments that are not active until they are activated in the body. They are called prodrugs. They have been with us in allopathic medicines and herbal tinctures: "A prodrug is a medication or compound that, after administration, is metabolized (i.e., converted within the body) into a pharmacologically active drug."[37]

New technologies are in place to provide a "platform" for developing and trying out vaccine candidates in the lab. For example, in 1967, scientists discovered that a virus stripped of some of its components would only replicate once in a cell. It became known as a pseudovirus.[38] Several decades later, private companies are offering to support research with this technology.[39]

Nothing is perfect, though a paper notes, "We show that some pseudotyped virus preparations give rise to low levels of replication-competent virus." In other words, some pseudoviruses do reproduce more than once.[40]

Let's turn to Objective 3.1 of the NIAID strategic plan. The first bullet under this objective is:

- Screen protease inhibitor[41] and nucleotide analogue class agents and other small molecules with documented activity against other coronaviruses.

These go by many names, "small molecules," "micro or a nanoparticle." These are small enough to enter a cell without the help and chemistry needed for a virus, such as SARS-CoV-2's need for a spike protein to muscle its way through the cell wall. Nanoparticles have various means to slip through biological barriers.[42] The first nanoparticle approved by the FDA as a cancer treatment was Lupron Depot® in 1989. Since then, this technology has been used to treat things like fungal infections, organ transplants, and HIV. By 2010, small particles were used for a vaccine, using virus-like particles for Human Papillomavirus (HPV), a sexually transmitted disease or skin-to-skin contact. Since about 80% of the population will get the disease at some point, most not even knowing they have it; some will get genital warts or cancer. So far, there have been no severe long-term side effects, even in the face of studies looking for correlations with some chronic conditions. The drug has been taken by about half of the middle-school children in the United States.[43]

Nano-drugs can have a variety of ways of working or mechanisms of action, as it's called in the literature. A COVID-19 nanoparticle treatment in the works will cause controversy among scientists when it comes up for approval in 2021.[44]

It is a complex system, the human body. We have declared an emergency, so the aggressive search for a cure and a vaccine is an emergency. The cautious steps of development, trials, approval, manufacturing, and distribution are being rearranged so that manufacturing setup and distribution logistics are being done in parallel with trials, which makes good sense, cutting a year or more off the typical timeline. But, to deliver on the promise of a vaccine in a year, the trials must be done in record time. Will we be able to "Make haste slowly?"

Despite the Governor's suspension of evictions, the plaintiff's law firm objects to the $300 sufficiency. So, Peter goes to a hearing in District Court on May 13th with me in tow. The court is filled with people waiting for their case to be called. The judge quickly gets to Peter and asks about the case. Peter tells the judge that he knows he has to leave the property, is not asking to retain possession, but needs to care for Laurel, who had been his tenant and still living at the property and cannot take care of herself. He only wants permission to be on the property to care for Laurel, which Judge Ray's order has forbidden.

It is over in about 5 minutes, over the objections of the plaintiff's lawyer, who is on Zoom.

The amount of the sufficiency never comes up. The District Judge dismisses the case and remands it back to Ray's court. We walk out around 2:15 to Big Creek to get coffee. After a short visit, Peter says he has some errands to run.

Moore voice mail, March 13th, 3:23 PM.
Jim it's an emergency. Laurel just called me the sheriff's at the back door with the court order to evict both of us immediately. Call me back please, soon as you can. Think we need you here too and I just missed a call from somebody too. Anyway give me a call please. Thank you bye.

Laurel Ferriter text:
HELP!! The law is at the back door.

Voice mail from Laurel Ferriter:
Hey Jim this is Laurel. The sheriff and somebody showed up at the back door saying we have to get out of here. The door is locked and I can't get up. I told him I had to wait till Peter came back. But please please come back with Peter, please. He's looking for you right now. They told me I had to call Peter to get home, so this is timely. I know he's probably already left you messages anyway. Please do get in touch with him thank you. Bye.

Laurel Ferriter text:
Peter doesn't know where you are...wants to arrive together.[45]

Not one to be tied to my phone, I finally hear it buzz and look at it. I can't get Peter on the phone. I get in the car and head over. I get to Peter and Laurel's house about 20 minutes later. Peter arrives a couple of minutes later. The deputy is gone. Laurel could hear someone talking to the deputy on the other side of the back door — she couldn't get up to open it — we assume it was the plaintiff. Whoever it was, the deputy had second thoughts.

I suggest to Peter this had gone too far for self-help; we need to raise money for an attorney. I talk to an attorney in Missoula whom I have worked with before. We come up with a $5,000 retainer, some from Peter, some from me, and some from Bill Good, an aptly named professor, and some small donations. The Missoula Attorney will end up deciding to do the work pro bono after we hire him — so we would get most of the money back when it was all over. Bless his heart.

Back in April, the plaintiff's lawyers had filed a complaint trying to get around the Governor's no-eviction order by claiming an exception,

> ... any action to evict a tenant based on damage or destruction to the premises, criminal activity on the premises, ...

They accused her of criminal trespass.[46] Laurel is very upset at being called a criminal and files a counterclaim with her response, saying they had knowingly misapplied the criminal statute.

With that situation, Laurel and Peter's lawyer calls into court. He told me that the clerk had indicated it was OK and given him the number to call. Laurel, Peter, and I show up in person on the 27th.

Judge Jennifer Ray opens with an admonition to their lawyer, saying he should have shown up in person and that she had not given him permission to call in. The Judge then asks the attorney if he is aware of the appeal. Of course, he is mindful of the appeal, but it seems that her point is not procedural but an admonishment.

Judge Ray goes over COVID-19 directions she from the Montana Supreme Court, seeming to have little interest in hearing about how the circumstances relate to the Governor's order or the relevant laws. She makes a statement to the effect that it is not her job to enforce a Governor's directives.

Judge Ray signs an eviction order for Laurel, despite her comorbidities, which had been included in briefs to the court. The attorney immediately appeals to the District Court, filed on the 28th, with the undertaking specified in the rule. Laurel's lawyer reports the next day that Ray's clerk told him that Judge Ray doesn't think an automatic stay is in order. I am confused; I see no exceptions in MCA § 25-33-204:

> Stay of execution when undertaking filed. If an execution is issued, on the filing of the undertaking, the justice or Judge shall direct the officer to stay all proceedings on the execution...[47]

Laurel has 72 hours to get out. After hashing it out, it didn't seem worth it to any of us. The remedy would be to ask the district court to intervene. We would simply be spending money. Peter and Laurel have to leave sometime — though if the Governor's Directive had been followed, it would have been a few months out; Peter and Laurel have always been willing to pay rent to the new owner. The lawyers agree to another week or so for Peter and Laurel to move out. The rental market is getting worse, not better.

The old Freightliner refrigerated 26,000 Gross Vehicle Weight straight truck that I had once used to deliver native plants, now waiting around to be sold, is chocked, dolly ramp rigged, ready to load, parked in front of the house and garages on North Burnt Fork Road, east of Stevi.[48] We have a dolly, ropes, and blankets — we need muscle.

I talked them into it: I hire my daughter and the oldest of her teenage daughters to help load up. Bill Good can come over for a while. Peter helps as best he can. There was no way everything would fit since there were two large garages full of stuff, some valuable mechanics tools. We start. Peter calls out what goes in the truck.

Just keep your body moving, don't stop contemplating what to do, just keep moving unless you are on a break; it will get done. A day and a half later, the truck is full. This 73-year-old body is tired. Bill Good is kind enough to allow me to park on his land, which is a few miles away.

The cats are still there, dozens of them, and most aren't that tame. So, now it's time for a cat roundup. I didn't know until now, but I am a decent cat wrangler. Well, Peter and I are decent cat wranglers. Peter holds a box or cat carrier at the ready. They repeat the same route as they scurry away from my grasp. The third, fourth, fifth, or sixth time around, with my work gloves on, I have a cat in my hands. The cat is not happy — claws out, wiggling, screeching, trying to bite. In the box, it goes. I get wounded — one latches on, though his teeth did not go through the leather work glove, I get punctured anyway.

Bill has an old bumper-hitch trailer and barn that Peter has set up as the cats' new home.

Peter and Laural go through a couple of motels in town until they land in Super 8. By now, we had come up with all kinds of potential schemes, such as an old RV and rentals in Butte, Anaconda, though it would be a hard sell with cats and no rental references. After paying full fare for a couple of days, the owners of Super 8 are kind enough to rent Peter and Laurel a room for $600 a month. Thank God for kindness: Peter and Laurel are not on the street. Laurel was not gurneyed being out under the watchful eye of a sheriff's deputy and driven to a hospital that doesn't want her.

A room at Super 8 will be their home for the rest of 2020. With COVID-flight well underway, Ravalli County is a favorite landing spot, and the lack of rentals is getting worse. Good people are there when Laurel and Peter need them.

A new group, Stand Together for Freedom, comes into being. Even though the lockdown is over, the uncertainty about disease, jobs, business, and protests in the streets on TV provide a ready audience for Alan Lackey and his wife when they decide to call a meeting.

Alan Lackey has been in the beef business "for a number of years" on leased land laying on the New Mexico high plains east of the Sangre de Cristo Mountain Range. Alan migrated to King County, Texas, for a while, a couple hours' drive to Corpus Christi, where I had gone to high school. Then, back to New Mexico, where it became too hard to hang onto "a historical job." By then, Alan had enough of a reputation to be hired as a consultant by a Wall Street "tycoon" to start an organic beef operation in upstate New York.[49]

He had been in the valley as a visitor for the first time in 1978 and, like many visitors, had it in the "back of his head" that it would be a good place to live. So, when he got a call in 2012 from his cousin in Corvallis wanting to start a wood-stove business, he came out to the valley. He had learned meat-cutting along the way. After working as a butcher for the market in Stevensville, he set up his own business selling custom meat.

As the pandemic hit, Alan was "like most people, minding my own business making a living, assuming that people in decision making positions thought like we did." When this seemed not to be the case, Alan thought, "I have not been a good citizen because I haven't been doing my duty which is staying involved in the process. So, I picked up the Constitution and started reading it."

They got a "gracious" offer from the Fort Owen Inn as a meeting place, just south of Super 1 at the Stevi-Y.[50] Everyone had been locked down and shut down, so when he put up bright pink fliers on bulletin boards that said, "Freedom Meeting," 35 people showed up. Through word of mouth, the membership quickly grew to 200 people.[51]

I got an email from someone telling me about the meeting. I am motivated to go, to see if I could get some traction on a First Amendment lawsuit — it had become moot when the Phase 1 order came out — but it might happen again. I called Alan to get a sense of what he was about. He says it is nonpartisan and "about our freedoms," and a Constitutional lawyer is ready to go — and that they intended to act.

While I am not sure where the group would go, maybe in some direction I wouldn't follow, it doesn't hurt to show up, but I demur.

I respond to the person who invited me:

LIBERTY TREE

Kareemov (shutterstock) Color change James R. Olsen

I am interested if it is really about the core values in the Bill of Rights and not anger-driven. But frankly, I am not sure it would be good for my health to jump into another issue with Peter and Laurel going on - so I am going to pass on the meeting.

I would later learn what I missed. The first few meetings become a discussion about how to keep Ravalli County reliant and free. The group ends up with four areas of interest:

- One is education; it has become more of an indoctrination system for socialism.
- Two is the Salish-Kootenai Water Compact, which the group saw as a threat to property rights — "and still is."
- Three is to get involved locally in politics.
- Four is to make sure we have a Constitutional Sheriff.[52]

One, two, and four have a long historical tail. At The Table, the discussions go something like this: socialism is the gateway drug to communism, the Soviet version, a dictatorship with an explicitly godless ethic that wipes away private property and freedom. But are we willing to watch a child die at the doors of the hospital emergency room because her parents can't pay? When we can get past the angst and partisan rhetoric, we do not, and never really have, lived in a free-market capitalist society — because no one was willing to see a child die at the doors to the hospital emergency room.

The wealth gap complained about by Karl Marx, the rich capitalists making all the rules and gathering in all the wealth, was not wiped away by "re-educating" the capitalists; the wealth-gathers were simply replaced with Communist party bosses.

Meanwhile, the worst aspects of free-market capitalism were mitigated by adding socialist underpinnings to our economy — the hospital has to treat the child, and the cost is spread across all the bill-paying patients; things like child labor went by the board, not based on some economic philosophy but because of Judeo-Christian ethics

with some secular Enlightenment thrown in. The real question is not if we have socialism — we decided that long ago — but how much?

I thought the Salish-Kootenai Water Compact was resolved a few years ago. I guess it's an old sore, as the tribe and the Federal Government filed hundreds of water claims, including off-reservation for water connected to the drainage on the reservation. The compact between the tribes, the United States, and the State of Montana put all those to rest. Not everyone was happy when the Montana legislature voted for the compact.[53]

This conflict is one of many. I live in the homeland of the Salish. My house lies on the river bottom, on land that the Salish once fished, hunted, and gathered bitterroots in the early spring. It lies within 5 miles of where Sacajawea walked on her travels with the Lewis and Clark expedition. Even though the Salish were living peacefully with white settlers, in 1855, the US Government insisted that they combine with other tribes and move to what is now called the Flathead Reservation.

Like many of these stories, to displace the people without a fight, the Hell Gate Treaty was made — it defined the tribes as a "nation," a dependent nation, that was given land for their exclusive use. Then, like in many other situations, the US Congress allowed plots to be homestead by non-tribal members 50 years later, over the objections of the nation.[54] This entangled legal history has resulted in a checkered history of attempts at assimilation and a return to the idea of sovereignty.[55] As to water rights, the fictional developer's claim that you can do anything with water on your private land in Montana in *Yellowstone* Season 1, Episode 1, is a myth that runs with the Jackalopes.[56]

I am not sure what Stand Together for Freedom means by Constitutional sheriff. Alan says a sheriff who believes in the Constitution; we both include Sheriff Holton in that camp.

The Constitutional Sheriff Movement that some people espouse at The Table is what I call an alternative theory of government. It is rooted in the question, "Who gets to decide if a law is Constitutional?" The theory of government expressed in the US and Montana Constitutions is that what is Constitutional is decided by their respective supreme courts. But people who don't trust the courts often put forth other ideas.

Some claim that a Constitutional Sheriff is the locally elected sheriff — the one who gets to decide what laws are Constitutional. Some aspirants claim the sheriff is above the authority of even a legislature if it comes to that. When this point is raised, I point out that the word "sheriff" is not used in the US or the Montana Constitution. In Montana, the sheriff's authorities and duties are specified in a law passed by the legislature.[57]

Nullification has a long history, reaching at least back to the Civil War. Nullification is the idea that a state can declare a federal law unconstitutional.[58] Nullification was in play in the 1960s, a term used in Martin Luther King's "I Have a Dream Speech."

Jury nullification goes further back to before the American Revolution: A jury can ignore the law based on moral principles. Examples include ignoring the Fugitive Slave Act by Northern juries before the Civil War and ignoring crimes against African Americans in the South after the Civil War.[59]

As Stand Together for Freedom grows, they begin a self-education program while seeing more and more flaws in America's response to COVID-19. "Most everybody is a Christian in the group — some aren't." Even so, Alan Lackey opens each meeting with,

> This is Stand Together for Freedom, where we have a heart for Jesus, America, and the Constitution.[60]

The group dives into the Constitution, educating themselves on the basic principles. They use videos from the Wall Builder Program that have a lot of quotes from the founders corresponding with each other, which Alan said "is good to know." Alan says there are flaws in the Wall Builders, but it is a good "historical perspective."[61] The other series is from the John Birch Society.[62] Alan notes,

> For the first few weeks, just like everybody else, [we wondered] how bad is this? But a lot of things just didn't make sense.

Medical professionals came out — good research and information about the protocols that didn't make sense. And they showed how it is more about money than health.

> The whole medical profession sold out... we are headed toward corporate feudalism. The problem is that politicians started making medical decisions.[63]

The very legitimacy of the government is being questioned more often by more people, some violently, some peacefully, in response to government mandates. Tumultuous waves from all directions are headed toward Montana.

WTO protests and Occupy coming from one direction. The Sagebrush Rebellion and Militia coming from the other. Now its Black Lives Matter and resisting unacceptable COVID-19 mandates. Is there a center, the place where one believes that the government can be fixed? There seems many who would still agree with the quote:

> My country, right or wrong;
> If right, to be kept right;
> And if wrong, to be set right.[64]

For the Great Divide, the Great Disagreement is, "What is right?"

On the 26th, the video of George Floyd flashes across screens everywhere. It is hard to justify — as more footage is released, it becomes impossible to justify.[65]

*31 May 2020: 107,674 COVID-19 Related Deaths Reported*

| 1 | 4 | 8 | 10 | 15 | June 2020 | 18 | 20 | 22 | 29 |
|---|---|---|---|---|---|---|---|---|---|

MT Traveler Quarantine Stops

data from private Surgisphere
authors could not vouch for
Lancet pulls 2 HCQ Studies,

recession since WW II
World Bank predicts worst global

2 Million US Cases

FDA suspends HCQ emergency
approval for COVID-19

UK approves Dexamethasone
for COVID-19

WHO ends HCQ study

NIH ends HCQ study

Study: 80% people with flu symptoms
may have actaully had COVID-19

Gilread set price for
Remdesivir at $3,120

### COVID-19 Treatment Guidelines (NIAID COVID-19 Treatment Guidelines Panel)

• Hospital: Dexamethasone Only for patients on oxygen (O2); Remdesivir: Ventilation, O2 < 94% or supplemental O2.
• Recommend against using Hydroxychloroquine • Consider Awake Prone Positition in lieu of Intubation when indicated

Image by STOCKIMAGE (Adobe Stock)

*Flag on a Wall. Protest*

# 9

# A KNEE ON A NECK, A CRACK IN THE WALL

*It is not enough for me to... condemn riots. It would be morally irresponsible for me to do that without, at the same time, condemning the... intolerable conditions that exist in our society.*[1]
— Reverend Martin Luther King, Jr

## JUNE 2020

I can't breathe. It has been less than a week since the image of the knee of a white police officer pressing into a helpless black man's neck swept the nation. The streets were already simmering with the killing of Breonna Taylor[2] in Louisville in March, adding to a long string of questionable deaths that go on year after year. Too many, for too many, Reverend Martin Luther King's dream is still unfulfilled. The word spreads.

10-4. The country explodes in protests, nearly all peaceful, but not all. There are always a few waiting in the wings for an excuse, saying violence is the answer, even if it's not. More people on the streets, more 12[*], more police on the beat. Angry people on the street march, chant, and scream, running, breaking, burning, and blazing bright on TV screens, computer monitors, and smartphones across the country — as the calendar flips from May to June.

III[**] percenters. Counter-protesters clash.

Police activate, separate, de-escalate. Some misbehave.[3] Black Lives Matter, I Can't Breathe, No Justice No Peace is scribed at the end of a sizzling iron rod, branded on the hide of the English Language. All Lives Matter quickly becomes a counter slogan. The protests and counter-protests burn like wildfires lit by a summer thunder-

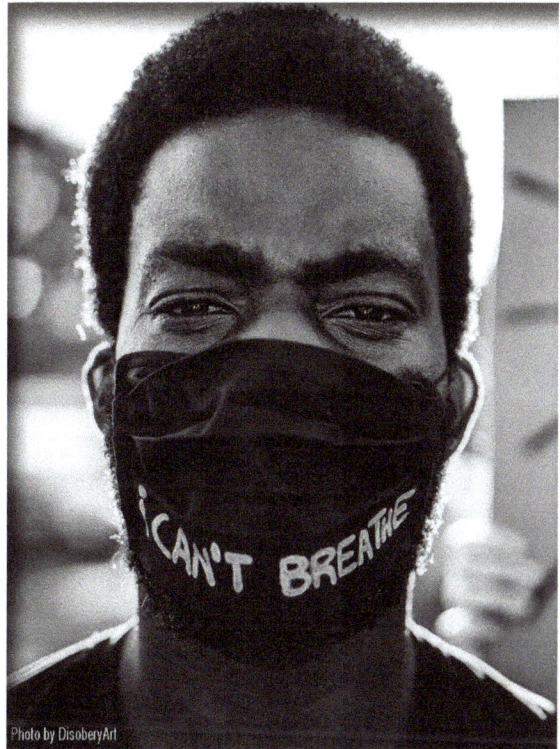

Photo by DisoberyArt

*I Can't Breath*

storm on a parched landscape driven by a howling wind. Peaceful demonstrations by day, a war zone at night is not uncommon.

A crack in the wall.

As the protests arrive at the gates of the White House, the President is briefly hustled to the "bunker." The Secret Service reports violent in-

---

\*    12 = Police present: Carson, Ken (performer), "Fuk 12 Lyrics," Produced by Bart how, Star boy & Outtatown., Track 15 on X., Released Jul. 8, 2022.

\*\*   III = 3-percenter. Symbol used in many 3 percenter patches. Wikipdia, Three Percenters: https://en.wikipedia.org/wiki/Three_Percenters

*Lafayette Park Floyd Protest*

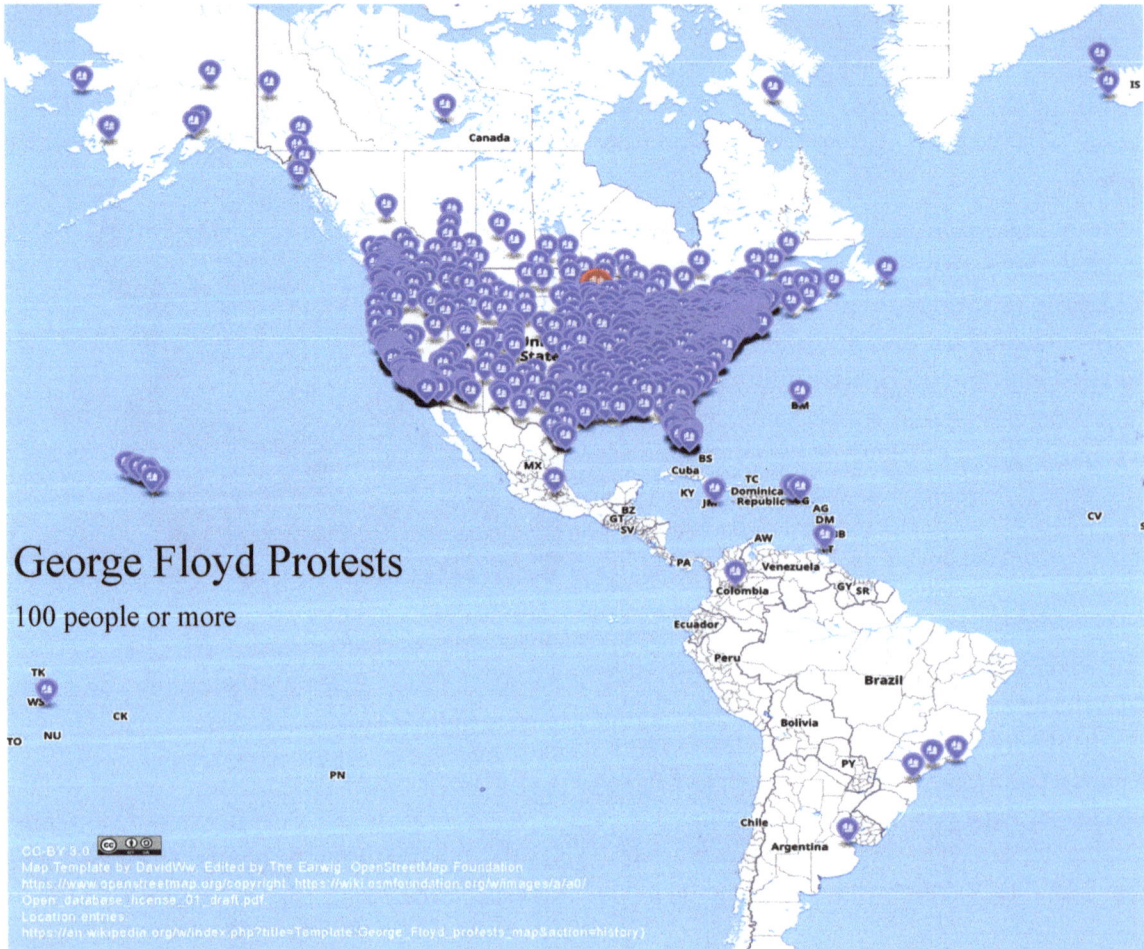

# George Floyd Protests

100 people or more

*Black Lives Matter Protest Map Worldwide*

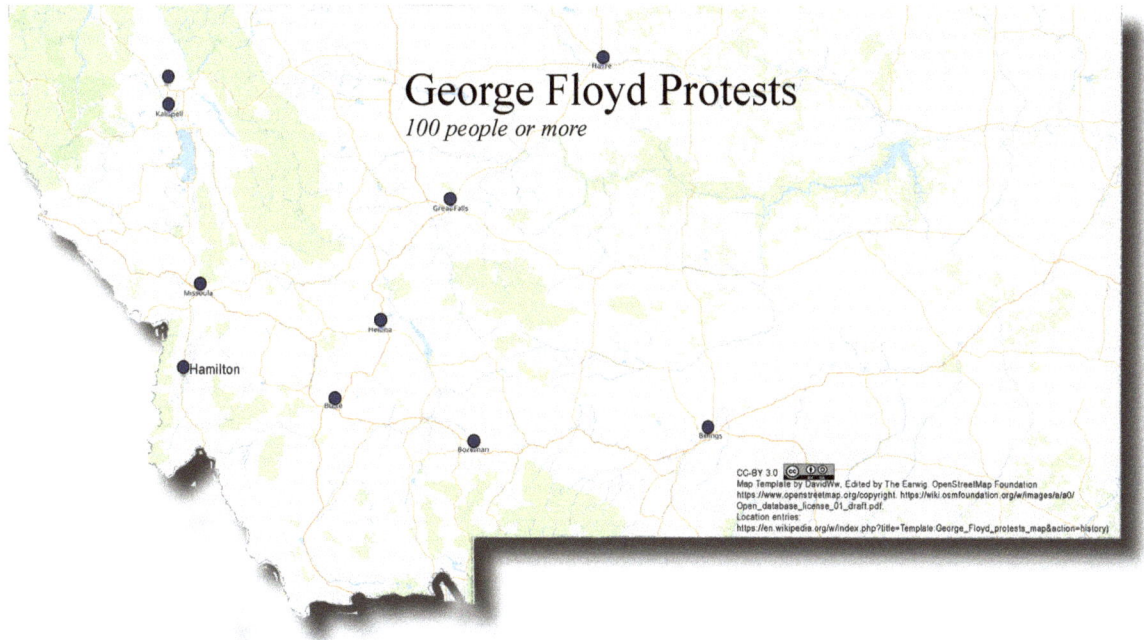

George Floyd Protests
*100 people or more*

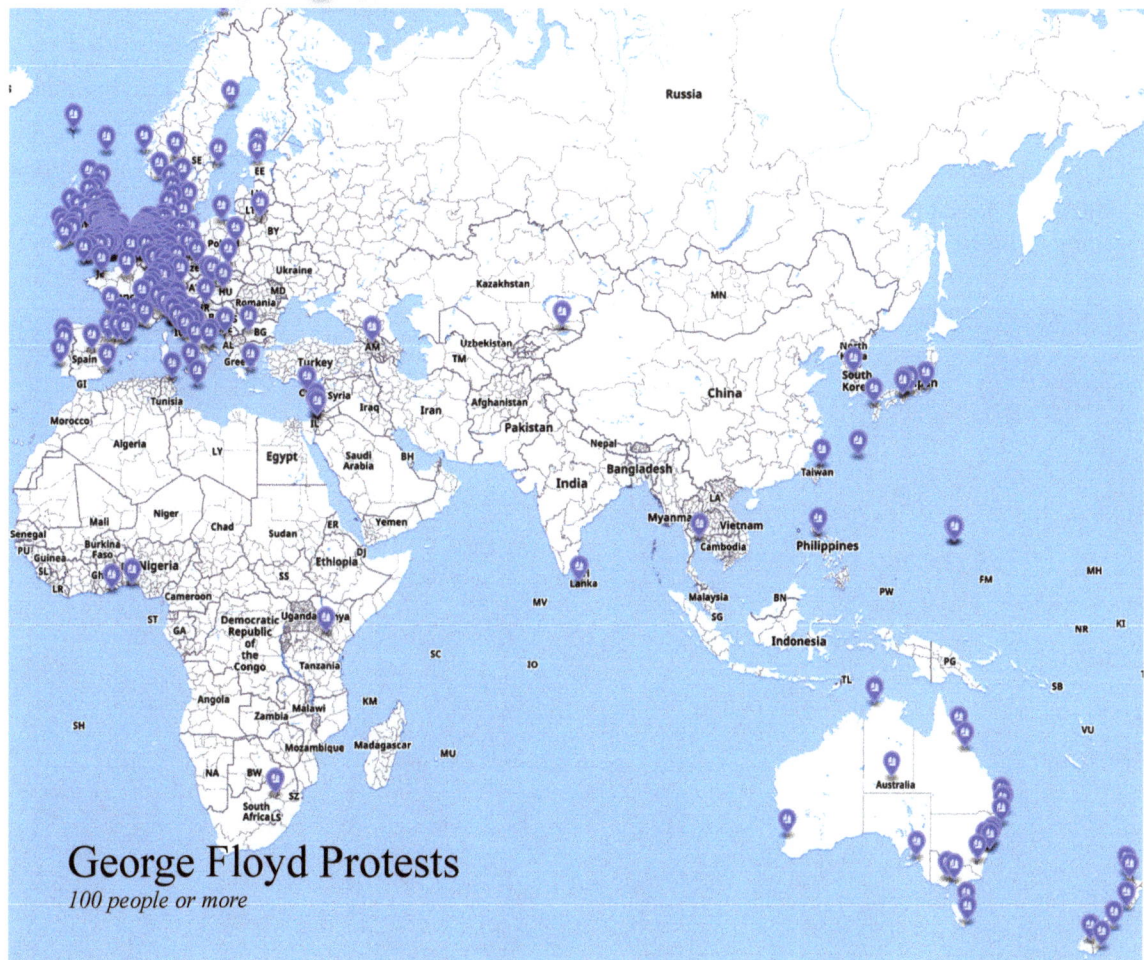

George Floyd Protests
*100 people or more*

cidents; the National Guard is called up in several cities across the nation.[4] On the 1st, President Trump walks, Bible in hand, from the White House to Lafayette Park to make a surprise speech in front of a church. Peaceful protesters are ordered out, pushed, rushed aside. This event would be cussed and discussed for over a year — it turns out the police were planning to put up a fence to keep protests further from the White House; Trump took advantage of the opportunity to make an appearance.[5]

Senator Tom Cotton's op-ed in the New York Times focuses on the violence during the protests, noting that businesses have been destroyed and police officers shot. His answer is "Send in the Troops," as was done in the '50s and '60s, invoking the Insurrection Act. While some have called on their state National Guard, no governor has asked for the active military.

On the other hand, the op-ed does represent the opinions of some portion of the population, so it seems fair that it should be aired. But the newspaper has a revolt in the newsroom. The Opinion Editor's sin is that he didn't vet well enough within the newspaper. However, if vetting means not allowing the essence to be published, it is a sin against balanced journalism.

The News Guild of New York condemns it, saying it "puts our Black staff members in danger, promotes hate, and is likely to encourage further violence." The editor, James Bennet, is forced to resign, although I cannot find "promotes hate" in the op-ed piece.

This seems to be a microcosm of the national divide. It can be challenging to support the cause but condemn violence no matter who perpetrates it. But that is needed.[6] Instead, it's a crack in the wall.

Black Lives Matter protests start in Montana cities, including Missoula, on the 2nd. Rumors that Antifa is coming to town — not true.[7]

How this will all play out is unclear, but Stand Together For Freedom will not just wait. Alan Lackey reports that they are doing some things to get ready:

> People were ill-prepared for this, so we put a lot of emphasis on preparation for a

Photo by Chris Owens (Shtterstock. Cropped James R. Olsen)

*George Floyd Protest Turns into Riot*

disaster of any kind. We have some minimum requirements for having food and water, and alternative way to heat the home. We have our own repeater so we have communications.

We formed neighborhood groups that meet. We have a lot of older people, a lot of widows. We have been able to help them.

Like minded people have come together. This has been a very dicey situation. We also developed a shooting range to let people become proficient with their firearms. We look at Portland that burned, and Kenosha, Wisconsin and all the lawlessness. We don't see the normal response you would expect from law enforcement.

We're seeing a breakdown of the social pattern and structure of society. And, people are frightened.

What we do, we do to protect ourselves in the Valley. We have people that are ready... It's not going to be a panic situation.[8]

After this conversation, we both agree that law enforcement in Ravalli County would provide an appropriate response.

This isn't militia rhetoric, but the group's earmarks will likely bring some scrutiny. Some of the members use ambiguous rhetoric. For example, a line from a letter on the Stand Together for Freedom website, "to actively resist anything, anyone, any group of people, and any other nation who want to take down America" after a dissertation about the "Marxist Left."[9] Of course, "active resistance" is very ambiguous as to the means since the term has been applied to everything from the non-violent resistance of Martin Luther King to the American Civil War.

Alan uses broad stripes for who is invited into the group when it started, so this does not necessarily represent the group's purpose or the sentiment of many of its members — I talked to some informally. Where the group ends up remains to be seen.

A crack in the wall.

It is almost a pop-up protest in Hamilton on the 7th — my wife says she is going to a Black Lives Matter demonstration on Main and Highway 93, a favorite corner for political rallies. Though I sympathize with the cause, I go primarily to be there with my wife if someone does something stupid.

I can't help myself. When I walk up to a group of people I know handing out little American flags, joking, "Is this the second amendment rally?" I can't get a laugh.

I take one and start to walk the few paces back to the corner of Highway 93 and Main. Someone pulls up at the stop light and shouts, "all lives matter." Someone nearby says how stupid he is. I react. "He's just trying to figure this out just like the rest of us."

It is a good bit of organizing; organized by Frances Carrasco, a high school student from Corvallis High, and her friends, over 200 people showed up. She says,

We can all remain peaceful in supporting a big movement that's much bigger than ourselves.

She goes on to say that she didn't think we have problems with policing that urban areas do.[10]

I learned later, watching TV news, that a block and a half away, a guy gets in an angry rant claiming that, in Black Lives Matter protests, "violent people use it for their gain." A female Hamilton police officer appears and asks him to disengage. When he refuses, she says he is being detained. He pushes her and says, no, he's not — backup arrives.

The amazing thing, what any good activist wants to happen, is that he came back after his talk with the police, apologized, and proceeded to talk about the issue[11] — standing in contrast to self-appointed militia counter-protesters arriving to "protect" businesses. In Missoula, a teenager is hunted down and detained by the so-called militia because he is afraid to show his face behind a mask.[12]

Black Lives Matter protests are televised from Salt Lake City, with plenty of peaceful protesters but infused with violent-minded agitators. It quickly gets out of hand. The Utah Governor calls out the National Guard.[13]

There is an old saying, "don't write anything you don't want to see in the newspaper." But, a school coach did write something on Facebook, and it did get in the newspaper. So, the Darby School Board gets to deal with Black Lives Matter. It is a microcosm of what the country is going through. The coach is reacting to news from Salt Lake City. Now, here it is on the front page of the *Ravalli Republic*:

> Replying to an initial Facebook post, he posted, "they should all be strung up and hanged in public like the old days. Lot less of that sh*t would go on."

The post has since been deleted.

> I'm deeply regretful that I posted that... That is my problem to bear.

The school board offers the following statement in a press release:

> Coach ... comments on social media were unacceptable. No effort to explain them... erases the fact that he chose those words and made the conscious decision to share them with the world.

> Those words do not represent this school, this board, our teachers and staff, or the community of Darby as a whole.[14]

He is suspended for a year. The two-hour meeting draws out-of-towners from as far away as Billings. A father-son team from the Blackfoot Reservation is organizing a boycott of the Darby football team, hoping no one will play them until the coach is fired permanently. Most speakers want the coach fired, but the coach's apology at the beginning of the meeting "really meant a lot to me," according to the School Board chair, adding, "It's bigger than Darby..."[15] True. The coach

and his family become the subjects of harassment and a death threat.

A crack in the wall.

Hollie is planning to go to Portland to look up an old friend. She asks how this compares with the 60s — she's not the first. In 1967, '68, '69, especially where I was going to school, the University of California at Berkeley, the world sometimes fell into violent chaos. Looking back at it, I say that many good things came out of it, such as steps forward in women's rights and getting rid of Jim Crow laws, though *James Crow*[16] is still with us.

I tell her about Jim Crow. When we moved from Detroit to Virginia in 1954, when I was seven, I remember asking my mom at a bus terminal why there were whites-only and blacks-only bathrooms. She could only say, "That's just how it is." I said nothing, but it didn't pass the kid logic test.

How could "how it is" be different? The answer was not as apparent as it seems now because no one had ever seen it, certainly not a seven-year-old. When I was in High School, one thing Martin Luther King did was answer the question. Though he wasn't winning any popularity polls at the time,[17] he planted the seeds of a truth that could not be long denied. Progress — obviously not enough.

I tell Hollie about my first vote ever. It was in the California primary in 1968 for Bobby Kennedy. I watched his assassination on TV the next morning.[18] A lot of hope and optimism died that day. I thought, "Well, they killed another one," as assassinations seemed to define this time in history — President John F. Kennedy, Martin Luther King, Malcolm X, three civil rights workers, little girls at a church, Medgar Evers, Vernon Dahmer, an unknown number of lynchings, leader of the American Nazi Party, and at least one politically motivated murder of Black Panther leaders involving the FBI.[19] Political assassinations were going on around the world, a decade of political killings: Indonesia, Iran, Iraq, Israel, Japan, Laos, Lebanon, Thailand, Vietnam, Argentina, Bolivia, Brazil, Dominican Republic, Guatemala, Haiti, Honduras, France, Germany, Greece, Italy, Portugal, Spain, Switzerland, Algeria, Congo, Ethiopia, Kenya, Mozambique, Nigeria, Senegal, Somalia, South Africa, Tanzania, Togo. It was like they were reporting the weather.

There was a war on and a draft on, both adding napalm to the already volatile mixture gathering on the streets in the '60s. I didn't have time to protest or party. I was working my way through school. I was at Berkeley because that's what I could afford. I got my draft notice after I graduated and signed up for the Air Force. My young wife pinned lieutenant's bars on my uniform in April 1970.

*For What It's Worth,*[20] I did see it unfold; I walked through it on my way to classes, on my way to work, protesters behaving and misbehaving, reporters sometimes describing what I saw and sometimes making it up. The heat was there for more than a field day.[21] The police finally killed someone "in cold blood" during the People's Park march as they sprayed unarmed people with buckshot.[22] When asked about the buckshot, California Governor Ronald Reagan, with an eye on running for President, replied,

> It's very naive to assume that you should send anyone into that kind of conflict with a flyswatter.[23]

"For all that, the world went on, things came out OK, history has always lurched along. And, the '60s certainly weren't as bad as World War II, which my parents endured." I tell Hollie, "This is probably no different — but I could be proved wrong."

I share my experience with how protests and riots work from what I have seen. When you see a protest on the news, the news camera zooms in with a telephoto lens on where the action is and where the violence is, filling the camera frame.

There are always a few people there to do "something stupid." By that, I mean to get their adrenaline up, to do some violence, often not even caring about the issue.

I go on to say that most people are peaceful. You have a wide-angle lens when you are in the midst of a protest or walking through one. When you see that flurry of stupid stuff at the end of the street, you can always walk the other way. When the excitement and emotion reaches out with octopus arms trying to suck you in, you can always walk the other way.

Hollie did go to *Portlandia,*[24] with its downtown walls cracking and windows broken, if you believe the TV. She came back unscathed, wisely avoiding downtown altogether.

Even as the news is filled with protests, the online sister to *Science Magazine, Science Insider*, shoots three more canon balls through the sails of the *SS Hydroxychloroquine,* "Three big studies dim hopes that hydroxychloroquine can treat or prevent COVID-19,"[25] the latest in a series of news pieces questioning the effectiveness of hydroxychloroquine in treating COVID-19 and complaining about President Trump's grandstanding for the drug. People in the FDA and CDC are concerned about over-using it.

The first study cited in the *Science Insider* article is an ongoing RECOVERY trial in the UK. "No Clinical Benefit from Use of Hydroxychloroquine in Hospitalised Patients with COVID-19,"[26] reads an early press release.[27] A study will come out in November in the *New England Journal of Medicine* showing a 27% mortality rate with HCQ and 25% with the Standard of Care *in a hospital setting.*[28] It seems consistent across the studies that HCQ was not doing much once you are in the hospital.

Peter Navarro and Doctor Fauci continue to butt heads on HCQ. Even so, Fauci arranges a Zoom meeting with the COVID-19 Treatment

Guidelines Committee for Navarro to make his case. He doesn't.

On June 15th, the FDA revokes the emergency use of HCQ for COVID-19. There have been statements that, even though widely prescribed for other conditions, HCQ increases the risk of serious heart issues. While the FDA found about 130 heart-related issues reported, the correlation of the drug is unknown. Clinical studies, including RECOVERY, found no increase in heart-related mortality as a side effect. HCQ is a safe drug when monitored by medical professionals.[29] I review the FDA's revocation and find it unjustified, sending my analysis to the County Commissions and Health Board; the effectiveness of HCQ for outpatients remains unknown.[30] This is quickly followed by the NIH and WHO stopping trials that are in progress. As a result, it will be awhile before we find out if hydroxychloroquine is effective for non-hospitalized people exposed to COVID-19 or who have mild symptoms.

The FDA may as well have climbed to the top of a bare ridge during a lightning storm. The ruling has the opposite effect than intended. Pulling the emergency authorization for HCQ makes the problem worse, not better, in Ravalli County. It is not clear that the number of people treated with HCQ is decreasing.

So, is HCQ useful as a prophylactic or taken early with mild symptoms? Millions of people have taken HCQ for malaria and other diseases, often for over a year.[31] A later Spanish open label trial, "A Cluster-Randomized Trial of Hydroxychloroquine for Prevention of Covid-19" is informative. The study will conclude that HCQ does not reduce infections for people exposed to COVID. But, when I look at Figure 2 in the Spanish study, a line stands out — for nursing home residents, HCQ did improve outcomes. This suggests a larger followup trial.[32]

Even as the news is filled with protests, Doctor Calderwood reports to the Health Board that the last COVID case has recovered. There are now zero cases. She and Jeff Burrows have a meeting scheduled with the hospital CEO tomorrow to discuss testing.

With a broad smile, she adds, "Other than COVID, there are other things than COVID in the world."[33] There are eight cases of Colorado Tick Fever.

The Public Health Nurse says they did 1,674 tests for COVID-19 last month. She's hoping to get back to their "regular stuff."[34]

Zero cases don't last. On the 26th. it's official, the first community-acquired case in Ravalli County.[35]

*30 Jun. 2020: 126,806 COVID-19 Related Deaths Reported*

❧

# Cumulative confirmed COVID-19 deaths per million people

Our World in Data

For some countries the number of confirmed deaths is much lower than the true number of deaths. This is because of limited testing and challenges in the attribution of the cause of death.

**Data source:** World Health Organization (2025); Population based on various sources (2024)

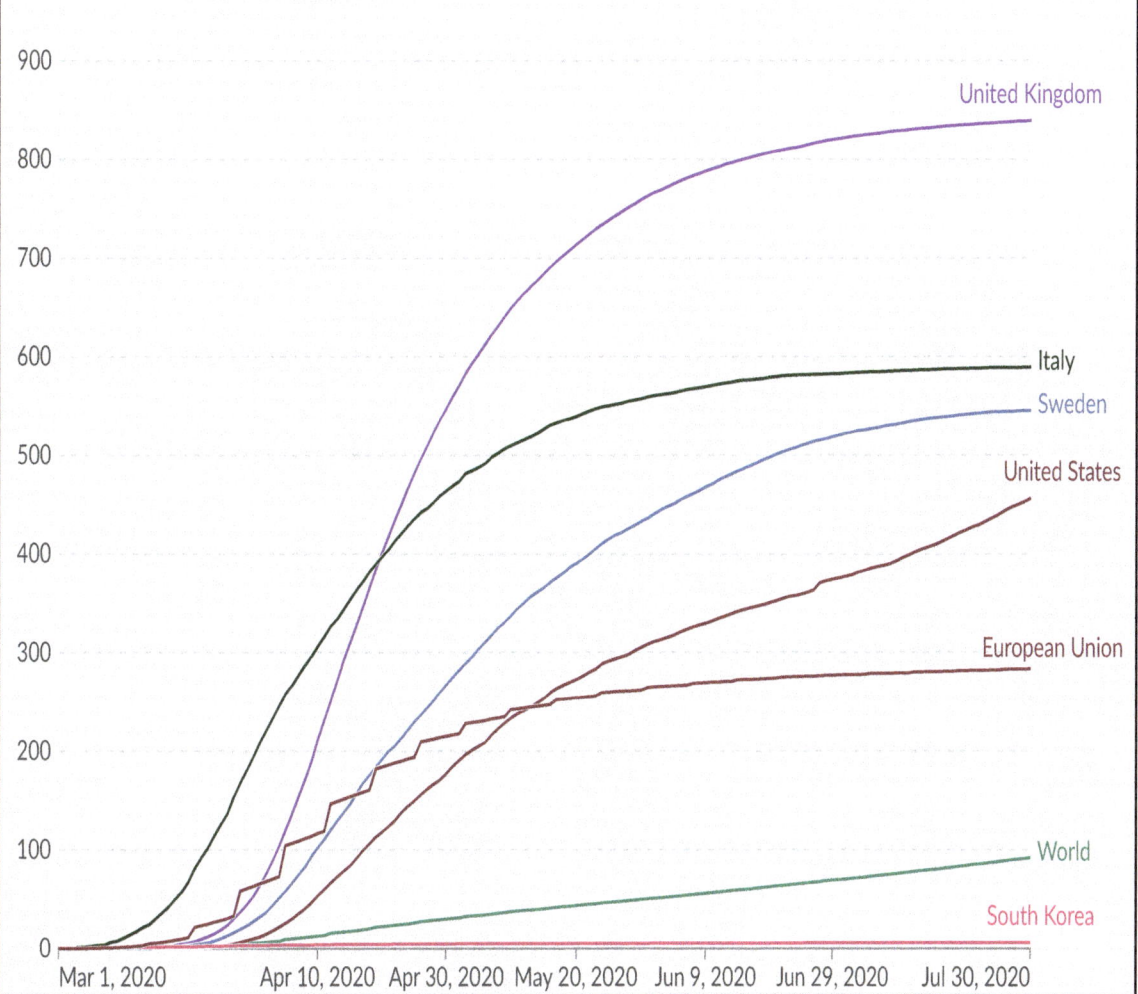

*Cumulative Fatalities per Capita June 2020*

2    4    6    7         14   15   July 2020   20   21   23      27      29

California, Indiana, and other states reverse reopening plans

INDEPENDANCE DAY

Scientists request WHO revise guidance re: airborne transmition

US exceeds 3 million cases

Early tests of Moderna vaccine show promise

Montana governor issues mask mandate

Diagnostic delays lead to cancer deaths

AsterZeneca & CanSino vaccines show promise

Report: antibody drop 3 mo. after infection

Moderna begins Phase 3 trail

Rapid antibody test approved

**COVID-19 Treatment Guidelines** (NIAID COVID-19 Treatment Guidelines Panel)
- Vitamin C, D, Zinc. Literature reviewed, not enough information to recommend for or against.
- Redesivir: Priority supplement O2 due to limited supplies.
- Dexamethasone substitutes when not available: hydrocortisone, methylprednisolone, prednisone.

*Mask Wars* Photo by Melina Nagy (Shutterstock)

# 10

# PULL UP YOUR BANDANNA

*What we are all pining for is freedom that imposes restraints upon itself for the sake of society.*[1]
— Gandhi

## JULY 2020

The month starts with 148 people in Ravalli County in home isolation. The *Ravalli Republic* announces, "After flattening the curve since March, the Ravalli County number of community spread coronavirus has increased over the weekend."[2] The administrators at Marcus Daly Hospital realize COVID-19 will "linger longer." A Viral Clinic is set up down the hall to the left from the Emergency Entrance, its operation an extension of the Walk-in Clinic. The Viral Clinic becomes a mainstay if you have symptoms.[3] Our Health Officer on her way to becoming an expert on the symptoms of COVID-19 because Doctor Calderwood runs the clinic.

July 1st is a big news day. Things are getting back to normal. Many people, including me, look forward to the County Fair every year, the 4-H[4] kids showing off their animals, local non-profit fund-raising booths, each with a traditional fair food offering, and a small carnival that draws plenty of teenagers. The blue-ribbon, yellow-ribbon, white-ribbon, best-of, award-winning pies, cakes, cookies, vegetables, eggs, chickens, roosters, rabbits, pigs, cows, sheep, goats, flowers, quilts, photos, poetry, the list goes on in the fair rule-book. And a rodeo. The Commissioners pass the budget and give the "go-ahead." The Fairgrounds Manager, Melissa Seville, says,

> Around here, people are very rebellious about wearing masks, but masks are not mandated. We're going to buy bandannas for people to use, but we are not going to police that... We're encouraging social distancing and have a huge property.[5]

The fairgrounds are right in Hamilton. They expect about 30,000 people to participate in the four-day fair. Melissa notes they had funding for extra sanitation and would love to have more exhibits. She notes that other counties have canceled their fairs, "but our public wants it," adding that she gets calls and messages daily saying, "Thank you for having the fair.."[6]

But everything isn't back to normal. The Hamilton Playhouse,[7] Hamilton's local community theater, is putting on *Steel Magnolias*[8]; masks required. The annual brew fest is postponed — a big fundraiser for the Chamber of Commerce, a chance for local brewers to show off their wares during a downtown street party.[9]

There is a new sense of impoliteness, creating an undercurrent of wariness, as one encounters slights over whether or not to wear a mask. There is a sense of uneasiness as images of street violence arrive daily through the news feeds and social media. The idea of an uprising gains traction in some quarters or even a civil war. They aren't acting on it but expecting it, some even preparing for it.

Meanwhile, Manzella says, at a Stand Together for Freedom meeting that Sheriff Holton has been invited to address,

If we have the threat of riots, he can raise a posse and he'll pull from his CWC (concealed weapon carry) list for the purpose.[10]

A couple of people who grew up here are worried, not about outside agitators nor Antifa invading Ravalli County, but a civil war of our own making. If they are worried, so am I. Do we really have to do this?

I relent, not for a civil war, of course. You won't need one for that, but something in between. I put in for a Concealed Carry Weapon (CCW) and go out to Whittecar Rifle and Pistol Range,[11] not to sight in my hunting rifle, the usual reason for the trip, but to brush up on pistols. A small arms expert ribbon on my DD-214[12] saves me from having to take the CCW course. It seems there is a run on everything, including Concealed CCW Permits. It takes a while to get an appointment for fingerprints.

It is not the first time our firebrand, the "Donald Trump of Ravalli County," State Representative Theresa Manzella, takes some political hay and lights it with a flame thrower. She is running for State Senate this cycle.

The tradition of militias that was evident as the patriots turned out during the Redcoat's march to Concord in 1775 traveled, in various forms, across the continent. In Montana, local militias were a fact of life as time marched toward the turn of the twentieth century — but, unlike the self-styled, self-appointed militias that threatened our Justice of the Peace in the 1990s, these militias did the bidding of the established government — Sheriff Holton makes it clear that if he enforced anything, it would be the law of the land as legislated.[13]

Theresa portrays herself as a cowgirl, and while I am not sure if she has ever pulled a calf or driven cattle to pasture, she does know her horses, has ridden well, and trained them well for decades. She moved to the Bitterroots a couple of years before I did, in 1991, from Michigan,

still in her 20s, settling in Sula for a while before moving toward Hamilton.[14]

In 2008, Montana had, along with much of the country, a bank crisis and economic downturn. Horses got caught in a domino effect, becoming a liability for many owners, resulting in hardships for many horses in the state because owners could not (or would not) care for them.

Theresa came to the rescue with a grassroots effort to find homes for abused and neglected horses, founding an organization called Willing Servants. The program included a "final act of kindness euthanasia clinic" for old and infirm horses and a search for burial sites for horses — and offered informal grief counseling.[15] She would later say that this led her to politics because she was picked on and threatened by the animal rights movement.[16]

She educated herself on the meaning of the US Constitution using one of the same video series from the John Birch Society[17] used by Stand Together for Freedom.

In 2016, Representative Manzella wrote,

I'm a Christian, Constitutional, Conservative, Republican with a 100% proven voting record over the last three sessions, from the Montana Shooting Sports Association, the NRA, the Montana Family Foundation, United Property Owners of Montana.

Elected to lead the state delegation to the RNC (Republican National Convention)... the first step in the election of President Trump.[18]

Since Willing Servants, Manzella is an adept opportunist, riding on controversy or movements that someone else has gotten going. She danced with the idea of turning over Forest Service lands to the state — this latter idea gaining no traction in a county where hiking, riding, and hunting on public lands is widely valued — and happy to have Federal money fight wildfires.[19]

Theresa Manzella embraces a brand that was forged and hammered by the Democrats. She pulls it out of the hot coals,

> I have been dubbed the "Donald Trump of Ravalli County" by Democrats for my candid, straight talk, which is something I embrace.[20]

Though I doubt the Dems meant "straight talk" — more likely matching her up with Trump for their similar wild and woolly politics.[21]

On the other hand, nearly every Democrat and more than a few Republicans wonder if she subscribes to the basic ideas of a representative democracy. A Letter to the Editor from Dave Bedey, who had been the Chair of the Hamilton School Board for years, calls her on it,

> ... what is more shocking is her lack of understanding of the Montana Constitution and the basic principles of representative democracy. Yet she continues to loudly insist that she is a "constitutional conservative."[22]

When we had a school walkout over school shootings in 2018, a person on Facebook said that the school district was planning to,

> Use our kids as organized pawns for the gun control lobby… It is my understanding that kids are being told to walk out against gun violence and to support victims of the recent school shooting. We've seen this all before. Of course, it's all carefully masked by supporting administrators and teachers as "student led."[23]

Manzella added that she was investigating teachers unions as part of the "radical left."

Someone was talking through their hat. The walkout was student-led; I knew an adult who was in contact with some of the students planning it. The scathing response from the school administrator noted that some individuals "had the courtesy" of calling the school. In contrast, others chose to "fire off their false, and often inflammatory, claims against the school in public and closed forums."[24]

o by Photo Contributor @Shutterstock

*Trump Supporters' Flags*

As far as posses go, the Sheriff agrees that the law allows him to call up citizens[25] but that it is unlikely, noting he would first call surrounding law enforcement agencies.[26] All that said, self-appointed militias are still around.

In Ravalli County, Republicans have not forgotten being called "a basket of deplorables."[27] You can buy those words on a T-shirt at the Republican Booth.

This characterization is not accurate for most people I meet in Ravalli County. Some people will say and write that Democrats, liberals, and the like are idiots, don't know the truth, and are lazy parasites who feed off people who work for a living. This is an example of the Great Divide, much of its ego-driven tribalism and partisan sound bites.

Republicans in Ravalli County sometimes have brutal primaries. The confrontation moved on to arguing about the very foundation of the United States. The Republican who won the state legislative seat in the district that includes Hamilton has a conservative mindset but serves as a contrast to Manzella. Firmly rooted in a mainstream view of the government, David Bedey says,

> Extremists at both ends of the political spectrum threaten the social cohesion necessary for our constitutional system to function.
>
> Those on the far Left have explicitly rejected our constitutional heritage and would divide us along racial lines within some sort of socialist utopia.
>
> This stance has predictably generated opposition not only from conservatives like me but also from those inhabiting the far-right.
>
> But many of these "patriots" are agitating for a second American revolution — a civil war. They have given up on America. I refuse to join them.[28]

Bedey is from here:

> ... born and raised in Hamilton. He studied civil engineering and economics at Montana State University...
>
> During a 30-year military career, Dave served in combat units—including deployment during the Persian Gulf War...
>
> Upon retiring from the Army as a colonel, Dave returned home, ... chief financial officer of a small engineering firm. ... served on the Hamilton School Board for nine years.[29]

Ravalli County politics is not just about Republicans, though. The founder of Hamilton, Marcus Daly, was a Democrat and was heavily engaged in party politics in the 1890s. After he died in 1900, the county often elected Democrats.[30]

When I moved to Ravalli County in 1993, the county attorney was a Democrat; the Ravalli County Board of Commissioners had a Democrat, Steve Powell, as a member. When I was, for reasons I still don't know, invited to the Democrat Central Committee, the members included ranchers, the kind of people who a hundred and fifty years earlier had created things like the Grange, whose halls still stand in Ravalli County.[31]

Like the Republican Party, the Democrat Central Committee has changed with in-migration. It is now in a challenging position in a county that has not elected a local Democrat for over a decade. The numbers have been consistent — the Democrats have about a 28% straight-ticket vote.[32] The Republicans have a single-ticket vote that has been creeping up — about 40%. There are still split tickets to chase after. In 2016, Trump got 67% percent of a record turnout vote. But, the math showed that 15% of Trump's voters also voted for a Democrat for Governor — Steve Bullock.[33]

The rule of thumb is that a statewide Democrat will not get a majority of votes in Ravalli County. Still, they need at least 38% of the county's votes to win statewide. This is why we have US Senate Democrat Candidates open campaign offices here. Other Dems will tag on — some-

**2016 Early Vote**

| Federal & State | State Auditor | State Senator Dist. 44 | State Rep. Dist. 86 |
|---|---|---|---|
| President | Laslovich - 4431 | Olsen - 1851 | Ehli - 1779 |
| Clinton - 3591 | Rosendale - 6600 | Thomas - 3411 | Neal - 1052 |
| De La Fuente - 29 | State Superintendent | State Rep. Dist. 85 | State Rep. Dist. 87 |
| Johnson - 472 | Arntzen - 6480 | Manzella - 2057 | Ballance - 19 |
| Stein - 117 | Romano - 4521 | Young - 902 | State Rep. Dist. 88 |
| Trump - 6583 | Public Service Comm. | | Gorski - 1104 |
| Representative | Gutsche - 3849 | County | Greef - 1756 |
| Breckenridge - 335 | Lake - 7157 | Clerk of Court | |
| Juneau - 4012 | Retain McGrath | Trautwein 9126 | Municipal |
| Zinke - 7048 | Yes - 7933 | Commissioner Dist 2 | Pinesdale Selectman |
| Governor | No - 2205 | Chilcott - 6713 | Gold |
| Bullock - 5194 | Supreme Ct Justice #3 | Smith - 4304 | |
| Dunlap - 361 | Juras - 4676 | Commissioner Dist. 3 | Stevi Elem Levy |
| Gianforte - 5849 | Sandefur - 5380 | Hoffman - 9380 | For |
| Sec. of State | Retain Shea | | Against |
| Lindeen - 4003 | Yes - 7697 | I 116    I 181 | |
| Roots - 280 | No - 2232 | Yes - 7066   Yes - 3608 | Stevi High Levy |
| Steckston 1980 | District Ct Judge | | |

*Ravalli County Early Vote* 2016 (Trump had a big election-day turn out of 7,929 compared to 2,632 for Clinton.)

times including presidential campaigns for a ground game sending a young out-of-town hired hand to recruit door knockers and organize GOT-V,[34] get-out-the-vote call banks. Bullock got 41% of the votes in Ravalli County in 2016 against Republican Greg Gianforte.

Mask mandates are controversial and are sweeping the country. When asked if he would enforce mask mandates if they show up, Sheriff Holton responds that the Health Board has already decided to leave COVID-19 mandates as a civil matter. Any enforcement would go to the County Attorney, who would take it to court as a civil matter. If the Board did call upon him to enforce social distancing, he would not have the manpower. If it is a choice between going to a domestic abuse call or enforcing mandates, he will deal with domestic abuse.

On the 8th, confusion reigns over the Health Board Waste Water Regulations. A hearing is scheduled at 2:00 PM, following the monthly 1:00 PM Health Board meeting. The wastewater revisions have been in the works for years, partly to keep up with the Montana Department of Environmental Quality (DEQ). A few Republicans and some developers, in addition to the Realtors® Association,[35] have been following the progress.

The regular meeting of the Health Board starts quietly enough at 1:00 PM on the 8th, with about a half dozen people in the audience, David Bedey among them, a few masked, most not. The Health Nurse reports that the National Guard is going around the state and will be doing free drive-through COVID-19 tests at the Hamilton High School Parking lot on July 15th — but the samples still have to be tested in Helena — which is still overwhelming. She will call anyone who tests positive but plans to mail negative notifications because of staffing issues.[36]

The tension in an otherwise routine meeting rises as people start lining up in the hall, waiting for the hearing to begin. As the crowd files in, they don't fit, leaving people peeking in through the door. Commissioner Jeff Burrows says that he didn't know if everyone saw the notice that they weren't making any decisions today. Several rumors and worries about what could happen are circulating.

Seeing the crowd, Jeff says,

> With the Health Officer here and with the rules, we'll try to disperse immediately. Doctor Calderwood says we have to have six-foot spacing.[37]

Excited muttering sweeps through the crowd, but Burrows keeps tamping it down. The crowd had turned out, Theresa Manzella among them, and they weren't going anywhere. They want to talk about COVID as well as wastewater regulations.

The Board tries to extricate itself. The board members discuss moving the meeting to a larger space to accommodate social distancing. The crowd is unmoved and unmoving. The heat starts rising. The pot is about to boil over when Burrows shifts the focus to the public process, saying the Board is ready to take everyone's comments and the large crowd will be accommodated, adding that there is a lot of misinformation and misunderstanding.

A person from the crowd raises their voice,

> We're fine, we're not going to spread out. It's not going to happen.

Then others chime in,

> We have the right to peaceably assemble; the right to be free of inspections.

The crowd starts talking all at once.

Burrows interrupts, amplified by his mic,

> Just a second, just a sec. We're not going to have a free-for-all.

The crowd starts discussing rights. Before a minute is up, Burrows interrupts again, raising his voice,

> We're not going to shout people down.

The Board decides to reschedule the meeting; the crowd keeps talking. Someone asks if Facebook is accurate. Burrows says, "It was embellished at best."

Then Burrows puts the crowd on the spot, "How many people have read the regulations?" Few hands.

After a lot of back and forth, lifting a notebook, pages open, facing the crowd, Burrows says,

> Before you come here with your pitchforks and your torches, pick up the regulations and just read them and know what you're talking about when you come here. That's all we're asking before you call it a Communist Manifesto.

Manzella tries to pin the Health Board down — a commitment to meet without social distancing. Burrows responds he is not scared of the disease, but some people are.

Another angry comment: "We are the people," and they "aren't going to social distance" and will show up and have their voices heard. "

The Board votes to adjourn.[38]

Then it happens — a death attributed to COVID-19. KPAX comes on the air with this news,

> Good morning, Ravalli County Residents. Montana is reporting another 127 new cases for a total of 1,593. There are 710 active cases, with 855 recovered. Montana is reporting 25 deaths. Ravalli regrets to inform its residents that we will be adding 1 to that total... Ravalli County has 12 active cases that Public Health is actively monitoring, along with numerous close contacts in quarantine.[39]

Costco has begun requiring masks to enter their store. Their customers feel safer shopping. More customers file in through the sliding doors — Wall Street votes with their money. Costco's stock price rises as the DOW falls. When large retail chains see the surge in customers, they follow suit. Where big business goes, politicians are usually not far behind.

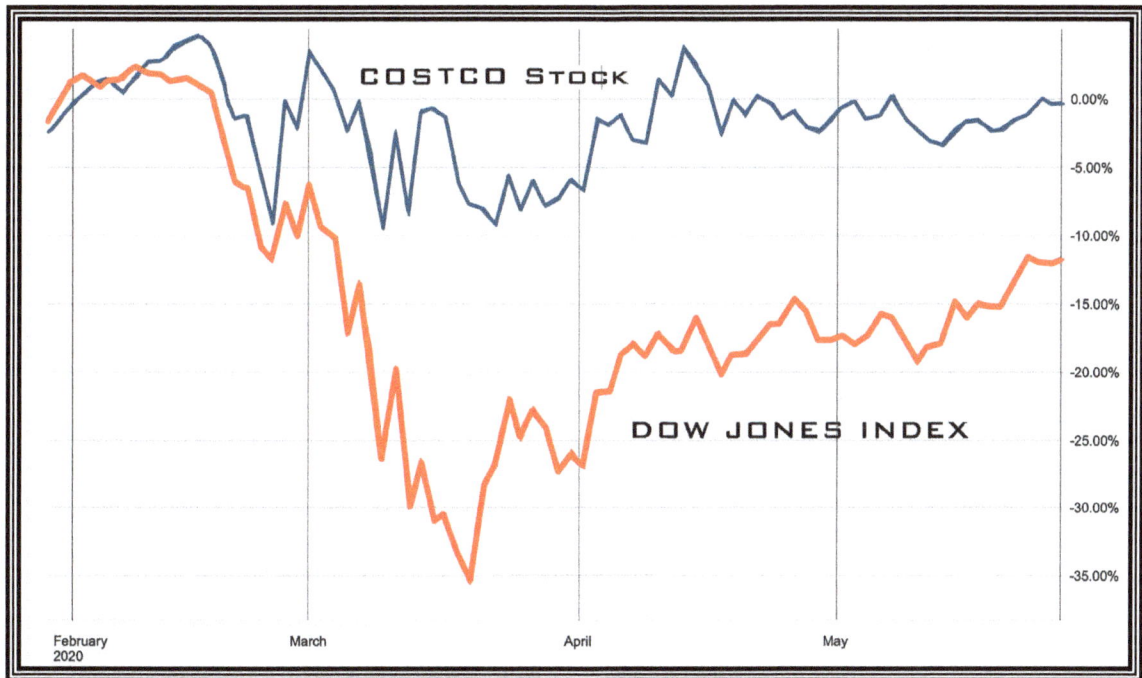

*Costco Stock Price vs DOW*

Do masks work? Yes and no. Marshall Bloom has a great visualization, explaining that masks, social distancing, and sanitation work like a stack of Swiss cheese, each slice with holes, but when you stack them together, the holes get filled — Bloom notes that, together, they work better.[40]

If the infectious virus is in the air, a mask becomes a layer between it and the immune system. While there are several modes of transmission from the infected to the uninfected, through the air is most common. Is it a sneeze and a cough or simply breathing? It is both.

An infectious person will exhale some aerosols that carry infectious virions. The 0.1-micrometer virions are not often hanging in the air on their own. They are crowded together in water vapor particles, the small ones called aerosols[41] — which is what is exhaled by someone infectious — and inhaled by those who are around them — possibly even after they leave the room. The aerosols can persist depending on size, gravity, evaporation, air movement, and ventilation.[42]

Wearing a mask is one way to interrupt the flow of virus-carrying aerosols for at least a short time.[43] Three kinds of masks are often compared:[44]

1. Cloth mask — a mask made of cloth such as cotton, thick enough to block the light when held up to a bright light.

2. A medical mask — called procedure masks by the CDC, the kind surgeons and nurses wear in hospitals.

3. N-95 or KN95 masks — also called respirators. N-95 masks are certified according to US law; KN95 masks are certified to similar standards under Chinese law. "N" means not resistant to oil, 95 means stopping 95% of particles 0.3 micrometers in size.[45]

The purpose of a mask is to keep the virus from being inhaled in the first place and from reaching others when breathed out. They work much like a HEPA filter. Having read through a paper years ago, a citation from the RML BSL-4 EIS on the HEPA filters they use,[46] I learned that HEPA filters are not like a tiny window screen, a grid stopping anything smaller than the holes. In-

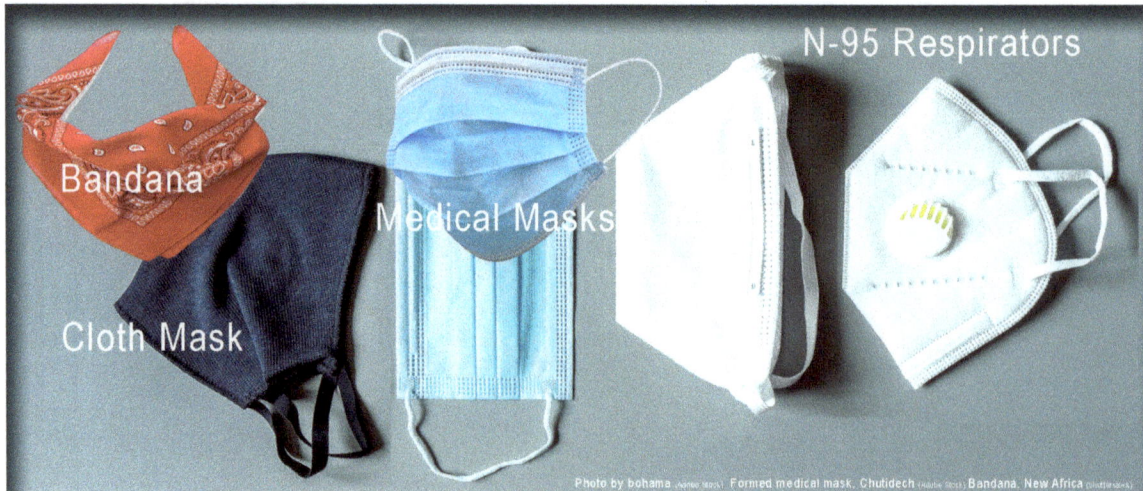

*Mask Types*

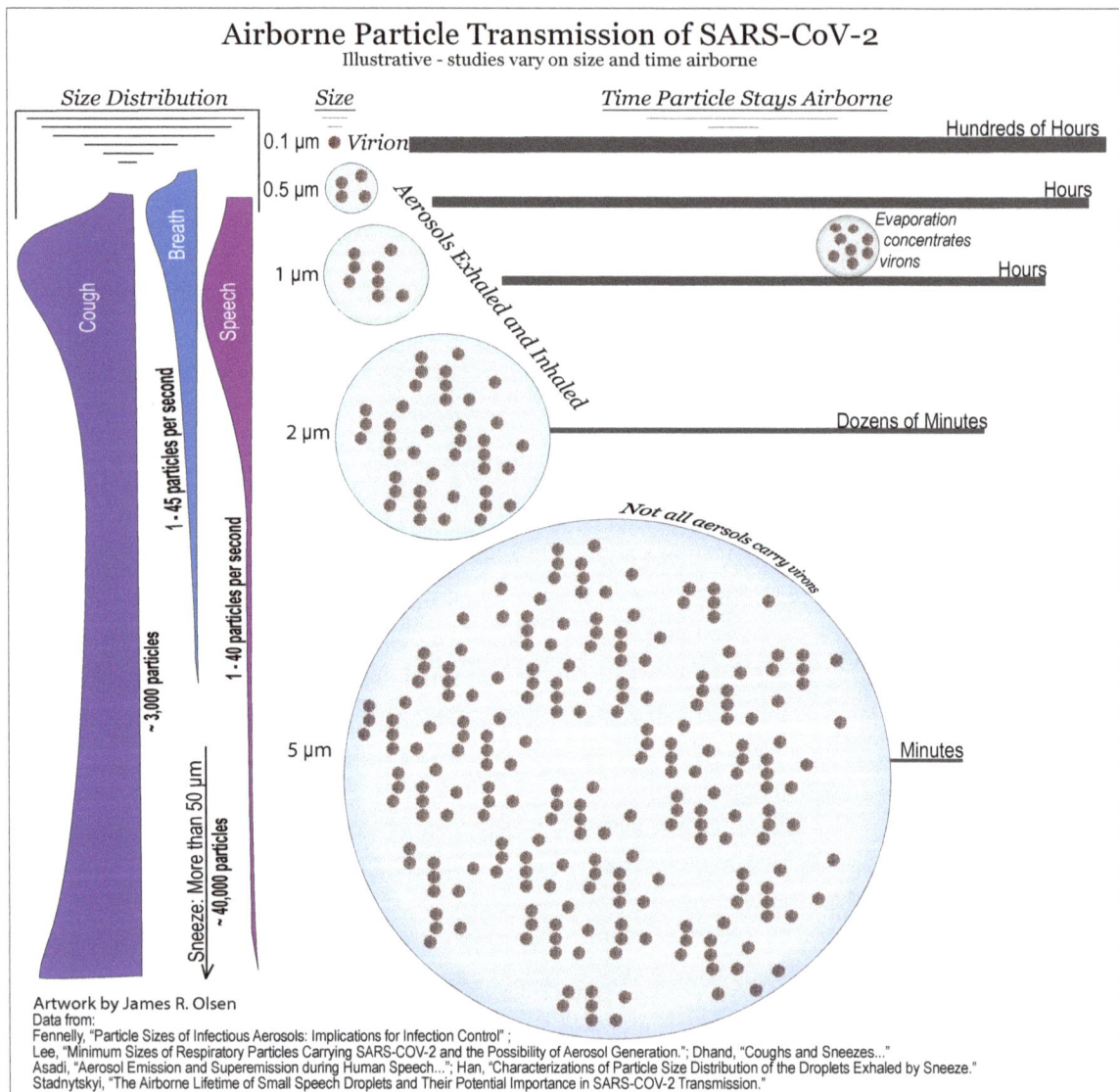

Airborne Particle Transmission of SARS-CoV-2, size, duration

stead, a HEPA filter and an N-95 mask are made of fibers to which the particles stick as they attempt to pass through the filter.[47] The rating is based on testing using various particle sizes. Even a well-sealed N-95 mask is not perfect.

While several mask tests use particles of various sizes, the University of Tokyo study gets right to the point with the real SARS-CoV-2 virus. It demonstrates that masks in the real world never stop all viral particles. They put actual SARS-CoV-2 virions in a nebulizer that makes cough-size particles.[48]

The infectious mannequin is spaced at an intimate talking distance away from a receiving mannequin to try out different masks. In one test, they duct-taped an N-95 on the receiving mannequin. Even then, some virions get through.

The experiment uses a high dose of the virus with particle sizes: <3 μm,[49] 20%; 3 to 5 μm, 40%; >5 to 8 m, 40%. With the infectious person unmasked, the virus that would be inhaled is,

- Reduced by 20% to 40% for cotton masks,
- 47% to 50% for medical masks,
- 57% to 86% for N-95 masks.[50]

When the mannequin wears an N-95 mask, 10% of the infected particles go through the mask.[51]

A limitation of the University of Tokyo study is that smaller particles exhaled while breathing or talking are not tested. On the other hand, the high viral load used may have exceed the masks capacity — the masks may more efficient at lower viral loads.

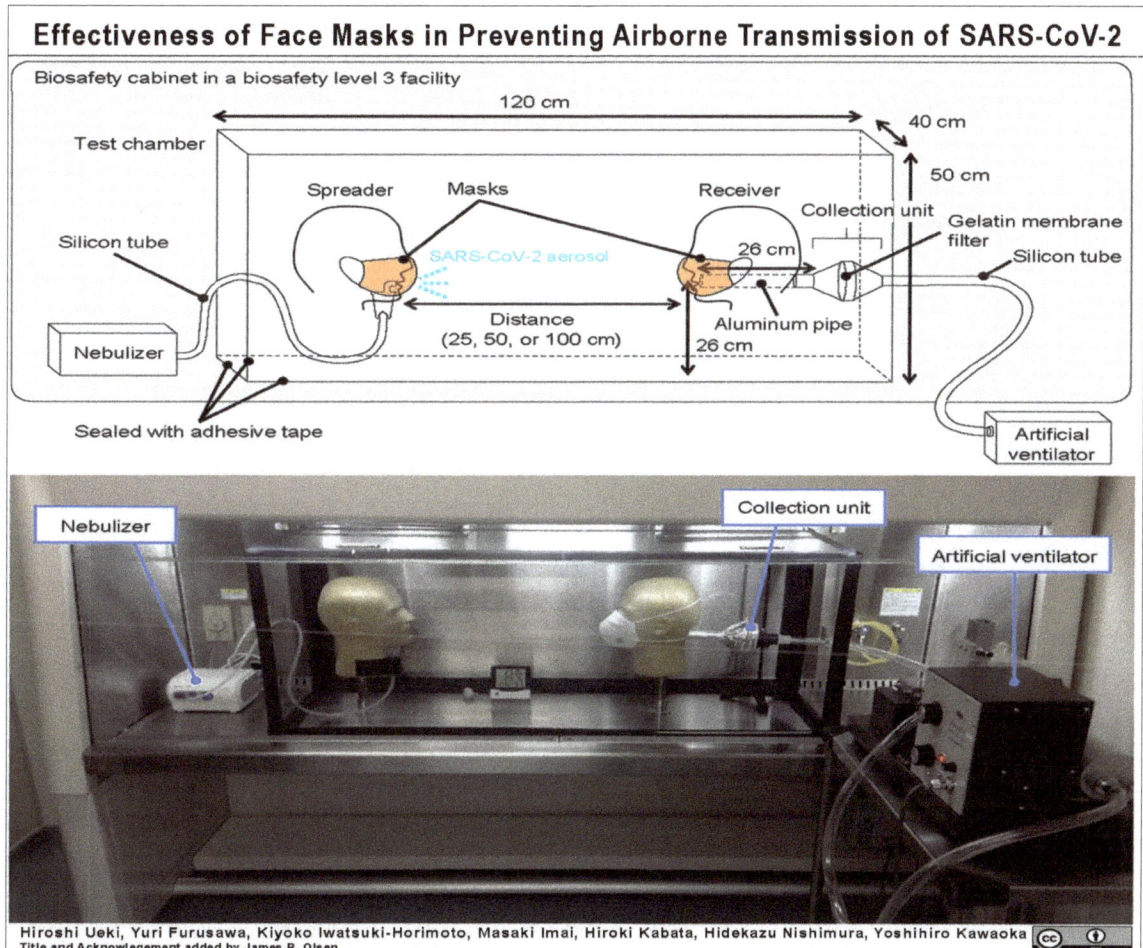

**Effectiveness of Face Masks in Preventing Airborne Transmission of SARS-CoV-2**

*Mask Test Using Actual SARS-CoV-2 Virus*

A study measured particles being expelled through masks during speech, admittedly with a small number of subjects and the limitation of the camera resolution. It shows,

- Reduced by 65% to 85% for cotton masks,
- 90% to 98% for medical masks.[52]

Even if the mask is perfect, it must fit on the face so nothing leaks around the edges. It should at least be as good as donning a snorkeling mask. The standard fit test is to hold the mask to your face with the straps dangling and breathe in through your nose. If the mask fits, it stays on your face when you let go. This is how you know fit is good.[53] If it doesn't, you can be sure water will leak in when you enter the water.

The balance of evidence is that masks are better at protecting from infecting someone else than protecting the wearer from breathing in the virus. So, how effective are masks in the real world? There are many studies.

The CDC will go so far as to issue a Science Brief "proving" cloth masks work. It is not science, however. Comparing the claims in the brief to the citations, one finds the citations are papers on masks of all kinds, not cloth masks.[54] When one adds American politics to the equation, there is a consequence for pushing the idea that cloth masks work with no evidence. There is a significant consequence for leading a good portion of the public to believe claims in the Science Brief are proven facts only to be later to be persuaded it is misleading by scrutinizers. It may not be an Army scrutinizers, but there is a Regiment, a regiment whose soldiers are scouring the data — and each soldier has a bull horn.

A meta-analysis by Jefferson et al., "Physical Interventions to Interrupt or Reduce the Spread of Respiratory Viruses," says the same thing I have found in the literature, "that there is a relative paucity [of Randomized Controlled Tests] given the importance of the question of masking and its relative effectiveness."[55] A review of the literature in 2015 on saving money on masks for surgeons by the Royal Society of Medicine failed to answer the question even in the operating room:

> However, overall there is a lack of substantial evidence to support claims that facemasks protect either patient or surgeon from infectious contamination.[56]

Mask mandates, mask politics, and mask studies go back to the Spanish flu, the Flu of 1918, which stormed through the country, sweeping through a city or town for a month or two, raising the all-cause mortality by a factor of 2 to 10.[57] The observational studies at the time, comparing community to community, concluded masking the general population did not work.[58]

Because small aerosols can carry the virus and are persistent, Marshall Bloom's Swiss cheese model would have us thinking about the room we walk into even before we don a mask. In a crowded, unventilated room where someone is infected, anyone would almost certainly be exposed to a reasonable dose if they stay long enough, mask or no mask. While I can find no good trials on ventilation, both common sense and modeling suggest that a ventilation system with HEPA filters and a reasonable rate of air exchange would probably reduce the risk of infection.[59] Air-mixing from many spaces and/or recirculating indoor air with little or no fresh air exchange probably increases the risk of spreading the disease. In other words, the less time in a risky indoor space, the better, mask or no mask.

Some narratives will cite the harmful effects of wearing a mask, including oxygen deprivation and reinfection when pathogens are breathed back in.[60] 90% of one cohort of healthcare workers reported the adverse effects of wearing medical and N-95 masks when caring for COVID-19 patients for an entire shift. The impact for short durations of 1 to 3 hours is much diminished.

There is no evidence of oxygen deprivation or air exchange for healthcare workers. But a mask

## Effects of Prolonged Mask Use among Healthcare Professionals

Figure 1: Adverse effects after wearing a mask for a full shift.

- Acne — 182 (53.1%)
- Headache — 245 (71.4%)
- Skin breakdown — 175 (51%)
- Impaired cognition — 81 (23.6%)
- I have none of these — 29 (8.5%)

- Within 1 hour of wearing my mask
- After 1-3 hours of wearing my mask
- Greater than 3 hours of wearing my mask
- I don't have headaches from wearing a mask

24.5% / 15.2% / 30.6% / 29.7%

Figure 2: After how many hours of wearing a mask does the survey respondent experience a headache.

Elisheva Rosner. Rosner, "Adverse Effects of Prolonged Mask Use among Healthcare Professionals during COVID-19."

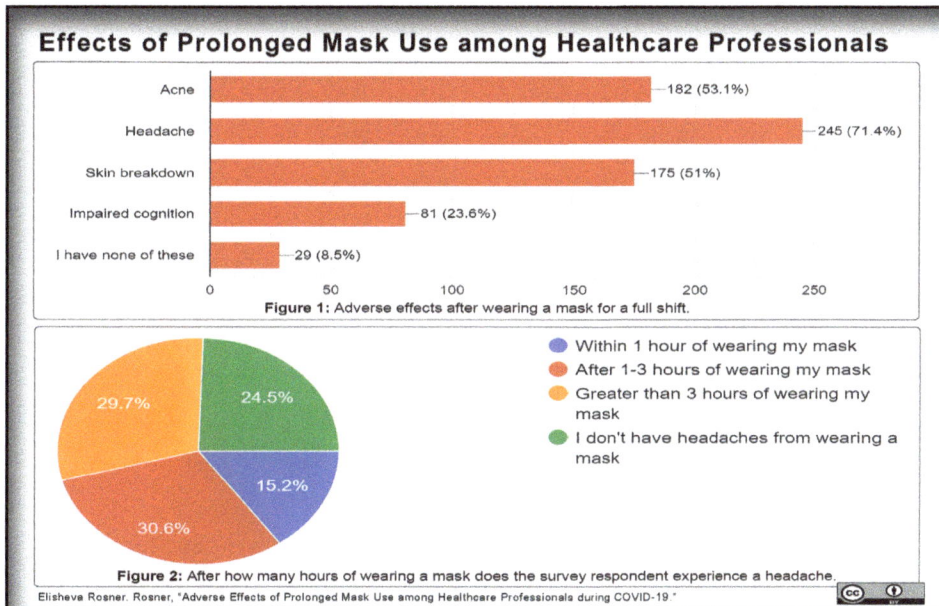

*Adverse Effects of Long Term Mask Health Workers*

impairs breathing patterns, so one may have to breathe differently and deeper. Science doesn't seem to know the answer for the general public, the older, or the impaired.[61] One wishes these questions could be answered with a large-scale controlled trial. A controlled trial of the effectiveness of a mask mandate versus none.

Most studies on the effectiveness of a mask mandate deal with case rates. Several of them show a correlation, like the study "The Effectiveness of Government Masking Mandates on COVID-19 County-Level Case Incidence across the United States, 2020."[62] A case, most often getting sick then recovering, has happened to most people and is something our society accepts as normal, the common cold being a prime example. A case can also be someone who tests positive but has no symptoms.

While the study recognizes its limits regarding potential factors such as crowd gathering behaviors may vary from county to county, it does not recognize in its limitations that case rates depend on the COVID-19 testing regime in the county. There is very likely a correlation between a county's policy and the population's attitudes, on average, regarding COVID-19. One might guess

that a county whose public officials put forth a mask mandate might have a correlated policy for testing and public messaging availability. This would suggest the possibility of a population and a mask-mandate county that, on average, is more likely to seek testing and medical care with mild symptoms.[63] That's what makes such studies so hard.

In observational studies, even in a hospital setting, no one knows to what extent the study subjects get exposed when they leave the hospital.[64] The bottom line is that science does not have a definitive answer as to how well masks work. Nor does the science provide a consistent answer for mask mandates, either in before-and-after observations or comparisons between locations or populations.

The mask question for the person on the street is, "What are my chances of getting severe symptoms, having to go to the hospital, or dying?" I have yet to find a study that definitively answers this question for masks. This complicating factor is that the decision to wear a mask cannot be held in isolation in the real world. For most people, a decision regarding masks is correlated to a range of choices to modify one's behavior.

Mandates versus fatalities is nearly always uncorrelated as we'll see later in this narrative — suggesting that, if masks do work, other factors are more important than a mandate in the real world. Regarding mortality, the balance of evidence is that even if masks work, mask mandates don't.[65]

Most papers attempting to prove masks work are either inclusive or poorly done. In the poorly done studies, predetermination bias is evident regarding masks.

Predetermination bias is not new. In 2018, a Harvard student presented a dissertation looking for a correlation between Type 2 Diabetics of different kinds of food and exercise and the risk of health setbacks such as heart disease. When the data is examined, foods like yogurt help a bit. But the big standout is ice cream.

That can't be right; ice cream is loaded with sugar and fat. It's terrible for you — look again, says the professor. But biology has no politics. Even though the dissertation slices and dices the data and possible explanations, low-fat dairy performs well, and low-fat ice cream wins out. The press release touts low-fat foods, denying ice cream its day in the sun.

The dissertation gathers dust where it was shelved, ice cream performs better than yogurt — few scientists are ready to bet their career on that observation — but it's what the science says.[66] Something else is happening besides the fat and sugar grams listed on the package label. It took five years, but now there are studies examining milk fat's effect on the risk of heart disease.

The CDC publishes a "Science Brief" in which they claim cloth masks work, citing studies about medical masks and N-95 masks and dismissing studies that did not fit the CDC narrative using a criteria such as a too small sample size, even though that same criteria would have thrown out studies the brief cited in support. This brief is selective persuasion, not science.[67]

The Counter-narrative sites do the same thing. The public has no source trustworthy information.

"Do masks actually work?"A good mask very likely reduces the risk of infection for short periods. Nevertheless, even those who wear a mask all the time can still get COVID-19.

I wear a mask where people expect me to or if I am in a poorly ventilated space for more than 10 to 15 minutes. But this is my take on approaching the real world, given that the science and data do not show that masks are effective for the general population.

But, mask or no mask has become a red herring; there is another way to control the spread of the disease, and it works. South Korea had prepared ahead of time to do classic epidemiology — the 3 T's: Test, Trace, and Treat:

1. Test, with near universal testing;
2. Trace, with comprehensive contact tracing, hot spots disclosed to the public, QR code records of high-risk venues such as bars, and self-isolation while maintaining contact with case officers, backed up by the possibility of a surprise visit.
3. Treatment prioritized on severity.[68]

While the Ravalli County Public Health Department tries this classic epidemiology, they do not have the resources. Public promotion campaigns from the CDC are a hindrance. There is little news of the CDC touting the three T's. Instead, it is masks, masks, masks.

Everyone will have to figure out mask-wearing for themselves.

On the 15th, Governor Bullock directs everyone in the State of Montana to wear masks in public places. Now it's unavoidable, like them or not, everyone has to think about masks.

Though touted to be science-based, no one cites peer-reviewed scientific literature. I suspect Governor Bullock took people's word for it. Politically, it was a way for the Governor to "do something," since saying he did not know what to do is a sure election loser in the United States.

Of all the COVID-19 directives issued in Montana, this is the most widely misunderstood, ineptly crafted, unnecessarily divisive, and rambunctiously argued. Governor Steve Bullock, Democrat, being termed out for Governor, is running for US Senate this year. After fading in the presidential primaries, Democrats begged him to help keep control of the Senate.[69]

A National Public Radio (NPR) news piece a couple of weeks ago captures the essence of Bullock's campaign — Montana has the lowest case rate in the nation while starting a massive testing campaign, opening the state for business, having acted quickly with the initial lockdown order.

He lifted the quarantine requirement for incoming visitors, saying, "there is ample contact tracing now." He's wrong. There isn't enough contact tracing capacity by a long shot.

Contact tracing capacity in Ravalli County has a razor-thin margin that would be unable to handle any bump in cases. And, ironically, more tests will mean more cases; more cases mean more contact tracing. The medical literature still poorly understands the relationship between symptoms, test results, viral load, and how it is passed on.[70]

With all of these uncertainties, Bullock is staking his campaign on COVID-19. Now he adds a mask mandate — it will cost him.

Bullock is a native son, born in Missoula and raised in Helena, a few blocks from the capital building. Going to Helena and back to comment to a subcommittee meeting or do a bit of citizen lobbying can be done in a long day trip from Hamilton. Helena still has a "small city" feel, growing from a mining camp called "Last Chance Gulch," which is still listed on Helena's street maps, home of a gold and silver strike.[71]

The City of Helena is Democratic, electing Democrats to the state legislature routinely from districts within the city limits, only to be surrounded by Republicans in the outlying rural areas — like most of Montana and the country. A strong union history and affinity underlies the working class mining heritage in the Democratic strongholds in Helena and Butte — still a strong pull in Montana Democrat convention hall speeches.

Bullock is the son of a teacher. He got into a small, competitive, well-rated liberal arts college in California[72] and then to law school at Columbia, ranking a close third behind Harvard and Yale.[73] After spending some time out of state in private practice, Bullock came back to Montana. He made a run for Montana Attorney General in 2000, losing, trying again in 2008 to win the election and becoming Governor in 2013 for two terms. Since we have term limits, Bullock cannot run for Governor again.

I first met Steve Bullock when he came to Emma's House. I was invited to the event because I had some involvement in helping it along. Back in the '90s, I was recruited to be on the Board of Supporters for an Abuse Free Environment (SAFE),[74] the local domestic abuse advocacy group, in the early days for reasons I still don't know. It was on the threshold of growing from a volunteer-driven, take-care-of-your-sisters outfit to its first hired staff in the 1990s. It was formed as a formal non-profit by the Soroptimists[75] who had started the idea in the 80s.[76] After a decade of one-on-one education with County Attorneys, detectives, sheriffs, police chiefs, and judges, it was finally accepted that you should actually arrest a domestic abuser. So when the director raised the money for transitional housing, the community was ready. As it became one of the most popular non-profits in the valley — it grew, becoming known as the "rock star" of advocacy groups in Montana.

At a SAFE Board Meeting, it became clear that something was missing, "You mean if she is 18, you can meet her in a parking lot and make her disappear until you get a restraining order, but if she is 17, you can't?" After that question, I got an education on how the system works and doesn't work for children. Long story short, the SAFE staff got a grant for a Victim Advocate. Valerie Widmar, who was working for SAFE then, picked up the torch with great compassion and enthusiasm.

Years later, Val Widmar learned about a new program for interviewing child victims of crime, which is better for the child. She asked me how to find a home setting instead of a commercial building. Long story short, my company recruited another investor. Val and I went house shopping in areas zoned as transition — zoned to become business but still residential. We found an estate sale for a place that was well-known for its beautiful garden. The heirs agreed we could use their mother's name, Emma. Val made Emma's House the shinning star of child abuse response in Montana.

Nearly every Ravalli County Attorney, Montana Attorney General, and an occasional governor has Emma's House as a must-visit. When Steve Bullock visited in 2009, he brought some staff from the Montana Justice Department. Val had set up an outside picnic table — it's the only time I have said more than a few polite words to Bullock. Of course, the press showed up. When I look back at my quote in the newspaper, I wonder if we have outgrown the idea:

> Emma's House was one of those grass-roots projects that work in rural communities because people haven't lost touch with one another. You get five or six people in a rural community together and

Left to Right:
George Corn, Ravalli County Attorney
Valerie Widmar, Emma's House Director
Steve Bullock, Montana Attorney General

Photo by James R. Olsen

*Attorney General Visit to Emma's House*

have everyone pull out their phone; it's remarkable how many people you can reach.[77]

Bullock is not a firebrand. He has a demeanor that says, at heart, this is a nice guy, both in private and when he is on the stump. Going into the 2016 election, I had more than one Republican tell me they believed he would win "if he didn't mess it up."

Everyone seemed to be looking for a protest candidate for President in 2016, a candidate who was not an insider. In June 2016, Bernie won the Democrat primary in Montana by a decent margin.[78] For Republicans, the protest candidate was Donald Trump. Of course, Trump would win the general election vote in Ravalli County simply because he was a Republican. While Trump got 67% in Ravalli County in 2016, it was not 80% or 90%. So, it's not all super-conservative. Bullock did win the state in 2016, carrying 41% of the votes cast for Governor in Ravalli County, a nice margin over the minimum needed, beating Greg Gianforte in the statewide vote for Governor.[79]

This year, after making a bid in the crowded Democratic Presidential Primary, after a lot of pressure, Bullock is running against Republican Incumbent Steve Daines.[80] Bullock is polling to win — with general approval of his administration's handling of COVID-19. A poll in April showed a hard core of 10% believing the Federal and State governments should not be involved in managing COVID-19, shrinking to 7% at the local level.[81] This confirmed that most people were willing to give up a lot for a short time when the pandemic started, and everything was uncertain. This changes as the months drag on.

Governor Bullock has resisted pressure to issue a mask mandate for over a month, fearing a backlash. As the pressure builds, including from the Montana Nurses Association,[82] Bullock makes a political calculation that he should do what other Dems are doing.[83] When Governor Bullock issues the mask mandate, the backlash he fears becomes real. The divide becomes a rift — instantly. The mandate becomes a cause célèbre. In Ravalli County, which has already gathered the firewood, stacked it, laid on some tinder, Bullock's mask mandate lights the match. It burns.

The mask mandate is something everyone anticipated, but no one understands; I don't. Neither do the business owners who call the Chamber of Commerce asking if they are required to provide masks for their employees nor the restaurant owners who ask if customers should wear masks before they get to their table, obviously needing to remove them to eat, nor the owner who wants to know if his cook who works alone in the kitchen needs a mask, nor does the wedding planner who whats to know what to do about people on the dance floor at a wedding.[84]

For Stand Together for Freedom, having already organized around mandates and masks, any nuances in the Directive were unlikely to matter much for most of their followers. Their reaction comes in short order.

But, firebrands also inhabit the pro-mask camp. The ink is barely dry on the mask directive when 911 is flooded with calls attempting to turn in people and businesses for not masking up.[85] At the Farmers Market, "a couple of people" are sneaking around taking pictures of people setting up who weren't masked — outside, uncrowded. "Give me a break," I tell the market master when she tells me that story.

This anonymous behavior casts a negative pall. "Don't call 911. This is reserved for emergencies," announces Sheriff Holton in the newspaper.[86]

I have to get Commissioner Chris Hoffman to walk me through the Directive to finally understand what Sheriff Holton had said out of the box — it is not something he will enforce because it is not something he can enforce given the way

*Ravalli County Fair*

the Governor's Directive is written. It is not up to him to enforce — it is a civil matter according to the Directive — not that the Sheriff will start arresting people for not masking if he wants to keep the peace. The vague wording leaves each county and city to sort out the chaos:

> Local public health agencies are directed to assist all officers and agencies of the state are directed to assist in the administration and enforcement of this Directive...[87]

There is no enforcement provision for the individual in the Directive:

> This Directive is enforceable only against businesses and other persons who are responsible for indoor spaces open to the public...[88]

So, the logic is that when someone walks into a store refusing to wear a mask — there is no consequence for that person unless confronted by

a staff member. Employees are now to police an angry populace, no combat pay included.

It is poor public policy. To make this policy worse, enforcing it for these businesses has no clear criteria.[89]

The bottom line is that the mask mandate is not, and never will be, enforced in Ravalli County. Officials release a statement,

> Sheriff Steve Holton, in conjunction with Ravalli County Commissioners:
>
> We encourage all residents to educate themselves and conduct their business as safely as possible during this pandemic. We ask our residents to always treat each other with compassion, and respect each others right to make decisions about their own health.
>
> Private business owners may choose to enforce the Governor's Directive. Everyone has the right and ability to shop or patronize businesses they feel com-

fortable in.

Criminal citations will not be issued for violations of the mask directive. The RCSO [Ravalli County Sheriff's Office] will however, investigate and enforce trespass and disorderly conduct complaints.[90]

Each business is left to their own devices. A few businesses post signs saying, "We proudly don't require masks." Others post a sign saying that they honor each person's choice. Many simply post nothing or hang the county's "wear mask" poster but do not worry over whether people are masked up or not. Some businesses decide to require a mask and ask anyone who doesn't comply to leave. There are a few anti-mask posters — reports of rude behavior to customers wearing masks at several businesses.

There were insults and arguments. People call the Sheriff, asking if they would be arrested if

they don't wear masks. "Holton says the behavior he's been seeing over this issue goes against the grain of how people usually treat each other in Ravalli County."[91]

County fairs are being canceled across Montana. It's unfortunate that Doctor Calderwood didn't consult with the Commissioners before issuing this order,[92]

---

I have made the decision to cancel most of the 2020 County Fair, with support of the Ravalli County Board of Health, solely due to an imminent threat of ... exponential spread with subsequent increases in morbidity and mortality...

I have pondered the fair extensively and cannot justify allowing a mass gathering at this critical time, with such potential for deleterious impact. The responsibility of making these hard decisions are mine and I cannot let the fair go forward now, with good conscience.

We are also required by law to uphold the Governor's Directives for reopening and are currently in phase 2. The Fair Board worked hard on plans, and I know outside activities are generally safer, but I do not believe in my heart that people would socially distance in this setting, and spread would be unmitigated. Tracing cases and contacts acquired at the fair would be impossible and attendees would unknowingly spread COVID more rapidly throughout the community.

Please also see BOH letters of support.

Sincerely,

Carol Calderwood, MD

---

Closing the fair pours gasoline on the fire.

We still have a great fair in this county. The 4H kids come out in force, the old two-stroke engine farm gear putters in the fairgrounds yard, surrounded by tents hosting vendors of all varieties, the local non-profits selling fair food. I even thought of entering a huckleberry pie with a raspberry glaze once. But my wife, who picks the huckleberries, feared they'd go to waste. Don't ask where she picks; that's like asking a hunter to name his favorite drainage. The carnival attracts the younger crowd when the sun gets lower in the sky. And we get a rodeo.

The fair cancellation notice appears in the newspaper the same day the mask order comes out. Even though other popular events were being canceled, such as the Brew Fest,[93] Celtic Games,[94] Shakespeare in the Parks,[95] and outdoor concerts,[96] many would call county fair sacred if it weren't a sacrilege to say so.

The Fair Board has to make a final decision, though, whether to obey or not. The Fair Board immediately meets after receiving the Health Officer's order that evening. Terri Lackey is on the Fair Board and loses no time protesting the shutdown order. It is "sad" and "emotional," as it takes a year to organize the county fair. Participants and vendors are lined up. The stock contractor tells the fair director that his business is going under if the fair isn't held.[97] People speak for and against it. Alan Lackey comments,

> What are you afraid of? Are we going to hunker down under our beds because of some invisible force that's out there? God gave us immune systems, and this isn't any different.[98]

But another person says, "If even one person gets sick or injured at the fair, it's one too many."[99] At least a couple of people from the Ravalli Democratic Central Committee attend — at least one stands to support cancellations.

Terri calls for a vote to open the fair with no restrictions, having gotten 608 letters in support saying, "the Board's hands weren't tied by the County Health Department's position." The decision was not on the agenda and would probably

have to be redone if someone made an issue of it, which is a pretty fair bet. Terri presses the point. The vote fails 4 to 2.[100]

It gets uglier in some counties. Health Officers had been threatened — in the Flathead, about a three-hour drive from here, a Health Officer had a call from the County Sheriff, "Do you know how to use a gun?" — because someone had threatened her. She resigns.[101]

Meanwhile, Doctor Calderwood undoubtedly feels the added stress of the Sheriff and Commissioner's statements. The reaction to shutting down the fair is more visceral than anything she had experienced.

On Saturday, she resigns.[102]

The Health Nurse, whose job title is "Ravalli County Public Health Director," follows Doctor Calderwood out the door, resigning, while not wanting to give any reason to the newspaper. Another employee of the Health Department leaves with her.

I am confused. The Sheriff and County Commissioners didn't "disobey the law," as far as I can tell; the Commissioners did not say they did not expect it to be obeyed. So, even though I think the mask order is a bad idea, Doctor Calderwood has the authority to initiate civil actions to enforce Governor Bullock's order no matter what the Commissioners or the Sheriff have to say. That said, it would be unreasonable for her to go it alone in this volatile environment.

The same journalist reporting the resignation reported on the reaction by the medical community to the resignation — which is uniformly sorrowful in the quotes from a half dozen medical professionals.[103] The main thread of the story is a narrative by Marshall Bloom. The *Ravalli Republic* reports,

> Rocky Mountain Laboratories Associate Director Marshall Bloom said he has worked with Calderwood since she took the office 13 years ago and was horrified to learn of her resignation...

Ravalli County Elected Officials and To Whom It May Concern:

Though I have been able to work affectively with people of many different perspectives during my 13 years as Health Officer, I am a point that I believe we must "agree to disagree." I did not plan to institute required masking locally. However, I feel that I have been put into another no-win situation by the locally elected officials' decision to disobey the Governor's Directives without my input. Masks are easy to use, safe, cheap, and (now studies show) effective. The way I see it, I now have a choice of 1) staying on but not doing what I believe is legal and best for the county, 2) directly confronting leaders whom I do respect, or 3) bowing down. I have regretfully decided to give my 2 week notice of resignation as Ravalli County Health Officer.

I would have liked to have stayed on as HO at least long enough to see us through this whole pandemic, but I am extremely proud of impacts I have had, independently and as part of teams we built in the county and the hospital, through the early phases of this crisis up to this point. I have always suspected that the fall or winter will likely be the time frame of the local peak. However, I believe that actions we have taken already, in the county, state, and nationally bought us critical time to gain plans, supplies, and team connections. Systems are now in place that can allow a reasonable way forward.

Though at times I have been frustrated with our differences of leadership opinions, I have mostly appreciated the chances I have been given to take responsibility for decisions, as the nonpolitical health expert. I am reluctant to quit, not due to any desire to retain any power, but rather because of loyalty to county residents and to the Public Health nurses, who have relied on my support and input.

Especially if a non-physician is tasked with the position of HO, it will be critical to have a physician medical advisor for local nursing consultation, especially regarding COVID case/contract tracing and management. Without this, I'm not even sure who would be responsible to orders and results of COVID tests on asymptomatic contacts each day. Therefore, I would be happy to make a move to something like "Public Health Medical Advisor." For me this would require, at a minimum, maintenance of my county insurance. In the future, I would also consider serving as Public Health Medical Director if and when Dr Ashcraft is ready to step down.

Sincerely and with continued best wishes for the health and safety
of Ravalli County,

Carol Calderwood, MD

*Health Officer Letter of Resignation*

She has served with distinction on our Institutional Bio-safety Committee.

I am aware that she's been under a lot of pressure from the public as well as local elected officials and realize she has had to take some very forceful stands all in the name of protecting the health of the citizens of Ravalli County. Unfortunately, Carol is not the only health officer in Montana or elsewhere in the US to resign because of unfair pressure.[104]

While Health Officers are resigning across Montana, Commissioner Chris Hoffman says he doesn't understand why she is frustrated enough to resign, saying,

> The commissioners have supported the Public Health Department throughout the pandemic.

> We've just about given carte blanche on resources they say they need to combat this thing."

There have been some disagreements. Hoffman says the health and safety section of Montana Code Annotated, Code 50, makes it nearly impossible to not have disagreements because the code "almost completely has the commissioners abdicate their responsibilities in a time of crisis."[105]

Tiffany Webber RN is asked to step up as acting Public Health Director. She demonstrates that she has a heart for our community, having grown up here, been a Hamilton Broncs cheerleader, and nursed here. Calling it a "battlefield promotion," her job is to "prevent, promote, and protect." She finds "it painful to think we have aspiration for political gain."[106] People accusing her of political gamesmanship are more likely to find it if they look in the mirror.

The resignation letter is on the agenda when the Health Board meets on the 28th. The Health Board members, all masked up except for Jeff Burrows, sit at the dais facing Doctor Carol Calderwood, in the hot seat, at a table that has been moved up to within a few inches from the dais. The large crowd sits behind her, murmuring, some ready for a fight.

Burrows adjusts his laptop, moves his coffee cup, adjusts his mic, while glancing at the crowd. Hands placed on the tabletop, he asks someone to "shut the fan noise because everyone can't hear. Is it 12:30?"

Katie answers, "It's 12:30."

Burrows says, "Let's do it."

After role is called, Jeff opens the meeting,

> Let's make sure everybody uses your mic so that everybody can hear. And,

> just to start the meeting, if you have public comment, work your way to the mic. We're not going to have discussions amongst each other... We'll try to keep it civil and orderly in here today.

> The first item on the agenda is discussion and possible decision on the Board of Health's recommendation for the Public Health Officer position.

> Carol, I don't know if you have anything to say to start the meeting.

Carol responds, almost with a sigh,

> No. I would just like to help if I can. My preference would be to do the medical side of it. I've had so much support in this last week... that I have had a change of heart. I could continue to help an MD or DO suitable to the Board if available. I did resign. In the interim, I see how much I am still relied upon by the hospital and the nurses. That's my passion.

Turner says,

> Is that the only side of it?[107]

Calderwood says that the Health Officer is a whole job, making administrative decisions and signing compliance orders. So, if it can't be split to a medical advisor and someone else doing the other parts of the job, "I can live with that. My passion is not making decisions about the fair." Her passion is taking care of cases daily, testing, isolation, and quarantine. She is the person that the test results come through. She deals with "gray zones" on quarantine issues.

Obviously, there have been off-stage discussions. She really wants to see us through COVID-19.

Working seven days a week has been wearing on her — like it would anyone. When she finally delegated and took a couple of days off; as luck would have it, the governor put out the mask mandate. Now she feels she can't even take a rest.

The board members clearly want Carol to stay. Turner asks what she would think if they held her resignation in abeyance for, say, three months. She thinks that it would be a good solution.

Jeff looks her in the eye. “I just want to make sure that if you are going to do it that you are all in.” We may disagree in the future, but we could not handle another resignation.

Doctor Calderwood admits it was a misunderstanding. Burrows acknowledges that they may have done better if they coordinated with her before putting out the press release.

Then, public comment starts. It goes on for over 2 hours, some people wanting Doctor Calderwood’s resignation to be accepted, most of them wanting someone who would not fall in lockstep with the CDC or the Montana Governor.

Some want her to stay on, and a few want her to enforce what the Governor is doing.

At least one angry doctor says that patients were being coded[108] for COVID-19 even though they were there for something else because the hospital gets more money.

People stream up to the mic, saying it’s about health, a healthy immune system, as much as a disease needing treatment.

Some people ask, “Why aren’t you using hydroxychloroquine?”

Others say PCR tests don’t work; it’s just a way to get the case counts up so the government can take away freedoms.

There are a calls for cooperation, including by Doctor Turner and myself, to work together to stay healthy and keep businesses thriving.

I warn against perpetual emergencies.

While the opinions and stories are wide-ranging, there is almost universal agreement the law as written needs fixing.

State Representative David Bedey says that the Board of Health and Health Officer are abiding by state law. He talks about changing the law, saying,

> A lot of the vitriol directed toward Doctor Calderwood is misplaced. She does not have the authority, nor would you want her to have the authority to flagrantly disregard the law. I think that is a bad idea.
>
> I’ve looked into this extensively as soon as the governor issued his emergency declaration and secured a lengthy legal analysis from the legislative staff...
>
> I walked away somewhat with the attitude as Mister Olsen has. That is, I am not happy giving the governor unbridled authority in a perpetual emergency situation. I have to tell you that there is no other way to read the statutes that he has the authority and the legislature gave it to him.[109]

Someone from the crowd asks what happens if a business owner insists on a mask and they don’t want to wear one. Burrows responds,

> Speaking of constitutional rights, the private business owner has the right to refuse service — tell your friends and go somewhere else.

Ultimately, the Health Board decides to defer accepting Doctor Calderwood’s resignation until a qualified, acceptable candidate is found.

Local businesses are on their own in Ravalli County to figure out how far to go to protect their employees and what demands to make on their customers. In “conservative” Ravalli County, most people will recognize the principle of a “private property right” of a business owner to decide the rules of their store, but that doesn’t mean they’ll like it. There are already stories of people having angry encounters with employees behind the counter who simply did what their boss asked them to do.

Alan Ford’s policy is to recommend masks but not insist on it:

> In the old days, you would have been tougher and tell all the clerks to wear masks or get lost. Now it doesn’t work so well, so we recommended it. Most of his employees wore masks for about six months; a couple of Trump supporters didn’t. None of his staff got COVID, as far as Alan knows. He left it up to his customers, some seeing most of the

Photo by James R. Olsen

*Ford's Department Store*

staff masked, would say they forgot their mask and go back to the car.[110]

The only problem he had with customers is that two or three times someone would call,

I hear you are not enforcing the mask mandate, so I'm not coming in.

Alan shrugs it off, thinking, "They weren't a customer anyway."[111]

Mara Lynn Luther enforces the mask mandate at Chapter One Bookstore. When the CDC changes its mind, Mara Lynn thinks that's good because they are learning more. After a while, it's not so much hand sanitizer but social distancing. The staff tracks the CDC data, "wanting to err on the side of safety." None of her staff got COVID.[112] "We were always reacting to information and opposing information," adding,

We felt a weird responsibility to protect

people, even from themselves because we didn't trust people to do it for themselves and we didn't want to get sick ourselves. We have older people.

We did have angry customers.

When you have a storefront, people can be angry and not talk to you.

People can be angry and anonymously talk to you.

People can be angry and approach you and talk about it. When that happened, it went peaceably. We would say we are a privately owned business and can make the decision. We can change it.

We're not medical professionals, so we're just going with what the CDC is telling us and what we feel comfortable with.

If people had a conversation with us, we could sneak back to that community feeling. There are a lot of people we know we don't agree with. Yet we go to each other's businesses. We see each other's kids at school. It's part of living in a small town.

*Big Creek Coffee Roasters* (River Rising sign at far end of block)

But, other people yelled at the front door saying, "I will never shop here again," even though they never did. We got a lot of that. They were angry people.

But there were others who said they only came in because of our decision to require masks.[113]

A woman who came in is a regular customer, typically a very warm person. She wore a mask, but this time she is very curt. She didn't want to wear the mask. Mara Lynn started talking to her. She said, "I hate these things."

"I don't think anyone wants to wear them," replied Mara Lynn.

After talking a little longer, the customer admitted being scared, seeing a lot of people in the grocery store wearing masks, which she found unsettling. "It reminded you of the bigger issue."

The staff figured out that there were a lot of people scared on both sides — that's when their defenses would come up. Once we figured that out, we could get through this.

I respond with, "Fear turns into anger."

"Yah," Mara Lynn agrees.

The Table, which has been gathering at Big Creek Coffee Roasters, is almost broken up over this fiasco. A couple of men from our coffee klatch confront the owner of Big Creek about enforcing masks. The owner insists he will.

Even though the owner of Big Creek Coffee decided to require masks, some staff are still very concerned about staying healthy at work; I bring a few N-95 masks and airline-grade sanitizers.

I'm agnostic. Pretty much everybody except me goes up the block to River Rising[114] because they leave mask-wearing up to their customers. Of course, I will wander by and see if anyone I know is at River Rising when I go on my break and go in to chat. Hollie has been gracious enough to have coffee at Big Creek when we arrange to meet to roll the I Ching, even though she thinks the mask mandate is wrong. But, some people will never be customers again.

Photo by James R. Olsen

*River Rising Bakery*

This story made the *New York Times*, written by a freelancer who is passing through. Noting that the County Commissioners had decided not to enforce the mask order, "It kind of took some pressure off of us, because we're not having that confrontation with our customers when they walk in," says the owner of River Rising. Though some patrons want her to enforce it, she can't please everyone.

The owner of Big Creek Coffee Roasters is hoping the governor's order would put a stop to mask conflicts but blames the commissioners' for "putting us at odds with our customers."[115]

Mask conflicts continue in the community. In Stevensville, Kodiak Jax[116] posts on its letter-board,

We live in the USA, not China, take back our gov.

A few blocks away, the owner of Morning Star Caffeine and Cuisine says, "she has had a small handful of complaints for not requiring masks, but dozens of customers who have thanked her for not pushing masks on people."[117]

I wear a red bandanna around my neck. In places where people expect masks, I pull it up over my nose. If I am in a crowded place and worried about it, I don an N-95 mask, breathe in to ensure the fit is halfway decent, and slide the bandanna over it. I join the group at River Rising while still patronizing Big Creek. When someone starts picking on me for wearing a mask, I say that I wear a mask to be polite, adding,

It's a waste of time to argue about masks.

*31 Jul. 2020: 153,828 COVID-19 Related Deaths Reported*

| 3 | 8 | 11 | August 2020 | 19 | 24 | 26 | 28 |
|---|---|---|---|---|---|---|---|

Warp Speed orders vaccines ahead of time $2B to GS and Sanofi to develop vaccine

Trump signs orders and memos to extend unemployment assistence, payroll tax holiday

Warp Speed: $1.5B order for Moderna Vaccine

CDC study shows native americans most severe outcome

Remdisiver benefit for hospitalized patients questioned in global study

Abbot rapid anigen test approved

First Reinfection US Case

**COVID-19 Treatment Guidelines** (NIAID COVID-19 Treatment Guidelines Panel)

• Recommend against Tocilizumab.
• Recommend against Ivermectin, works *in vitro* reach plasma concentrations is much higher and approved dose.

Photo by CustomPhotographyDesigns iShutterstock

*Riding in the Rodeo*

# 11

# WHAT A RODEO

*A bucking bronc just wants to throw the cowboy off.*
*A barrel racing quarter horse responds to the slightest press of the her knees or a tug of her hand on the reins, while leaning 45 degrees and galloping 30 miles an hour.*

## AUGUST 2020

Everything seems split. While the 4-H show was the exception to the Health Officer's order, which could still be held with restrictions, the response by 4-H isn't much of a surprise. Instead of having the auction at the fairgrounds, Triple-D Cattle Feedlot and Sales comes forward. It is to be the "Freedom Show."[1]

The Hamilton Players split the stage actors from the audience with a live-stream presentation of *The Steel Magnolias*, not wanting to stack an audience in a crowded theater.[2] But, Darby rodeos on. At the Darby Arena,[3] the annual Wild West Rodeo draws 25 competitors from around the region for a breakaway roping contest — how fast can you get that lasso around a running calf's neck.[4] With the County Fair Rodeo canceled, another rodeo is organized in Darby, a bareback buckin' rodeo, with the best in the country invited from the Pro Cowboys Association.

They did come. A Canadian, Orin Larsen,[5] took the top prizes for $11,000. Sixth-ranked in world standings, he rides in front of a crowd that includes Hollywood actors and crew from the *Yellowstone* Series, which had put up sponsorship money — thus tagline, Yellowstone Riggin' Rally. It sold out quickly. Sold out of everything but masks.[6] "It was definitely cool to ride in front of Hollywood celebrities, and having the town come out was a cool event," said Orin.[7]

Mother nature seems to be giving us some breathing room this year, when a nasty fire season, also known as smoke season,[8] might have pushed us over the edge — clear skies and no sprawling fire camps. We have some wildfires every year, living in a fire-adapted landscape; it's just a matter of how much. We started with a snowpack that was worrisome in January. Still, the snowfall rebounded quickly, so we are probably not due for a bad fire season.[9]

Fire crews put out most fires quickly, with six still burning. Just as the Cinnabar Fire started to blow past the firelines, burning 2,000 acres, rain damped it down.[10] Our own Health Board member, Roger DeHaan, and his wife hiking in the wildlands may have saved us from another blow-up. They met some hikers coming out as they went up the trail. A little further on, smoke caught their eye. The smoldering coals had gotten under the duff, into the root system, from a campfire that was left too hot. Luckily, it hadn't spread. After several hours running back and forth from a nearby creek, they put it out with their only fire extinguishers — water bottles.[11]

Unfortunately, the Bitterroot River acts as if it is hosting a pagan god demanding a sacrifice, taking a life every year or two — a tragedy. On the surface, it often seems like it couldn't be dangerous. Someone dies in what is the most common cause, a pin — someone finding themselves in water as the current presses them against underwater branches that act like a sieve.[12]

The results from the National Guard drive-through snapshot tests all came in negative; maybe we can breathe easy.[13] But it isn't going to last long.

Doctor Calderwood is not the only one worried about crowds. The mask debate shifts to schools, as Governor Bullock leaves it up to each school district. Each school district is being pulled by parents with intense feelings in a tug-of-war, to go one way or the other — mask or no masks.

After Stevensville and Darby decide to make masks optional, the Stevensville mayor asks them to reconsider.[14] But the Stevi Mayor has his own problems. In addition to a recall election coming up, he has to answer to protesters with signs saying that virtual meetings are locking the public out of the process.[15]

The mask mandate is still in place for the rest of us. The Health Board has to navigate that red-hot-potato in the face of uncertain science.[16] Board Member Katie Scholl recommends a letter that is on the agenda. Katie suggests that the public needs to hear that the Health Board supports the Health Officer and Public Health Department. Doctor Calderwood adds, "I don't see that the Board or Commissioners have embraced the Health Department and Health Officer that we are your experts and you trust us."

Burrows says he can support someone without always agreeing with them. Obviously struggling with the dilemma, he understands what Calderwood said, but he feels in his gut that we are overreacting. He can't get past that. "The solution is worse than the problem," closing businesses.

I call in. While a fan of Doctor Calderwood, I agree with Burrows. At our last Farmers Market board meeting, I told the director,

> Next time somebody with a gun on their hip starts arguing with you about masks, call the cops. The mask arguments have not gone away."

The CDC certainly adds to the confusion, with Alan Lackey citing studies on their website that masks don't work and Katie citing a study from the CDC that masks do work.[17] Katie's letter is edited, toned down, passed on a vote, with Jeff voting no and doesn't sign it.

---

We understand there is much uncertainty with what appears, at times, to be contradictory information. The COVID-19 virus is a new virus so worldwide scientists are discovering new information daily and as this happens information is updated. One particular area of information updating is the use of masks....

We concur with the CDC that indoor use of face coverings and barriers are beneficial in reducing the spread of COVID-19...

We share the concern of everyone about the devastating effect this pandemic has had... We respectfully encourage everyone to wear a face covering... We do not believe that civil or criminal penalties are necessary at this time.

We know the people of Ravalli County hold true to Montana values, are fiercely independent and possess a desire to improve quality of life. In light of this, we ask the help of all Ravalli County residents to help preserve these values of life, equality and opportunity by respecting your family, friends and neighbors to help slow the spread of this virus by utilizing an appropriate face covering according to current recommendations.

Respectfully,

Ravalli County Board of Health Members:...[58]

Political silly season is well underway in Montana. Every few years, it seems some political hack gets the idea of promoting a spoiler for the competition rather than supporting their own candidate. Instead of building up their own or tearing down their opponent — which is ugly enough — they go to the even bigger lie and support a third party, with whom they have no other sympathy than to have it strip away votes from an opponent. The target is a group of voters who will not vote for your guy or gal but get persuaded to vote even further, right or left — the direction depends on who's doing the scheming.

This may be hard to follow since only a political operative could come up with this one. In 2012, there was a well-funded attempt to push a Democrat in the US Senate race after the Citizens United Supreme Court decision.[18] With $51 million in dark money and independent groups spent to influence a half million voters in Montana, a group called the Montana Hunters and Anglers funded ads for a Libertarian Candidate, submitting all of the required forms for expenditures and donations sources. They weren't hiding the fact that much of it came through other organizations.[19] Why? — to pull votes from the Republican challenger.

> Montana Republicans blamed Montana Hunters and Anglers, made up of a super PAC and a sister dark money nonprofit, for tipping the race. Even though super PACs have to report their donors, the Montana Hunters and Anglers super PAC functioned almost like a dark money group. Records show its major donors included an environmentalist group that didn't report its donors and two super PACs that in turn raised the bulk of their money from the environmentalist group, other dark money groups and unions.[20]

This year, it's the Republican's turn. While we have our share of Libertarians, we also have a Green Party. They both have to work to get enough signatures to get their party on the ballot. In Montana, the parties don't officially pick who runs for their party.

This year, Gary Marbut, president of the Montana Shooting Sports Association, an outfit that makes the National Rifle Association look like a bunch of left-wing liberals by comparison, filed to run as a Green Party candidate. Why? — to pull votes from Democrat Governor Steve Bullock, who is running for US Senate.

The Montana Republican Party puts up about $100,000 to fund signature gatherers for the Greens. When 568 people who had signed the petition to get the Green Party on the ballot found out about this, they filed to have their signatures removed — which is legal in Montana — so the party would fall short of the requirement. The Republican State Attorney General filed with the Montana Supreme Court to keep the signatures; the state court turned him down; the US Supreme Court let it stand. Bullock will stand on his record, after all. But will it be enough?[21]

Democrats do better if voting is easy, and Republicans do better if everyone has to show up at the polls and register ahead of time; at least, that is the working assumption of the political operatives.[22] The political hacks have looked at the polls showing a preponderance of Democrats planning to vote by mail.[23]

So the politician's view will become their party's view, voting access (Democrat) or election security (Republican), each marching as if on a holy crusade. They rile up their base so that, depending on which party is doing the talking — be it voter suppression or election fraud.

It is bound to be close in any case. I have now changed my view about Bullock's chances for the US Senate Seat. The Governor is a home-grown Montanan, a nice guy, who retained the Governor's seat even as the state leaned further right. At the Democrat booth next to the Farmers Market, I tell Central Committee members that Bullock will lose because of his mask mandate.

The Trump campaign seems to be hedging its bets as the polls get close. The hedge is the long-running eroding trust in government in the conservative movement, now aimed with red-dot-laser accuracy at the voting process. Lawsuits were already being filed to try to force local election officials to put more restrictions on voting in key, competitive precincts back east — it's in the news — it's passed down to the party machinery down to the county level.[24] Upping the ante, President Trump assures his supporters that he has overwhelming support.

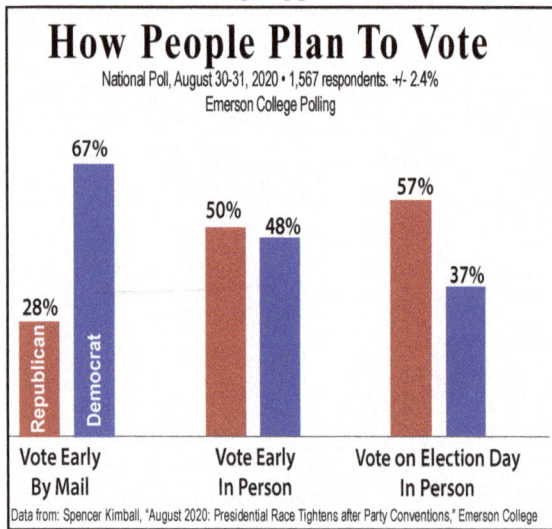

## How People Plan To Vote
National Poll, August 30-31, 2020 • 1,567 respondents. +/- 2.4%
Emerson College Polling

| | | | | | |
|---|---|---|---|---|---|
| 28% Republican | 67% Democrat | 50% | 48% | 57% | 37% |

**Vote Early By Mail**    **Vote Early In Person**    **Vote on Election Day In Person**

Data from: Spencer Kimball, "August 2020: Presidential Race Tightens after Party Conventions," Emerson College

*Poll: Mail Ballot or In Person Planned 2020*

And so, the voting security issue arrives in Ravalli County. The problem is that Doctor Calderwood and the school boards are not the only ones worried about crowds. Faced with calls for in-person voting from her party and fear of voter fraud being stoked by the usual suspects, Regina Plettenberg, Republican, Clerk and Recorder for Ravalli County, in charge of elections, is struggling because she can't find enough election judges to staff the polling places while trying to figure out safety protocols for the polling places. It is the straw that breaks the camel's back.

While pressure politics is the order of the day for the election-security crowd, they'll have a problem getting our Clerk and Recorder to bend to their will, even though the office holder, Regina Plettenberg, is a Republican. Her reputation is statewide.

It's a voting day tradition. People gather in the basement of the courthouse a little bit before poll-closing time, waiting to get the first glimpse of the early votes written on a whiteboard, sitting in one of the chairs just inside the door of the election counting room, temporary plexiglass walls separating us from the counters, ballots moving through the room with the grace of a ballet troupe dancing to a live orchestra, with Regina conducting. This orchestra has played for over two decades. When the polls close, the early and absentees are usually already counted. Local reporters show up, familiar faces every year, along with some candidates, usually those in contested races, with a couple of supporters in tow.

A few years ago, I said that there is one Republican who you could always trust to do her job, Regina Plettenberg. Two journalists piped up, "She is the most trusted official in Montana." As usual, Regina came out and stood at the whiteboard to give an informal interview and take a few questions. The early results were written out on the board, comments were made on any surprises, and reporters with a mic, camera, or notepad took candidates aside for an interview. The last time anyone ran against Regina, Republican or Democrat, was in 2006 — it wasn't close.

It is her turn in the spotlight, the Governor's directive for the general election, having expanded voting deadlines and allowing a local option for mail-in voting only.

Montana already has a well-tried process for mail-in.[25] In Ravalli County, special elections are already done by mail-in only. Even though every one of the 56 counties in Montana had the option to open polling places, all 56 had opted for mail-in-only in the June primary. For the general, the thought is to do mail-in-only again. But, before she submits the plan to the County Commissioners, she is savvy enough to plan a public meeting to take comments on the 17th, holding it in the big hall at the Fair Grounds.[26]

On the 17th, the big hall is not all that full. Facing the crowd, along with the County Commissioners, Plettenberg tells the crowd, "My problem is, that's what I rely on, are my experienced judges," as they edged toward a mail-in ballot for the general election.

A few think it is a good idea, but several had brought their torches and pitchforks. It is an attempt to throw the election —

> We are Americans. And we have the absolutely 100 percent right to vote in person! — those of us who... really care about our liberty, and freedom, need to be among the 3 percent who say, this is the line and I'm just not going to cross it anymore.[27]

Terry Nelson, the chair of the Ravalli County Republican Central Committee, makes the same point with less inflammatory language.

I have known of Terry Nelson for years. An Electrical Engineer who ended up as a licensed surveyor. There was some controversy when he, a well-known member of the Republican Central Committee, was hired as a county planner at a time when planning and zoning had become as politicized as COVID is today. But, he is qualified for the job; the political anti-planning rhetoric feared by some doesn't happen.

We became better acquainted when we dealt with the AR-15 lottery prize at the Republican booth a few years ago. He came to the Farmers Market Board meeting. He was great to work with to dampen the ardor of his closed Facebook group and some people staffing the Republican booth. Like many local politicians, he seems to have gotten into it for the right reasons, not just self-interest. But, my observation has been that it is hard to be Central Committee chair for a political party without becoming a creature of your party's partisan politics.

It is just too hard with most of her experienced election judges getting along in years, with most opting out of a crowded polling place.

After announcing the start of the meeting and a pause for the pledge of allegiance, Commission Chris Hoffman announces the first item on the agenda, "Designate Polling place for November 3, 2020, General Election," calling "madam clerk."

Sitting at the table, Regina Plettenberg says, "I need to make one last-ditch effort to talk to you guys about mail ballots. With every fiber of my being I believe it is a mistake to not go with mail-in ballot. You guys haven't seen the ballot yet, but its long," nodding her head to emphasize the point,

> It will be even harder to keep the flow going; people will become hung up....social distancing at public events... It's a monumental challenge to get this done in one day.

> ....mail-in-ballot allows people to do in person voting. What's the difference between going over there (to the fairgrounds) or coming here and spreading it out over a month.

Regina worries about voter confusion because we get our TV news from Missoula, which is doing mail-in. She knows voting is a Constitutional Right but cannot find anything about how it is done in the Constitution. Regina notes that mail-in gets more turnout. She addresses rumors of the US Post Office slowing things down, but she works closely with the Post Office. "They are people like us just doing their job."

She finishes with an apology, "I'm sorry." It's an administrative decision, not political.[28]

The law gives the commissioners the right to object to the plan — no action is taken. We will have a mail-in-only election in Ravalli County. People trust Regina, even though Regina applied for and received a grant from the Center for Tech and Civic Life Foundation, a grant to help defray election costs during the pandemic, a foundation funded primarily by Facebook owners the Zukerbergs, seen as liberally biased by the likes of Ted Cruz.[29]

Commissioner Jeff Burrows and Chris Hoffman, Chilcott being absent, both say the feedback they got is that people trust Regina. "Even those who don't trust mail-in-ballots say they trust Regina."

Terry Nelson approaches the mic with only one other person in the audience. He had advocated for in-person voting a week ago. Now, he has second thoughts. He is ready to accept mail-in-only, saying, "If the commissioners do, he will post this letter in the newspaper," reading it out loud:

---

Let me start by saying I am against Mail Ballot Only elections. I believe they provide many more opportunities for fraud, error and coercion on many levels. That being said, I am also adamant for FAIR elections.

Our current Governor, who is also running for a US Senate seat, has completely abused his powers during this COVID panic. For example, he received $1.2 Billion to IMMEDIATELY help Montanans back in March. As of an article from July 2nd, he had spent just over 10% of it, saving the bulk of it, I'm sure, to use in the month or two prior to this fall election.

On August 6th, he also unilaterally decided that all counties could decide whether they want traditional elections or mail only. This again was an abusive stroke of genius and gameplay by the Governor.... Even if he wasn't running for office, this would be a great breach of law and public trust, but the fact he is a candidate in one of the most important elections in the country makes this one of the biggest conflicts of interest I've seen...

Why do we need to allow this? Let me explain.

In the 2016 presidential primary, Ravalli County had 48% of registered voters vote. Missoula had 44%. In the 2020 presidential primary, because of mail out only ballots, Ravalli had 58% and Missoula had 53%... That means almost 7000 more votes for Democrats in Missoula. If Ravalli does mail in only, that should increase Republican votes by over 3000.

The evil genius of the Governor's order to "allow" counties to make up their own minds about the mail only ballots is that he knows conservative counties like Ravalli will choose traditional elections while liberal counties like Missoula will do mail only. This creates a completely unfair election warped toward the Democrats.

That is why I will continue to fight for a traditional election, while also be asking you to join me in supporting our Republican Commissioners and Clerk and Recorder in Mail Only ballots as long as the Governor's edict is in place.

Terry Nelson.[59]

---

Organizations are still adjusting to COVID. Mayor Dominic Farrenkopf reads poetry most weeks on radio station KLYQ[30] after giving a usually light-hearted report of some going's on in Hamilton. The Poetry contest at the County Fair is canceled. So, KLYQ resurrects the event, using its studio on Highway 93 in Hamilton to have the winners read their rhymes over the airways.[31]

KLYQ started in 1961 and has been bought and sold a couple of times. Now, a News Talk station, look for your favorite conservative on the dial. KLYQ still features local news from the Bitterroot, along with a farm report early in the morning. If you have a local event you want to pitch, you can get on the air live on Friday morning with a call to Steve Fullerton. Known as the "Voice of the Bitterroot," he is often recruited as a master of ceremonies or moderator for a political debate.

Two big annual fundraisers, the Chamber of Commerce Brewfest and Emma's House annual gala, have had to give up their crowd. The Mayor of Hamilton is always willing to be dunked on Daly Days. Not this year. A front page in the *Republic* picture shows Sheriff Holton, in an old west Sheriff's outfit, has Mayor Dominic Farrenkopf's head and hands in stocks for a scene in Emma's House silent film Western.[32]

The series *Yellowstone*, starring Kevin Costner, is staged around here. Like the Dallas TV series decades ago, the theme is a family legacy — in the *Yellowstone* series, the fictional ranch is the biggest ranch in the country, homesteaded in the 1880s.

The set for the Yellowstone Dutton Ranch is on the Chief Joseph Ranch, just south of Darby. Many of its buildings are now adorned with the *Yellowstone* logo. While staged here, the storyline is set in the Paradise Valley north of Yellowstone Park and Bozeman.[33] The *Yellowstone* production company must have returned to work after a good time at the rodeo. They show up at Main and 2nd for a bombing and a gunfight. As one commentator said, the body count in *Yellowstone* exceeds the population of Montana.

The reality in Ravalli County is that much of the wealth was brought from the East, as in the case of Marcus Daly; now a lot of it comes from the West Coast. The Chief Joseph Ranch house was built in a flourishing apple orchard in 1914 with money from a glass factory tycoon and a judge from the back east, renaming it for themselves the Ford-Hollister Ranch. Shifting from apples to dairy, then cattle, it became one of the first and more well-regarded dude ranches in the Rockies by the 1930s.[34]

The reality is a quieter but surer reality — no gangs with guns are trying to take over ranches; it's sub-dividers with a bag full of money. For the last few decades, the land has been more valuable as a subdivision than a ranch. But people still farm and ranch — everyone I know who farms here has a passion for it. But, all agree that farming is financially challenging and will consider selling off small pieces of land, often reluctantly, to raise capital to make improvements.[35]

The crew and actors of *Yellowstone* brings some wealth to Ravalli County. *Montana Business Quarterly* claims that 527 permanent jobs have been created, $25.3 million in "household income," and $85.8 million in business spending was done for Montana — though there was no citation for the source of these numbers.[36] According to the article, the population of Western Montana increased by 233 people, probably many in Ravalli County. The State of Montana gave $16.5 million in tax credits.

The article threw me when it said the product "increases exposure and awareness to our physical landscape, culture, and history to audiences that are potentially global." — it does indeed show off our landscape. Still, it has almost nothing to do with the culture or history.

Some of the *Yellowstone* crew brought their family and children. Hence, Mara Lynn gets to help several crew members, actors, and family get their favorite book. She met some interesting people and had a few laughs. When the Chamber of Commerce asked to hand out Yellowstone fliers, one staff member insisted it would be too controversial — no, no, they are fliers for Yellowstone Park.[37]

*Yellowstone* didn't shoot in front of Chapter One, but they did in front of Ford's. "It was real casual," said Alan Ford when the production company paid $4,000 a week before the shoot. It was to compensate for a day and a half business impact of being on the streets next to his building. It turned out that, even though they blocked

Permit in hand, the scene on Main Street in Hamilton is filmed and edited.[60]

---

*After a shoot-out filmed in an office setting, we move to the corner of Main and 2nd to see emergency vehicles in a long telephoto shot down 2nd street looking north toward the Roxy Theater. The scene cuts to inside a dark office.* Then:

### YELLOWSTONE SEASON 4 - EPISODE 1 — HALF THE MONEY
...

**8 INT. SCHWARTZ AND MEYER - DAY** 8

Beth in closeup foreground.

Beth pushes open office door of Schwatz and Meyer leading to street.

SFX: High pitch ear ringing, low rumble.

                                                                    CUT TO:

**9 EXT. SIDEWALK NEXT TO WALL SCHWARTZ AND MEYER - DAY** 9

Beth in closeup foreground.

Beth. bloody. torn dress. dazed. next to brick wall looking toward street.

SFX: car horn 3 sec cadence, sirens. helicopter whirring.

STUMBLING MAN, holding cigarette. dazed. walking toward Beth.

Aftermath of explosion. fire truck. crashed cars. smoke. flashing lights. emergency responders.

Beth walks slowly to curb, looks slowly around. Sits on curb.

P.O.V. Beth.

Stumbling man, cigarette hand shaking.

SFX: High pitch ear ringing, low rumble.

**STUMBLING MAN**
                    Are you all right? (No sound. mouths words)

                    BACK TO SCENE

---

the parking, he did good business with the crew. He ordered some *Yellowstone* shirts. They sell well at Ford's.

Fictional Beth is standing next to a brick wall that is a real Ravalli County multi-generational legacy — Ford's Department Store. The legacy began when John Ford was sent up to Hamilton from Salt Lake City to open a JCPenney® store in 1929, having been taught the retail business by J. C. Penney himself.

His son, Tom, born in 1925, moved with the family to be raised in the Bitterroots. Tom remembered fewer fences during the Great Depression, "people knew you and would lend a hand if they could." His favorite place was on the back of a horse, ready to "go to the hills."[38] Interested in the wild hills, Tom was a member of the Ravalli County Fish and Wildlife Association,[39] an outfit that goes back to the late 1700s and is still active today, a conservation organiza-

Photo by James R. Olsen

*Ford's Department Store* and Old Roxy Theater — *Yellowstone, Half the Money Scene was shot a about 20 feet behind the stoplight.*

he was just the right age to join the war. Rural America sends more than their fair share of their sons and daughters to war. He signed up for airborne and got the Army Air Corps instead, "all the guys going to airborne got the B-17, and it was the reverse; they got airborne." Catch-22. In fact, he ended up in the Italian air campaign that is the subject of *Catch-22*,[42] in the Fifteenth Air Force, commanded by Jimmy Doolittle, the man who led the *30 Seconds Over Tokyo*[43] raid.

Tom was stationed at one of the airfields around Foggia, Italy.[44] A Sergent, Tom was a bombardier, then a waist gunner. As a bombardier, he would take control of the airplane on the final bombing run while peering through the eyepiece of a Norden Mk XV Bombsight.[45] His skill as a bird hunter served him well as a gunner.[46]

The number of missions in a tour went up as the air campaign progressed, from 25 missions to 30 to the 35 Tom flew. The odds of coming back were about 50%.[47] When Alan, still a kid, asked about a picture his father had of a bomber crew, Tom simply said, "They didn't come back," a day when Tom did not go with them. Probably sick, but Alan never asked. Tom did beat the odds even though he was shot down. Tom and his crew bailed out behind enemy lines a bit north of Trieste, Italy. Hiding until night to hike out — they made it.

tion with a long history of projects to conserve fish and game.

The old JCPenney® store was on 2ⁿᵈ Street, in the same block where Ford's sits today.[40]

Tom was just the age to enter High School when the German Blitzkrieg swept over Poland. Instead of attending Hamilton High like his friends, he attended Saint Thomas Military Academy in Minnesota.[41] When he graduated in 1943,

When Tom came home from the war, J. C. Penney was retiring, and the JCPenney® was in turmoil. So John Ford set up his own business, Ford's Federated, opening to a congratulatory ad from Valley Furniture:[48]

The war over, Tom went to the University of Montana, probably on the GI Bill. Alan would later meet his fraternity brothers; they said all Tom could talk about was work — the "loyal son" would come down from Missoula as often as he could.[50]

In the 80s, two of his children, Alan and his sister, who you will still find behind the counter, began to run it. Tom stayed on into his 90s to help out in the shoe department. Tom picked out more than one pair of shoes for me. His son grew up on the same landscape, hunting and fishing, learning to be a retailer, who, in his younger days, would sometimes have to hustle over the fire station and suit up — a volunteer firefighter. There is a strength of character behind the unassuming manner of the Fords. A local family legacy. While *Yellowstone* is fictional, the multi-generational operations in the valley are dealing with real issues.

However, you can find some excitement in the woods around here. Within a month, a young man, bow hunting, will be making elk calls at Bass Creek Overlook. He draws a grizzly bear instead of a bull elk. Having hunted bears and seen a lot of bears, this one is the biggest he'd ever seen. Man and bear stare at each other, man and bear 10 to 15 yards apart. Telling the bear to go away, the bear stands its ground. The hunter fires a few 7mm rounds into the ground. The hulking grizzly retreats a few yards, continuing to stare as the man backs away.[51]

There is irony in the story. The sighting of a grizzly bear in the Bitterroots two decades ago would have been national news. It was a time of bitter arguments, collaborations, manipulation, and politics over a plan to reintroduce the grizzly bear to our area.[52]

The *Grizzly Wars*[53] wasn't just about bears, though; it is about the landscape, how people use it, and how some people want to keep it wild. This conflict over land, what it means, and how to use it is mostly managed by Federal government agencies.

Of course, people want to make money off the land by extracting its resources. As was often done back in the day when our own Copper King simply ignored the Federal Government as his timber company cut at least 21 million board feet off the national forests, shipping it to the mines around Butte, improving his profit margins while shoring up his mine shafts.[54]

The trees grow, the animals live, the rivers run, and tension breathes in the beautiful landscape and the resources within it. People like Dale Burk, who passed away this week, who, at 83, was planning a hunting trip with his son, even though he was in "a lot of pain" the day before he passed on. Another Montana legacy, he had grown up wandering from place to place in the woods, his father gyppo-logging. A reporter for the *Missoulian* in the 60s and 70s, shining a bad light on clearcuts and terracing, he had much to do with improving the laws governing forest management on public lands in the face of a heated debate in Ravalli County.[55]

The question of whether the Federal Government can legitimately manage public lands sometimes sprouts violent outbreaks such as the Bundy Standoff. America's troubled love affair with Federal Agencies is a *Heart of Glass*,[56] the CDC and the FDA being the latest targets of much mistrust.

The politics of COVID-19 is the latest chapter in the ongoing story of conflicts in Ravalli County. By the end of the month, the fire danger is rising as the political temperature is nearing the ignition threshold. President Trump lights off his drip torch,[57] getting ready to question the validity of the election if he doesn't win, repeating a new slogan: stop the steal.

If only we had a top notch rodeo clown to save us. Are we going to end up laying in the dirt, having just bitten the dust, stunned, hoping a hoof does not find a vulnerable piece of our flesh, as we turn to look up at that raging bucking bronc?

What a rodeo — trying to get back to normal, as we see our country fumble through a pandemic, as the fatality rate for the United States, accelerating, catches up with Sweden, as election politics gets meaner by the day, as both business and the work force struggle to stay in the saddle.

*31 Aug. 2020: 188,789 COVID-19 Related Deaths Reported*

Photo by Margo Harrison (Shutterstock)

**4**   **7**   **14**   **16**   September 2020   **22 23**   **24**   **28**

100th day of rioting on Portland

LABOR DAY

US Airports stop screening international travelers

Trump releases vaccine distribution plan using HHS & DoD

US Deaths from COVID over 200,000

More infectious Alpha variant in Huston

Trump does not commit to a peaceful transfer of power when asked by reporter

Trump to ship 150 milliion tests

Worldwide Deaths 1,000,000 [1]

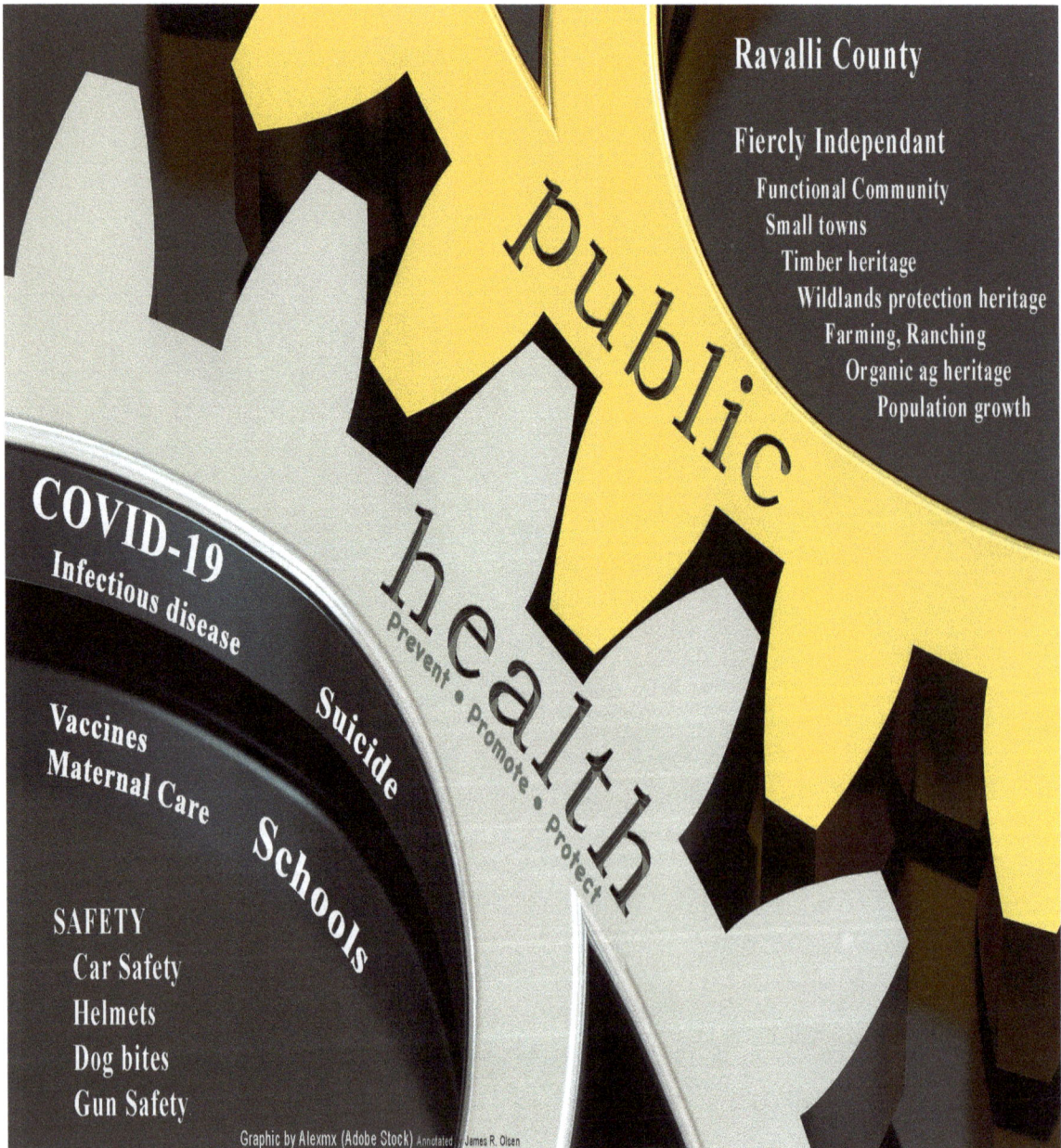

Ravalli County

Fiercly Independant

Functional Community
Small towns
Timber heritage
Wildlands protection heritage
Farming, Ranching
Organic ag heritage
Population growth

public health
Prevent • Promote • Protect

COVID-19
Infectious disease

Vaccines
Maternal Care

Suicide

Schools

SAFETY
Car Safety
Helmets
Dog bites
Gun Safety

Graphic by Alexmx (Adobe Stock) Annotated by James R. Olsen

*Public Health Ravalli County*

# 12

# TIME OUT FOR PUBLIC HEALTH

*People have their own ideas, so I have to work within that to bring you to the table.*[2]
— Tiffany Webber, Public Health Nurse/Director

## SEPTEMBER 2020

The fire hazard is lower, though still high; the fires are few.[1] The only people who attend the Health Board meeting were there to talk about septic systems, allowing the discordant beat of last month's meeting to slow to a waltz.[2] Cases are manageable, dipping last month — 35 cases and 2 hospitalizations.[3] Let's enjoy it while it lasts.

On the 16th, as the newspaper announces increasing fire danger, the Commissioners meet to interview Tiffany Webber for the permanent job of Ravalli County Public Health Director — the Health Nurse.

The meeting begins with the Pledge of Allegiance. Health Board members Roger DeHaan and Doctor Turner join the three County Commissioners. Tiffany is well dressed in a black blouse with a scarf draped around her neck, giving a splash of color, a simple arm clasp bracelet reflecting the light on her arm. Her glasses, pulled up on her head, gives the more formal-than-usual outfit an air of informality.

Answering the first question about who she is, with an engaging smile, Tiffany Webber explains that she was born and raised here, giving her family name as Trollope, "so you might recognize that." After graduating from Hamilton High in 1990, she got her RN, married, had three boys, then went to Ireland in 2014 for five years.

An Irish newspaper has her picture in a news story about a job fair for the Irish Health Ministry. She chatted with Health Minister Simon Harris.[4]

When Tiffany returned home from Ireland, she had done a lot of clinical nursing, including some time at Marcus Daly and private practice. Then, someone from "HR" recruited her to work in Public Health in 2018. She had great mentors in the Public Health Department.

Asked, "What is mission of public health?" You could use her answers as a training video:

> I look at my community, at the needs of the community, the community as a whole, not the individual like in clinical nursing, focused on the individual and the individual's needs. You have to span out and look generally at the population. What does the population need?

In her spare time, Tiffany goes around to different places that might have a good idea, a flow of people from the community. She adds,

> Before I could dive into what the needs are, COVID hit. Now, we have a need on our hands at the moment.

> But, the core mission is to be independent, to see the needs of the community, to liaison with other health people.

> We need to be able to coordinate with the hospital and other providers, to bring all of our stakeholders in for a conver-

sation so we can get to the heart of what we need.

Doctor Turner asks her the follow-up question,

> How do you see the public health in Ravalli County compared to other places?

Tiffany answers,

> It is unique to us that we have a seriously independent population of people.

Turner pops in with a note of irony,

> Oh really.

Tiffany goes on,

> That's really helped me. I do understand that; I come from that. It's really important we are assessing the needs of the community. You have to take that into consideration, because without it we can't bring people to the table to have the conversation.

Moving her hands as if tilting a balance,

> It's are real fine balance I find. How am I going to meet this person's need if the person's not even going to listen to what I'm saying.

> What I have to sell to y'all is that its for the greater good for everybody, not just yourself.

> It's hard for us, because we are all fiercely independent. We all have opinions, and we've all done the research, and we all know everything. We're privy to all the conspiracy theories in this county. Everybody has their own ideas, so I have to work within that to bring them to the table.[5]

Tiffany gives a list of what she thinks the key issues are at this point:

• Communicable disease surveillance.
• Vaccinations — always.
• Health and Safety — we're always reassessing.
• Mothers and children will always be an issue.
• Gun safety • Helmets • Seat belts.
• Suicide is a big one.

Asked about her relationship with Marcus Daly, she says she worked there and knows a lot of people there. She has developed great relationships with the supervisors there. She is trying "desperately" to convince the hospital that they can't be the only ones who work with viral illnesses. There's still this teeny bit of "us and them. We're keeping this private, don't tell public health."

Doctor Turner recalled that once, they argued about who would pay for rabies shots. The commissioners thank Tiffany for her work. Hoffman adds,

> Thanks for stepping up through this whole thing. Its been very clear who the workers were and the work piled on and trying to get the job done through this difficult time.

"Don't make me cry," Tiffany says with a laugh.[6]

Nurse Webber leaves the meeting to face the new disease. The local schools announce their first case.[7]

The Bitterroot Affordable Housing Coalition[8] consists of both providers and anyone else who shows up — which I have attended off and on for the last several years. I have done much except learned a lot. Everyone is so busy this year that I volunteer to make a presentation[9] to the County Commissioners. Laural and Peter did not get counted in the HUD Point-in-Time homeless count because they slept in a place "designed for or ordinarily used as a regular sleeping accommodation for human beings."[10]

Driven by COVID-flight, people are coming to the Bitterroot Valley like animals fleeing a wildfire. They are flowing through the passes, bringing too much cash, driving housing prices up, creating a building boom, pushing rental property owners to sell out, driving renters into RVs, tents, a friend's couch. The pressure on the Bitterroot culture, the new already outnumbering the old, the culture is threatening to become a cartoon of itself.[11]

**BITTERROOT AFFORDABLE HOUSING COALITION**

PRESENTED BY
**JAMES R. OLSEN**

# Too Much Rent • Zero Rentals

## Housing Security and Homelessness in the Ravalli County

### Bitterroot Affordable Housing Coalition
### Membership Criteria: *Show up*

#### MEMBERS
- Bitterroot Family Shelter •
- Bitterroot RC&D •
- District IX Human Resources Council •
- First Presbyterian Church •
- Ravalli Head Start •
- Ravalli Services •
- SAFE •
- Summit Independent Living •

other organizations and concerned individuals ...

# Airbnb
# VBRO

Graphics from websites Deleted

# Rentals Disappear

## And can we house a production company

YELLOWSTONE

- Air Bnb, vacation rentals
- Single family zoning in Hamilton

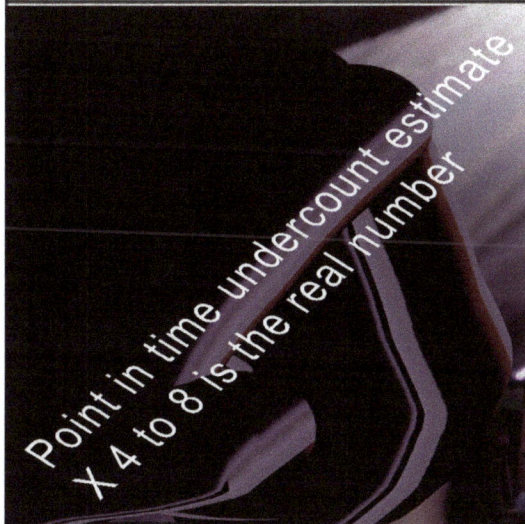

Point in time undercount estimate X 4 to 8 is the real number

# People without a roof over their head one day in January

COUNTING THE HOMELESS

# How Many?

## Estimate: 175 to 300

- Point in Time - 100 people - who
  - 66 Adults • 34 Children
  - 29 Were first time homeless
  - Half over 6 months, 17 over a year
- HOUSEHOLDS - 48 - family groups
  - 24 Single • 24 Families, 17 w/Children
- HOUSEHOLDS - 48 - how long
  - 35 in the County over 5 years
  - 14 of those, over 20 years

*Homelessness and Housing Security in the Bitterroots*

**800 sq ft condo**

Redondo Beach, Los Angeles

IT ONLY TAKES A FEW THERE
LOS ANGELES BASIN

TO HAVE A BIG IMPACT HERE
RAVALLI COUNTY

20 MILLION

42 THOUSAND

Photo SciePro (Shutterstock)

**Match Up
$1,000,000**

**Ravalli County**

Photo by Vacclav (shutterstock)

**COVID FLIGHT
Brings Real Estate Boom to
Ravalli County**

Photo changed from presentation

**Out of luck?**

Photo by pikselstock (Shutterstock)

RENTALS TODAY - MAYBE
# RENTALS NEAR ZERO

- Cardinal Properties
  No residential rentals available listed on their website
- Greener Montana Properties
  No residential rentals available listed on their website

- On trulia.com and zillow.com - may not be available -
  have to provide phone #
  $1,200/mo: 1 br
  $1,200/mo: 1 br
  $750/mo: 1 br

- Craigslist (two of these are also on realtor.com)
  $500/mo: Studio apt
  $600/mo: 2 br apt (listing 3 days old)
  $750/mo: 1 br apt
  $1,200/mo: 2 br cottage
  $1,000/mo: 2 br apt (7 months must be paid in advance)

Photo by JopsStock (shutterstiock)

THE FEW
VOLUNTARY
LIFE STYLE
CHANGERS

THE MANY
IN VOLUNTARY
WAGER EARNERS • SOCIAL SECURITY • BROKE
LONG TERM RESIDENTS • RESIDENTS

THE REST OF THE STORY

## Strategies for Living

- Couch sharing
- A camping trailer or RV
  - Permanent spot in RV camp
  - Forest Service Campgrounds • Or other site
  - Private land
  - On the street
- A Tent
- A Car
- Office rent is cheaper
- Homespun shelter in the woods or under a bridge

*Homelessness and Housing Security in the Bitterroots*

The coalition starts an initiative to get Hamilton to allow mother in law apartments on single family lots to relieve some of the pressure.

I was surprised by how involved SAFE was in the group, as housing security is becoming more and more of an issue for their clients. Stacey Umhey, the Executive Director, often leads the coalition meetings.[12] The Commissioners would probably approve any good idea, if the coalition bring the money to solve it. But the providers have little cash to spare.

The most common strategy in the Bitterroot Valley is trailer living or in the car. The texture of living in *Nomadland*[13] is no surprise to many in this valley, even though the life-changing premise is the opposite — not a shuttered plant, but no house to live in.

*30 Sep. 2020: 205,872 COVID-19 Related Deaths Reported*

2    5    7    October 2020    19              28    31

HALLOWEEN

President Trump tests positive of Coronavirus. Hospitalized.

Trump leaves hospital. COVID-19 spread to Whitehouse aides

New Zealand virus free after shutting down incoming travel

Global cases over 40 million

New rules make vaccines free

**COVID-19 Treatment Guidelines** (NIAID COVID-19 Treatment Guidelines Panel)
• Hospital: Severe: Remdesivir; Supplement O2: Remdesivir + Dexamethasone; Intubation: Dexamethasone
• Mild or Moderate: Insuffient data to recommend for or against Remdesivir • Long COVID desrciption

Iceberg photo by Alone. Titanic drawing by Aman Artist (shuttersock) Combined. Ship and iceberg distorted by James R. Olsen

*Tip of the Iceberg and Titanic*

# 13

# TIP OF THE ICEBERG

*It is the tip of the iceberg. You don't know how many people have symptoms and just stay home.[1]*

— Doctor Carol Calderwood

**OCTOBER 2020**

Every word swept over the crowd with a resonance reminiscent of a bass guitar at a rock concert,

> The only way we're going to lose this election is if it's rigged. Remember that.[2]

> And we're going to win four more years.

The crowd chants,

> Four more years, four more years, four more years, four more years.

President Trump goes on,

> After that, we'll have another four more years because they spied on my campaign. We should get a redo, four more years.

The crowd cheers, and cheers, and cheers.[3]

That speech is in Wisconsin. It resonates with many here, who already harbor a mistrust of the government, especially the Federal Government.

It's hard to believe that Governor Bullock's campaign opened his bid for the Senate with a political ad in the *Ravalli Republic* — a front page fold-over, presenting himself with a mask on.[4] It's like no one in the Democrat Party is paying attention to what is happening here.

On that same page, Tiffany Webber is doing rumor control again — announcing "they are not approving or disapproving events" — event organizers are deciding on their own. Of course, with partisan politics now in full throat, with pickup trucks driving down Main Street with Trump 2020 flags a-flying, the rumor could have been started by a local candidate spreading some drama, trying to put some breath in a local campaign when Trump is using most of the air — but there is no way to tell.[5]

Both Montana and Federal Courts rejected lawsuits trying to keep all counties from doing mostly vote-by-mail. Mail-in voting will start on the 9th. By direction of the courts, ballots must be received by 8 PM on election day, November 3rd — there will be no delayed mail-in ballots in Montana. Regina Plettenberg asks people to watch their mailbox and not mistake the ballot for junk mail, asking that they be mailed a week ahead of time — or put them in a drop box at the election office, or on election day, at the Florence Fire Hall, Hamilton Fairgrounds Premium Building, and Darby Clubhouse. Voters can also vote in a booth at the election office.[6]

Regina warns everyone to get their mail-in ballots in early, marking, sealing, signing, and putting a stamp on it, and drop it in the mail or bring it down to the County Building and put it in a locked drop box. Or come vote in a booth — but don't hand it to a stranger.[7]

As the voters make their final choice, Trump's man, Giuliani, tries an "October Surprise," as we all learn that Joe Biden has a son, Hunter, who is trading on his father's name.[8]

We are also on the last-minute-political-stump-circuit. It's not that rare — you could have seen Bill Clinton when he showed up one year in the county, stumping for his wife, or a drive to Missoula could have gotten a live speech from Obama or Bernie. Trump and family tried to unseat Democrat Senator Jon Tester in 2018 because Jon had "Poked the Bear" to sink the President's VA Secretary nomination.[9] At one of Trump's rallies in Missoula, the Rolling Stones' *Sympathy for the Devil*[10] screamed out of the loudspeakers, the lyrics of which a lot of Democrats would say fit all too well.

We see little of the Dems this election cycle, as we get visits from Republican Gubernatorial candidate Gianforte, followed by Ted Cruz. Republican Daines stumps in Ravalli County, running against Governor Bullock for US Senate — a campaign with cash flowing in from all quarters at a record-setting pace.[11]

Trump signs, Trump flags, Trump talk, Trump dissed and dismissed. Political tension, its rope pulled taut, is ready to snap.

Everybody's coming to Ravalli County it seems, for license plates, for votes, for good, for no good. With Missoula county offices still shut down and other counties shut down as well, with a slow-moving through-the-mail vehicle title process, the lines get longer in the basement of the County Admin Building for in-person title and licensing. People drive from other counties only to be turned away because they are not residents of Ravalli County. The Treasurer tries to stem the tide with a headline that reads, "Treasurer Turns Away out of County Requests."[12]

Anthony Fauci is now a full-up political figure whether he wants to be or not. In Hamilton, "Thank You, Fauci" and "Fauci should be arrested," sprout like dandelions in yards in Hamilton.[13]

Cases are rising, the number of people who need to see a doctor is growing, and hospital beds are almost full — as are all the hospitals in the area. The CEO of Marcus Daly is concerned about "the potential of the spike to exceed the community's health care resources and capacity." Adding, "As of this morning, we have the capacity for a couple more patients, but that doesn't mean COVID cases," adding that they are "struggling" to find hospitals to transfer patients to.[14]

It is about 10 days from the day a COVID-19 case is diagnosed to the need to go to the hospital. Hospital stays vary greatly, from less than a day to months.

Public Health Director Webber is not surprised:

> I think it's always been coming. You can't have the level of openness... and the influx of people from all over that we've seen this year and not see this happen.

In a pointed statement to virus deniers, she adds,

> If the virus wasn't real, we wouldn't be here, would we?[15]

> The health department was never the big bad wolf that was going to come in and tell businesses they are going to be closed down.

In a note of frustration, she reports that some businesses are taking her advice as if it is punitive. Tiffany says that Public Health is just trying to help a business avoid having an employee get sick and pass it around. She goes on,

> To add to the challenge, some people haven't been cooperative with public health in the contact tracing process. Others who tested positive refused to remain quarantined and went on to spread it to others.

> People who have the attitude that since they didn't get sick... it's no big deal.[16]

But some residents are in the hospital, and some businesses have closed. Hamilton and Pinesdale schools close due to COVID-19. Stevensville schools close because they cannot get substitute teachers.

"We've had two deaths. I can tell you that one of them probably had no idea going into this summer that they weren't coming out the other side," says Webber.[17]

This will prove to be the tip of the iceberg. Pockets of lamentation will spread through the valley. The dead will become sufficient to fill the ranks of a platoon, casualties of the COVID.

Still, science marches on as RML gets a 40,000 square-foot, $70 million upgrade to the 1963-era animal facility.[18] Most people go about their business, giving little thought to the struggling people in public health, the Viral Clinic, the hospital, or the other clinics.

However, healthcare providers probably feel transported to the 4077th Mobile Army Surgical Hospital (M.A.S.H.)[19] rather than working a day job. The first sniffle comes, achy muscles come, a positive test result; call the clinic; go to the clinic to see overworked doctors and nurses. The inhabitants of M.A.S.H., with their workload rising, have two more deaths to lament.[20]

The cooling snow has come in the mountains, falling lightly in the valley bottom, a white dusting of the streets of Hamilton, a portent of good tracking on the opening day of rifle season, a week and a half hence, when the orange army takes to the woods,[21] the anticipation pulling peo-

ple through the double doors of Bob Ward's[22] for hunting gear. However, there is a metric of fear: the pistol-ammo shelves are empty.

There is not enough snow to keep the heat from rising — emails, phone calls, visits from maskers and anti-maskers flooding into county officials.

October starts on a Thursday this year, so the second Wednesday, the day for the Health Board meeting, comes on the 14th. Though less than two dozen people are in the audience, most are from Stand Together for Freedom and the ever-present Representative Dave Bedey.

After a quick approval of the minutes, Doctor Carol Calderwood, still acting as Health Officer, kicks off the meeting, sitting at the table, notes spread out, mask on, reading from notes on a yellow legal pad,

> I don't really have anything except COVID. As you know, the cases have surged in the last week or so. I anticipate of the up slope that we have been pre-empting from the beginning. Both health care and health resources are already stressed.
>
> I am seeing firsthand in the Viral Clinic lots of people who are sick, whether they have COVID or not... clusters from businesses who are still having people go to work if they have been identified

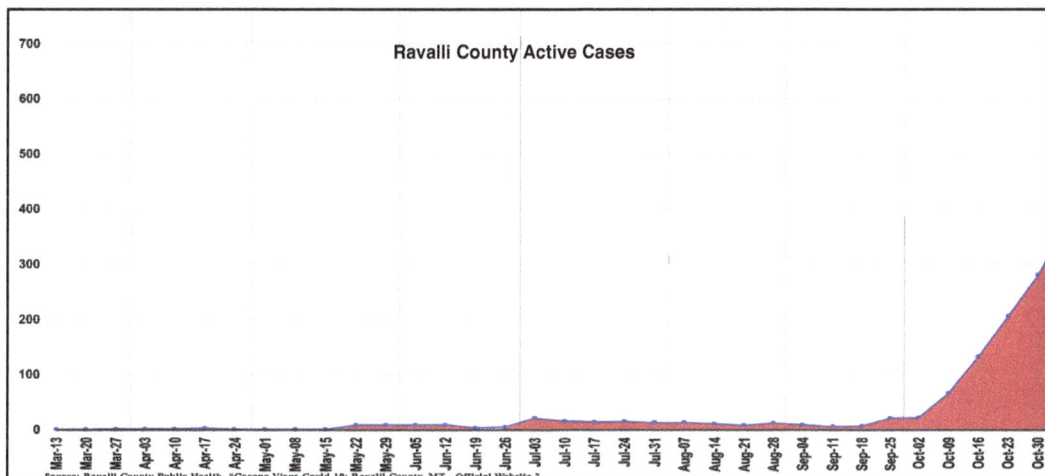

Active Cases mid-October

as a contact.[23]

Carol says, "Now is the time." She understands personal choice and does not want to do anything mandatory, but we need to put a plea out for people to take personal responsibility. She sweeps her hand up as she says,

> We don't know the shape of the curve, but it looks like the upslope... If we don't do something, we are not going to have access to health care. I see it already.

> I, and another person who came into to help, saw a 100 patients in two days. People with pneumonia, broken bones, and severe pain waited two hours to see me. That's already. I don't know what it's going to be at the peak.

> There have been days when all of the hospitals in Montana have been closed to transfer. We're going to have sick people with other conditions who won't have access to be transferred to a higher level of care facility.

She is challenged by the audience: the hospital was not full. "That's true," Doctor Calderwood responds.[24] It is not the beds, not the ICU units, but the staff. As usual, the hospital does not share actual numbers — a continuous and growing problem, never knowing how close we were to running out of capacity.

Jeff Burrows is getting "pounded" with messages.

> The commissioners and the health board are not doing their job, not requiring masking, Ravalli County has one of the highest numbers...

Jeff, out of curiosity, said that he ran the numbers. For actual cases, he notes that Missoula, which has strictly enforced masking, has twice the cases per capita as Ravalli County. Silver Bow, home of Butte, the Governor's poster child, 0.35% of the population is a COVID-19 case. It's 0.16% for Ravalli County.

Of course, the number of cases depends on a list of rules and personal decisions that are, at best, loosely related to actual infections. But Burrows makes the point that the masking message he got from someone in the county had it backward.

Carol responds simply to what she observes: cases are increasing, and we are running out of capacity. She notes that case count is people getting tested, but once you have a community spread,

> It is the tip of the iceberg. You don't know how many people have symptoms and just stay home. The numbers [cases] will not be as legitimate as when you trace them all.

Burrows replies that he is simply going by the numbers given to him, with Ravalli County having the lower infection rate of any urban county, why we are not doing our job, enforcing masks, closing businesses.

We are the "redheaded step child" in the eyes of the Governor, but our rates are lower than those of the "gung-ho aggressive counties." Yellowstone County (home of Billings) has nearly a 1% infection rate.

Doctor Calderwood jumps in,

> I think we are going there. We have a lot of people coming in, a lot of tourists. I'm not asking to enforce anything, just to do the right thing.

Her worried words fill the room,

> It's a different year. If you have a cold, stay home. I don't think people anywhere are taking it seriously, young people thinking they're not a risk. They're probably not to get serious illness, but then it spreads.

Doctor Calderwood simply wants the leaders to unite and say the right thing,

> Yesterday, I tested 20 patients, and 3 were positive; that wasn't happening before.

She depends on her experience now, having been in the trenches firsthand, and the science, as she predicts,

> We will see more sick people. A lot of

people will get it; we won't know how many, but people will come out of the woodwork and be really sick or die.[25]

Unfortunately, her instinct will be proven prescient before the year is out.

There are fears that the upcoming holidays will be filled with closely packed gatherings of family and friends. Doctor Turner asked, "Where is our surge capacity?"

Doctor Calderwood doesn't know. She only knows that it is harder to do transfers to Missoula. With 100 cases now, "What if it's 500?"

Alan Lackey comes up to the mic:

> For once, I actually agree with the Health Board about a statement. How about a statement of facts? You're talking about a positive test result as a death sentence.

> How many false positives are there? Let's state the facts and let people make their own decisions... The numbers just don't pan out... We're justifying what we're doing without facts ...

> It's almost like a cult... It doesn't justify the interruption of our lives...

> I agree you ought to have a statement: what's the truth, and what are the facts."

Bedey gets to the mic. He is a little bit skeptical about a public relations campaign,

> People have made up their minds about the severity of this particular disease. Of course, that's fed by the so-called facts and truth you can find on the Internet one way or another.

> The problem is that citizens here or anywhere else in America are trying to winnow through all of the competing facts that we see so that they can make informed decisions.

He goes on to say that it would be useful to step away from the national issues. "Define what's an infection, once we agree on that, simply report them." Focus on the local hospitalization rate — "that is a measure of the strain on our medical capacity. Right now there are just too many places to look, and there are a lot of them

that have an agenda."

As Nurse Tiffany Webber walks up to the mic, Burrows says, "One more thing to add to your plate."

Looking at Burrows, she says,

> I appreciate your numbers, it gives me perspective. I get asked about those every day, but I am so busy dealing with the people that I don't keep track of them. I do feel a little bit better that we are doing well.

She shifts slightly in her chair, her hand moving to place the palm over her heart,

> I'm overwhelmed right now. It's hard to hear that these long hours I put in, trying to call people back to give them the information they need, is not valid. That's a hard thing for me to hear. The work I'm doing is not valid work.

> I disagree. I am dealing with people who are sick. I can't call them back.

She can't share some information but does give an example with the punchline, "Don't go to work if you have the sniffles."

It's too much now; she can't call everyone like before. She has written down what people can do for themselves because they are not going to go through to public health.

> We have other diseases. I don't know why, but we have seven cases of gonorrhea. It may not sound like a lot, but there were none when I started two years ago.

She's trying to set up a testing service because the hospital only tests symptomatic people. She retests people who have no symptoms who test positive.

Roger asks if she could use other help.

Tiffany responds that people want to talk to a nurse. She is "not comfortable" having a non-nurse because they would practice under her license. She is talking to a couple of EMT providers to step in.

Burrows says money is not the issue; if she

needs anything, just do it. Looking at the crowd, he says,

> It's a lack of people. While we are sleeping, these people are contact tracing, so don't fire at them. Fire at us.

Tiffany goes on with her report,

> The number of contacts ranges from family members, say five, some others have 30 contacts. We have four dead.

All that said, Doctor Calderwood confirms she will stay on as long as needed.

The Commissioners have gotten a hundred emails in 5 days, mostly about masks, for enforcement, for the Commissioners themselves being an example to wear them — countered by other emails about freedom.[26] So the County Commissioners, Greg Chilcott, Chris Hoffman, and Jeff Burrows, take up a statement about COVID-19. There are 25 people in the audience.

The letter is posted on the video screen, discussed, and edited by the Commissioners. Then Chris Hoffman asks for public comment.[27]

Alan gets up first, asking for the statement to have more about health, asking again why hospitalizations and those recovered are not reported. The death rate is the same as driving down Highway 93. The whole thing kabuki theater, "we're loving this lie."[28]

The next speaker suggests that masks are causing as many health problems as they are solving. Over the next hour, other speakers repeat the themes, questioning the utility of masks and PCR test accuracy.

One speaker walks to the mic, mask on, with a different message, asking for signage on businesses saying they must wear masks.

> This [letter] has zero teeth in it. It's really just for show unless you add requirements for people.

Chris Hoffman, having announced his last term as Commissioner, everyone knowing he had been Sheriff, closes public comment with this postscript:

> The most important thing is, I do believe the worst thing to come out of this will be the division.
>
> I believe it is coming from both sides of the spectrum — I know it is for a fact; I've witnessed it.
>
> We have private businesses that are being harassed because they're trying to be compliant with what the governor put out, whether we agree with what the governor came out with or we don't. It doesn't matter.
>
> Individuals who run those businesses believe it, are afraid of it, and are complying with it because, like many of us in the room, they don't know what else to do.
>
> That is where enforcement has a place. If you go into a private business and harangue the employees in either direction, whether for wearing a mask or refusing to wear a mask on someone else's private property, you can be ejected.
>
> And if it gets bad enough, you can be charged with disorderly conduct. Not as threat, *but good God people, can we please let people be who and what they are.*

Burrows chides people from both sides who claim the other side didn't care about the health and welfare of their neighbors. "If someone on either side says that the person who disagrees with them 'doesn't care,' that it's wrong." He ignores the email.[29]

Hoffman chimes in, saying,

> No one in the county government is interested in being the mask police or making adult grown America citizens undergo prosecution. We want people to be good neighbors, to be what we have been famous for generations.

There are seconds from the crowd. The Commissioners sign the letter.

Having missed the meeting, I watch the next day. Nurse Webber seems to be in a double bind

**COUNTY OF RAVALLI**

STATE OF MONTANA

The Bitterroot Valley

Hamilton, Montana 59840

**The Ravalli County Commissioners are releasing the following statement:**

To the Residents of Ravalli County,

Because of the recent substantial increase in COVID-19 cases in Ravalli County, the Board of County Commissioners is encouraging all citizens to follow the latest guidelines, which include:

- Wear a clean mask in public (especially where social distancing is not possible or practical)
- Wash hands and frequently used surfaces regularly
- Practice social distancing and avoid large gatherings (where distancing is not possible)
- Stay home if you are sick (including work, school and gatherings)
- Respect and follow guidelines set by private businesses and public entities including schools

We also believe it important to point out that people should be employing healthy life style choices, including nutrition, exercise, and rest.

In order not to overwhelm Public Health and our community's medical resources, please follow the above recommendations. Now is the critical time for every Ravalli County resident to do whatever they can to promote the health of our community.

While it has never been within the purview of the County Commission to enforce public health directives, and our opinion has not changed regarding strict enforcement, we are encouraging all residents to educate and conduct themselves as safely as possible during this current situation. We ask our residents to always treat each other with compassion, and to respect each other's right to make decisions about their own health. We thank our neighbors, the citizens of Ravalli County, for their continued efforts to help keep our community safe.

Chris Hoffman, BCC Chairman

Greg Chilcott, BCC Vice-Chair

Jeff Burrows, BCC Member

Commissioners Statement on COVID-19.

— two conflicting demands that cannot be met — and needs more contract tracers. A contract tracer must be a nurse, but there are no nurses. However, a contract tracer does not have to be a nurse. New York City set up a Test and Trace Corps, albeit belatedly, after the worst of it. Applicants from all walks of life are eligible to be trained. But Ravalli County Public Health is probably too small to implement such a program on their own.[30]

The *New England Journal of Medicine* editors issue an opinion piece lambasting the administration concluding on the handing of the pandemic, "We should not abet them and enable the deaths of thousands more Americans by allowing them to keep their jobs." The critique rightly cites South Korea as having done it right, with early testing and contract tracing. What is missing is that the South Koreans had prepared ahead of time to do nationwide contract tracing and the United States had not.[31]

The Bitterroot Valley makes music, but a little less today. It's reminiscent of *Bye Bye American Pie.*[32] After many ups and downs, Chip Jasmin passes away, a beloved performer and teacher of music.[33] I remember being amazed by a high school student performing at the Banque Club years ago. It was Chip's son Cove.[34]

The homegrown rock and blues reaches at least back to the '70s with the *Big Sky Mudflaps*,[35] a dance band that brings the music of the Grateful Dead to the Bitterroot. I used to see Pam Small on double bass strumming some bluegrass. With more homemade music studios per capita than Nashville — maybe. Our valley delivers homemade talent, folk, bluegrass, jazz, county, classical, and rock and roll. Jenn Adams[36] spent some time in Nashville and brings a bit of class as a *Berklee School of Music* graduate. You take some music lessons, or, when Jenn turns professor, take a class on music theory. The music lives on with the Hardtimes Bluegrass Festival.[37]

You never know. A couple of decades ago, I was having dinner with my wife at the Spice of Life Café on 2nd Street, which often had some live piano. That night, a singer was singing excellent jazz. I turn to my wife and say, "she is really good."

My wife says, "That is Eden Atwood," in a tone that suggests I should have known.[38] I may not know my musicians, but I like the music.

We've had our share of famous musicians settle here, including Hoyt Axton,[39] Huey Lewis, and some of his band members.[40] Huey would sometimes appear in local venues with little warning — I can testify that he plays a mean harp.

The Bitterroot was not always kind, though. The then sheriff Jay Printz arrested Hoyt for marijuana possession in the late '90s, as Jay was riding high as a Second Amendment hero after winning a Supreme Court decision saying the Feds cannot demand that he enforce the Brady Bill.[41] Even so, the gun shop *will* still send your paperwork to the FBI.

Huey Lewis got stuck in running a legal fight over Mitchell Slough, which runs through his land. Since the Bitterroot River meanders quite a bit, the question is if the slough is a ditch or a river braid. If it's a river, the public can fish it. I know someone who fished that stretch as a kid in the '50s — but in those days, ranch owners did not worry too much about people passing through their pastures. Huey got tired of the slough being trashed; lawyers were hired; old-timers pulled out old maps. It took years.[42]

The thermometer reads zero out in the woods as rifle season opens. I skip it this year; I am no help to the Orange Army as it struggles to make the quota in 2020.[43] I spent my time studying for an amateur radio license to keep my mind active.

It does remind me of what I had forgotten from my Electrical Engineering education back in the day. I am unsure if I will build a rig to "chew the fat." Still, my grandfather had been a radioman in the early days, getting call sign 6AB from the State of California before the FCC existed in 1910. With local clubs shut down, I go to Spokane for the test.[44]

I put in for a vanity call sign that ends with "6AB." I buy a handheld, but there is not much traffic at the valley bottom. Momentum carries me to study for the final level, Amateur Extra.

On the 30th, I had gotten word of the Witches Brigade, an informal tradition started by a couple of dozen women in 2013, witches in full costume riding bikes through downtown Hamilton. The Brigade grew, recruiting a couple of hundred Witch Riders last year.[45]

As the sun sets, I head out on my bike to 3rd and Main to watch. The majority are AWOL[46] from the Brigade this year; Witch Riders reporting — about two dozen to brave the cold fog and mist for some cackling fun.

The storm clouds have settled in Ravalli County, leaving a kin to *The Mist*,[47] though it doesn't hold monsters; it holds pestilence.

*31 Oct. 2020: 230,195 COVID-19 Related Deaths Reported*

# Cumulative confirmed COVID-19 deaths per million people

For some countries the number of confirmed deaths is much lower than the true number of deaths.
This is because of limited testing and challenges in the attribution of the cause of death.

**Data source:** World Health Organization (2025); Population based on various sources (2024)

Our World in Data

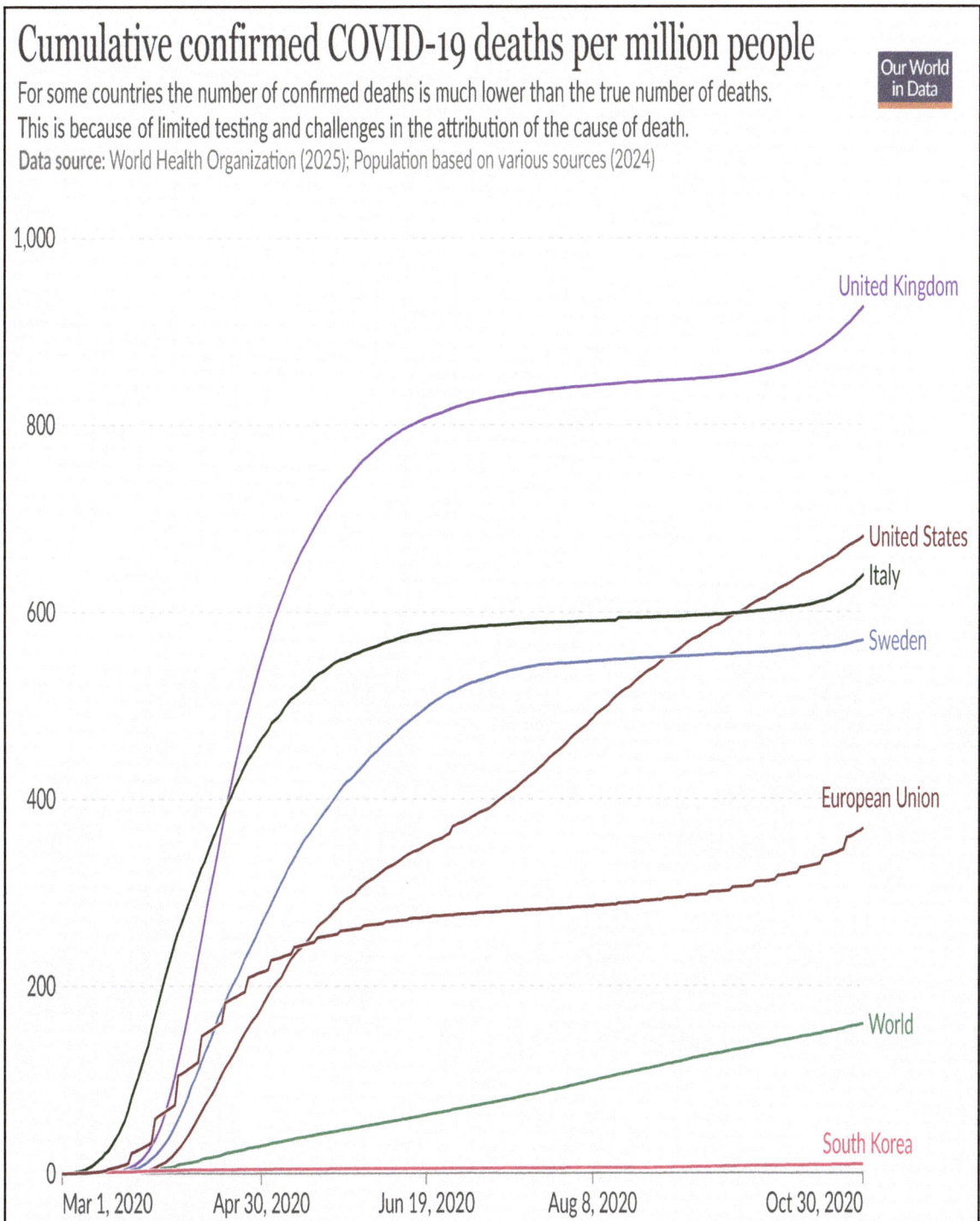

*Cumulative Deaths per Capita Mar-Oct 2020*

November 2020 timeline entries:

- CDC ends cruiseship no sail order
- National Election Day. Biden wins presidency according to reported votes
- FDA approves Bamlanivimab monoclonal (lab made) antibodies for non-hospitalized high risk patients
- Biden announces transition team
- Moderna vaccine reports 94.5% reduction in risk of getting COVID-19
- Paper - genetic structure does not rule out lab origin for COVID-19
- MT Directive close bars & resturants 10 PM
- Pfizer vaccine reported 95% effective
- Pfizer asks FDA for vaccine approval
- AstraZebeca vaccine 90% effective easier to distribute
- Regeneron cocktail appoved
- THANKSGIVING

Counties added the following number of new cases on the state website Thursday:[1]

- Gallatin with 164 (1,110 active)
- Yellowstone with 126 (2,437 active)
- Flathead with 105 (760 active)
- Missoula with 63 (971 active)
- Rosebud with 45 (135 active)
- Lewis and Clark with 34 (811 active)
- Cascade with 35 (1,255 active)
- Phillips with four (37 active)
- Ravalli with 23 (339 active)
- Toole with 20 (17 active)
- Blaine with 19 (107 active)
- Hill with 17 (397 active)
- Custer with 15 (108 active)
- Stillwater (68 active)
- Dawson with 13 (87 active)

- Park with seven (66 active)
- Daniels with six (23 active)
- Deer Lodge with five (176 active)
- Sweet Grass with five (34 active)
- Broadwater with four (66 active)
- Garfield with four (eight active)
- Jefferson with four (68 active)
- Glacier with eight (114 active)
- Richland with four (51 active)
- Beaverhead with three (33 active)
- Chouteau with three (59 active)
- Granite with three (30 active)
- Lake with three (165 active)
- McCone with two (10 active)

- Valley with 11 (76 active)
- Sheridan two (26 active)
- Wheatland with two (23 active)
- Lincoln nine (93 active)
- Mineral with one (one active)
- Carbon eight (48 active)
- Roosevelt 26 (389 active)
- Musselshell with one (32 active)
- Pondera with one (29 active).
- Sheridan with two (26 active)
- Sanders with 11 (16 active)
- Powder River two (10 active)
- Fergus with 14 (67 active)
- Madison with 11 (39 active)

Moment of truth

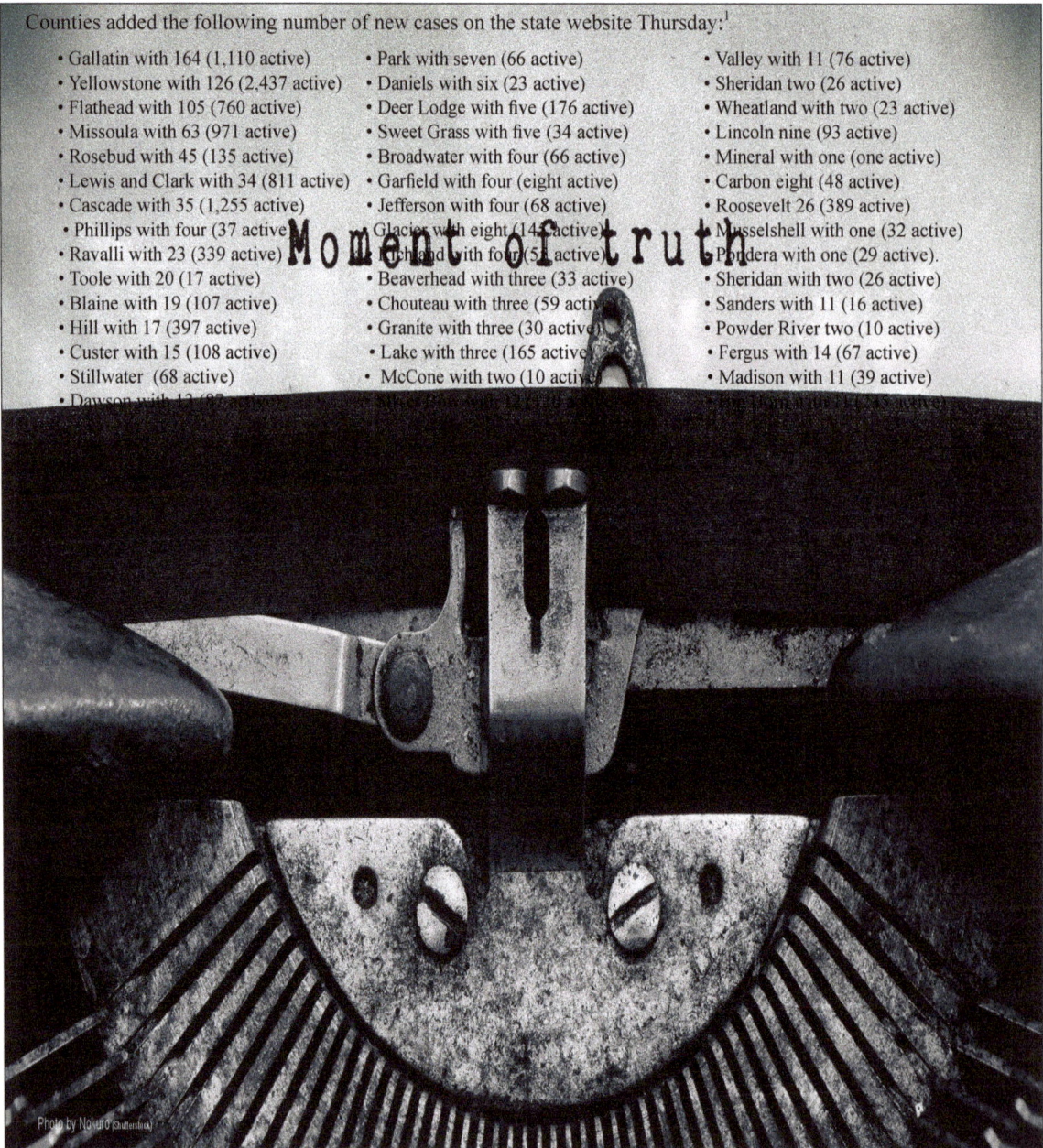

*Moment of truth on a vintage typewriter.*

# 14

# TRYING TIMES IN RAVALLI COUNTY

*We've got to come up with a plan before it gets any worse.*[2]
— Tiffany Webber, Ravalli County Public Health Nurse.

**NOVEMBER 2020**

The weather doesn't match the mood: a frozen morning warming up to the mid-40s, snow-covered mountains brightened by the almost clear sky, defining the boundary between us in our street-lined world from the wildlands beyond the ridges. We seem to chafe at the idyllic day, our mood better suited for the sound of a howling wind blowing against the waves of a hard-running river, whipping spray into a chaotic swirl.

However, the weather is not entirely out of sync as it sounds a more subtle note, bringing forth a bit of morning fog that fits the reality of it all, the uncertainty of it all. No one knows where this was going, not COVID-19, not election day, not the economy. But, as Nurse Webber had said in her interview, "We all have an opinion; it is Ravalli County, after all."

Another death in the news today in Ravalli County. "Fifth Ravalli County resident dies," cries the *Ravalli Republic*.[3] But there's more, as nearly every newspaper in the state picks up the latest COVID-19 counts from around the state, ensuring we know that there is no safe harbor if we were to sail around the state.

Counties with deaths added to the state website Saturday include:

- Yellowstone with two (94 people total)
- Gallatin with one (eight people total)
- Carbon with one (one person total)
- Lake with one (six total)
- Lincoln with two (five people total)
- Missoula with two (18 people total)
- Teton with one (one person total)
- Valley with two (five people total)[4]

Another death today: "The name's Bond, James Bond," the line made famous by Sean Connery, who, at 90, passed away, according to the paper today.[5] They're retro now, the James Bond series. They were new when I was in high school, one of the many textures of things about to change in the mid-60s, when these movies, which would be rated PG when Hollywood got around to a rating system later in the decade, were considered too "racy" for my girlfriend's father — we went anyway.

But, *The Times Are A-changin'*.[6] What is obscene today is the money being spent for my vote. Governor Steve Bullock is in a race for control of the Senate. With about 750,000 registered voters in Montana, the money spent on the campaign comes to $217 each, the most expensive in Montana History, with almost $100 million coming from outside groups, Political Action Committees, and the like.[7] They could have sent us each a check for $217 instead of ad buys, but from what I can see, it wouldn't make any difference in this county. After subtracting the straight-ticket voters, it will be a referendum on Bullock's response to COVID-19 in Ravalli County.

Anticipation modulates every line we speak as we go through the day. The line is not all that long at the county building for last-minute voters as Regina and crew begin the count of early votes, with everyone rooting for a winner.[8]

There is one upbeat beat in this divisive chorus: Laura Garber decides to cook apple pies. She tells a reporter for the *Ravalli Republic*,

> I think apple pie is about as American as you can get. Of course, we should be making apple pie for each other right now. We have to start somewhere.[9]

Seeing all of the division, she gets Cultivating Connections[10] into gear to have an "American-as-Apple-Pie Party, Hopes to Cross Party Lines" at her farm, Homestead Organics,[11] just south of town. Driving south from Hamilton, if you look east off Highway 93 just after you go by the turnoff to Skalkaho, just after a row of trees, you'll spot the corner of some tilled land — that's her and Henry's farm. Our grandparents would have called it a Truck Farm.

Henry, her second husband, was raised in Eastern Germany. He talks about the self-sufficiency of farmers in the postwar communist regime, bringing some of those methods to Homestead Organics.

I've known Laura since her now-grown children were still in strollers. A farmer from the toes of her boots to the sun hat on her head. The soil under her fingernails announces her calling as a farmer and an advocate of sustainable farming. Her soft-spoken voice can sneak up on you until you realize you are talking to a fierce heart willing to try one more time, with one more idea, to show the essential connection between food and community. So, along with the CSAs,[12] organic veggies, and organic chickens comes a teen mentoring program set in farming — and a run at introducing industrial hemp, local food in schools, and organizing community-building sessions. While partisan politicians try to divide us, she tries to unite us — let's hope Laura prevails.

The morning of November 3rd, election day, begins with a tweet from President Trump:

> To all of our supporters: thank you from the bottom of my heart. You have been there from the beginning, and I will never let you down. Your hopes are my hopes, your dreams are my dreams, and your future is what I am fighting for every single day![13]

Our polls haven't closed yet when states begin to be called. A seesaw affair, partisan hopes rise to great heights; disappointments sink to the dark depths of a black sea. It begins.

5:02 pm: Vermont, 3 electoral votes for Biden.

5:34 pm: 5 electoral votes for Trump. **Trump 5**, Biden 3 — West Virginia,

5:42 pm: Trump 13, **Biden 16** — Kentucky and Virginia.

6:08 pm: **Trump 24**, Biden 16 — Indiana

6:10 pm: **Trump 33**, **Biden 63** — Alabama, Delaware, D.C., Maryland, Illinois, Massachusetts.

6:16 pm: **Trump 80**, Biden 63 — Oklahoma, Florida, Tennessee.

6:50 pm: **Trump 96**, Biden 77 — Missouri, Arkansas, New Jersey.

7:38 pm: **Trump 126**, Biden 115 — Mississippi, Nebraska (3 districts), New York, South Dakota, Wyoming, Louisiana, Colorado, South Carolina.[14]

Trump is ahead.

A record turnout yields a Republican sweep in the county, 25,000 ballots counted of the more than 28,000 mailed by the Ravalli County Clerk and Recorder — early votes are written out by hand on the whiteboard in the basement of the courthouse at 8:00 pm when the polls close. Once again, Trump gets 2/3 of the vote, with 31% for Biden and 2% for the Libertarian candidate. However, Bullock does not pull 15% of Trump's vote like he did in 2016 — only 2%. With an 86% turnout in Ravalli County, Bullock gets only 35% versus the Republican Daines with 65%. It isn't even close state-wide, with Bullock getting 45% of the vote in a state he won four years earlier.[15]

The Presidential race dominates the news.

8:36 pm: **Trump** **144**, Biden 115 — Kansas, North Dakota.

It switches again, back to Biden.

9:10 pm: Trump 144, **Biden** **205** — New Hampshire, Connecticut, New Mexico, Oregon. California, Washington.

President Trump tweets:

WE ARE LOOKING REALLY GOOD ALL OVER THE COUNTRY. THANK YOU![16]

It will get close, but President Trump will never catch up in the count.

10:26 pm: Trump 210 **Biden** **214** — Ohio, Maine Dist 1, Idaho, Rhode Island, Texas, Hawaii, Iowa.

11:54 pm: Trump 213 **Biden** **225** — Nebraska (final district), Montana, Minnesota.

After two days of election maps, commentary, and a county-by-county drama in swing states, "Stop the Steel" takes social media by storm.[17]

Pennsylvania pushes Biden's count over the magic number, 270 electoral votes.

Nov. 6, Trump 214 **Biden** **273** — Maine, Main (1 dist), Wisconsin, Michigan, Pennsylvania.

A week later, the final tally puts Georgia and Arizona in the Biden column, making the litigation and maneuvering harder for the attempt to prove that it was really a Trump win.

Nov. 14, Trump 232 **Biden** **306** — Nevada, North Carolina, Alaska, Arizona, Georgia.

Lawsuits follow, 62 filings, mainly in six swing states that went to Biden: Arizona, Georgia, Nevada, Michigan, Pennsylvania, and Wisconsin. All but Nevada were won by Trump in 2016. Some lawsuits are dismissed for lack of standing, some based on attempts to change the procedures, and some on the merits of the evidence presented to the court.

It will go on and on. Even the most desperate Bronc rider will think of quitting when he gets thrown 61 times in a row. Lawsuits are lost, and many appealed only to fail again. Two go to the US Supreme Court, which is not convinced either time. There was one win, a Pennsylvania ruling that one could not "cure" a vote by going back after the election with a proper ID — it didn't change the outcome.[18]

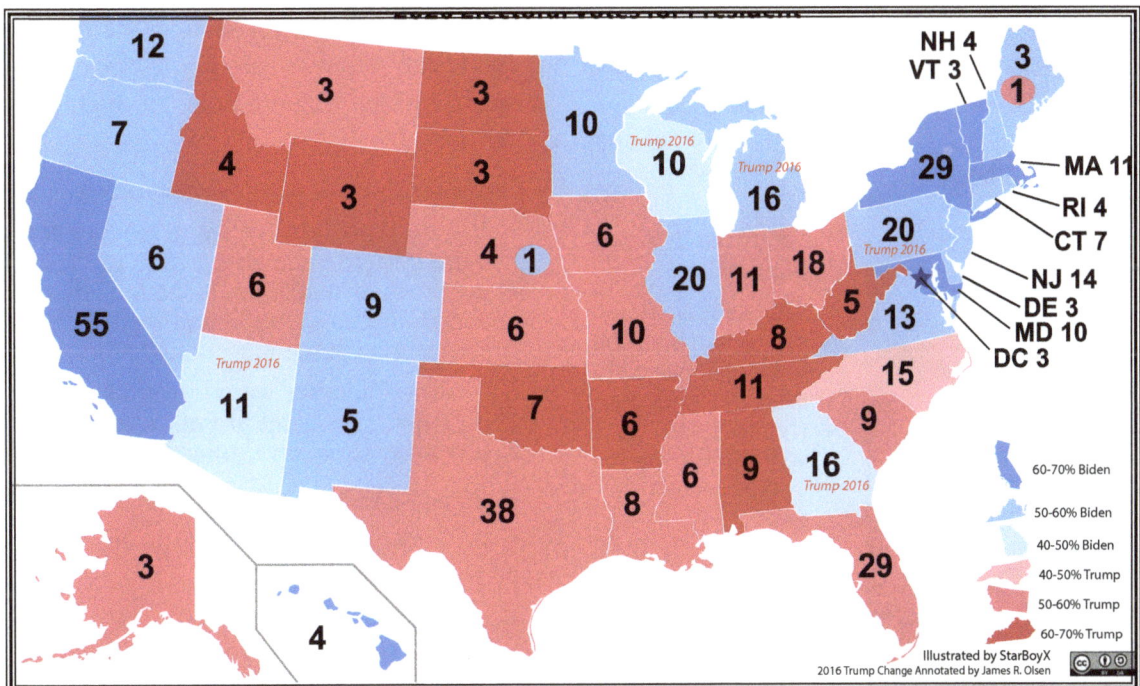

*Electoral Collage 2020*

Face masks and "Stop-the-Steal" start appearing in the same sentence on the street.

On the 11th, Emergency physicians from Marcus Daly Hospital warn of an impending disaster, already in progress, as patients with life-threatening conditions "such as heart attacks" will no place to go because of overcapacity in Montana's hospitals.

Winter sets in. A local food bank can't keep up with demand, though hunting recovered this year after the Fish Wildlife and Parks extended the elk season — some of the game ends up at the food bank.[19]

But, all of this did not halt the annual Veteran's Day Ceremony at the Dough Boy Memorial,[20] standing in front of the Museum, etched with the boys lost in the Great War, a war that ended with an armistice on 11/11 on the 11th hour in 1918, today decorated with 60 flags, including one for each of the 50 states, and one for each military service. I go, salute the colors, and bow my head during the prayer.

Bob Hafer, "United States Army Master Sargent E8, retired," setting up flags notes:

> I think it will be a good symbol for our times —
>
> unity and yet diversification in one big block.
>
> The veteran is the one who takes the oath that basically says, "If my country asks me to, I will die for service to that country."

The Museum Director, Tamar Stanley, adds,

> When times have been a bit divisive,

Dough Boy Memorial, Hamilton

unity is a strength of 50 state flags.[21]

Maria's Restaurant[22] steps up to give away 25 baskets of Thanksgiving dinner, "something we do every year for people in need. I wish I could do more, but 25 is all I can do this year," says Maria.[23]

The crowd almost fills the room. Most of the crowd is from Stand Together for Freedom.[24] This Health Board meeting is going to be a long one.

The room seems too poorly ventilated for that size of the crowd, something I had never thought about before. I have been in this room many times with even bigger standing-room-only crowds. The little virus has changed how we think about day-to-day life.

I can't wait to say my piece and get out, thinking I could rush home and see the rest on Zoom, submitting more comments if I wanted to as the meeting goes on.

First, Public Health reports, then public comments. The meeting opens with a report from Doctor Carol Calderwood.

When she starts her report from the back of the room, Jeff Burrows asks her to come up to the speaker at the table up front, "It might pick you up better."

Carol sits down, leaving her coat on.

> Just as far as COVID, we're seeing a lot more symptomatic people who are testing positive in addition to the total number. Both public health and the hospital are hanging in there, working hard.
>
> I just feel like we want to get that message out: stay home if you're sick, and stay away from large groups.

Someone else talks but can't be heard. Burrows asks people to speak close to the microphone so people can hear.

Carol concludes, "That's about it."

Doctor Turner asks about contact tracing. Doctor Calderwood defers to the Health Nurse, saying, "there is definitely stress in public health."

Then, Doctor Calderwood is questioned about ICU beds. She isn't "privy" to the numbers, but the hospital is getting more spaces. She adds that it is "always people," pulling people from other departments if they have ICU experience, adding,

> They're forecasting less availability to transfer patients...

Doctor Turner finishes her sentence, "because there is no place to transfer."

"Exactly," says Doctor Calderwood.

Asked about how the Viral Clinic is doing, "It's hard, it's doing OK," adding that it wasn't like other years, where if you had a cold, you would push through, but now you are more likely to come into the clinic. A lot of people are worried, rightly so, if they are symptomatic.

The doctor reports that the hospital has expanded the testing criteria to include people who had to go home for care. "There are so many cases and so many contacts."

I know from past meetings she's already been through a lot; it is not letting up.

The Health Nurse, Tiffany Webber, comes up to the mic. "I have a lot of stuff." She takes off her mask so "you can hear me better." She walks through the other activities they must do on top of COVID.

"We've gotten a couple of additional contact tracers in the medical profession," says Nurse Weber, "some retired, one PhD, not nurses."

Tiffany has been trained in the principles of contact tracing. Nevertheless, she has to invent a contact tracing program and communication plan on the fly. She has three contract tracers "going strong," but the number of cases we can get to is "too much." They are four days behind.

> If you think about people who go in to get seen, they're usually 3 or 4 days into their illness. Then it's another two days, possibly, to get the result. Then it's taking me 3 or 4 days to get a hold of you.
>
> Sometimes, by the time I get to them, they're ready to come out of isolation.
>
> Most are out in 10 days. Others are still too sick to go back to work.
>
> Unfortunately, we are seeing that people who are sick and being asked to come back to work, knowing they're infectious, knowing they are sick.

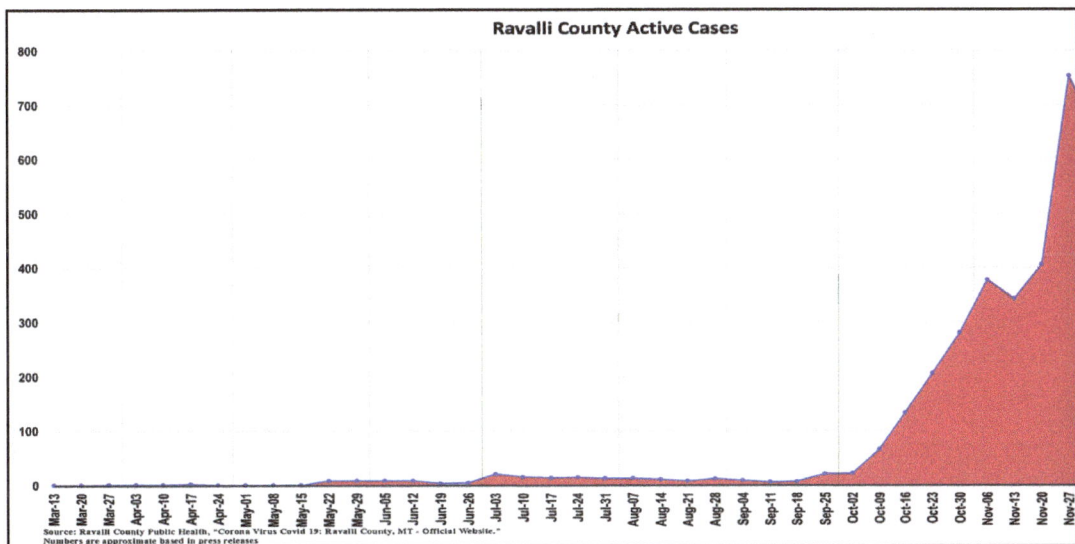

*Active Cases Ravalli County mid-Nov.*

People who are concerned often "reach out to us." Nurse Webber has to get in the middle of many situations that are "very uncomfortable." We must take their word that they are sick; we are not policemen and cannot send them back to work.

The contact tracing means that they are not swabbing (for tests), and there are not a lot of places to get tested. "The county needs more places to get tested," Tiffany says.

Tiffany also has to follow up on complaints to a portal set up by the state for complaints against businesses who are not complying with the mask mandate — and, of course, Ravalli County is one of the leading counties.

> There have been 475 cases for October. When they get a positive test, it is not automatically a case; we have to call them. Maybe they are not symptomatic, and it's a false positive or other factors... we aren't getting the information to the public that needs it. I'm very concerned about that.
>
> At night, I have to tear myself away. Where do I draw the line? I say, "OK, these 50 people I am not going to call tonight. I just can't." That's not good enough for this county.
>
> What I can't tell from the list is who wants to be called and who doesn't. If I could just look at it and say these people really don't want to hear from me. They're probably going to hang up or say a few choice words about who I am or what I'm doing.

Shrugging, she goes on,

> I'd love to know who they are. That's OK. Just move on.
>
> We've got to come up with a plan... it's getting worse.

There will be no plan.

Tiffany says the worst thing is that when she calls four days later, some people are angry and don't know they should be isolated from their spouse and kids.

Asked about nurses. She has one, two newly graduated nurses coming, and three part-time retired nurses.

Nurse Webber explains that the new cases reported every week (and reported by the press) are different than the active cases she has to deal with.

In response to a question, Tiffany says the cases were nearly all symptomatic because they ran out of capacity and had to prioritize testing asymptomatic close contacts, which the CDC wants her to do. Marcus Daly Hospital has given them a block of time, but they only test symptomatic people. That is why she wants to do their own swabbing. They are swabbing about 80 to 100 people a day.

If they test positive and have symptoms, she assumes they have COVID. They only retest for a positive that is asymptotic. But most of the people she gets are symptomatic.

Then Doctor Calderwood breaks in, saying that people who come in with COVID can have diarrhea or a cough. She has gotten to the point where she can pick out about 80% of the people who will test positive because of weird symptoms they often have trouble describing — loss of smell, chest heaviness, and pressure in their nose. She hears it 10 to 15 times a day.

I sit in the audience with my red bandanna, the veteran in me thinking, "we should pin a medal on her." Here we are, with over 40,000 people in this valley, and the Health Nurse had a staff of what? Six.

The Hamilton City Council requested a joint meeting, which is brought up on Zoom.

Jeff Burrows had a table and a podium set up with mics. After almost an hour of presentations from the Health Officer and Public Health Nurse, Jeff opens the meeting for public comments, prompting me to hustle up to the podium.

Alan Lackey walks from the back of the room, holding an already opened spiral notebook, held level, reminding me how one would carry a Bible open to the page that will be read to the congregation. He sits down, carefully laying the open notebook on the table.

After he positions the tabletop microphone, Alan pauses, glancing at me, thinking I might be first. Jeff says, "Jim, before you go, we'll go to Alan, then we'll go to you."

Obviously, the result of a lot of preparation, Alan delivers his presentation in an even voice with frequent glances at notes in a spiral notebook,

> It's become a routine here to get some truth into the situation and actually what's happening. I know [County Commission] Board Member Chilcott criticizes people for coming up with things off the Internet. CDC is a source you all probably use.

> Six percent of the deaths by COVID were supposedly just for COVID, meaning ninety-four percent had co-morbidities. Ninety percent were in the upper age group. But, just to get back to reality here.[25]

Alan continues quoting percentages of cases and deaths, "six one thousands for a percent" in Ravalli County (We all know both are lower than the national average.) He asks again what the hospitalization rate is for people who tested positive. Alan admits that hospitals are full, but from his perspective, using the percentages he quoted, it is not due to COVID-19.

> It is a disease and we need to protect the most vulnerable people with comorbidities, people in advanced age, but it is a disease obviously most everybody recovers from.[26]

Alan questions the cause, citing a lack of compelling evidence of the existence of a SARS-CoV-2 virus, expressing doubt there could be a successful vaccine, adding that it would contain things "we probably don't want in our bodies."

As I watch from the podium to his left, I think it is good to hear at least some "substantive" comments. Even though I don't agree with everything said, it is a relief from meetings that too often degenerated into angry rhetoric.

Then Alan says something concerning that needs a response, something I need to add to what I am prepared to say.

Alan continues. Looking at his notes, he asks the board if they took an oath. The answer is no.

We will get to that later, he says, going on,

> The board is intimidating when they enforce their mandates, and there is a severe loss of revenue, deciding who is an essential business and who isn't. You are severely restricting constitutional rights here.

> I believe the board and the decisions they make should be held personally responsible for the loss attributed to the mandate. 99% of the people trust the medical community, but by misrepresenting it, you are betraying our trust.

He adds that the board says they tailored their mandates based on research. Not so. His group made a FOIA request and found that the board depended on sources such as CNN.

> It is your personal and professional responsibility to find out what the truth is and not foster an atmosphere of fear, and that is what I see going on.

Turning the page in his notes, Alan talks about oaths, noting that one Health Board Member, a veteran, had taken an oath to defend the constitution and that it did not expire — adding "thank you for your service."

> It's a clear violation of our civil rights. The County Attorney can back me up on that.

Then he cites the Federal Code, Title 18-241, "Conspiracy Against Rights," 18-242, "Deprivation of Rights Under the Color of Law," 18-1038, "False Information and Hoaxes."

After a few more words, he says,

We don't need to destroy our community, our culture, and our economy as result of this overreaction. So, I want to put everyone on notice. Thank you.

I have heard this logic before, the idea that public officials are personally liable — but my reading of the US Code and Montana law says it is not valid except in minimal circumstances.[27] I don't know if Alan used his research or depended on a lawyer. I've walked through this before in conversations at the coffee shop. A critical federal law (28 USC 2679) is omitted — it protects public officials from liability when they act within the scope of their duties. In any case, as Alan noted, these are not Federal Officials but state officials.

I am next. As I start, Burrows has to remind me to give my name for the record. I first say I have handouts and passed a few out. I had provided copies earlier to the Health Board. I am nervous, not having prepared beyond a list of points to make.

> So, first of all, I served in the US Air Force for 12 years as an officer." Raising my hand as if taking an oath, I said, "I took the oath.
>
> You don't have to thank me for my service. It was my duty. So, I take it seriously.
>
> People that serve in the public, so long as they do the best they can and are not purposely lying — they're OK.
>
> To try to hold someone [personally] accountable for making mistakes when they are doing a public service for free is not what I swore an oath to uphold.

Turning to the crowd,

> Let's draw the line where it belongs.

I go on,

The sad thing is that the mask thing is kind of a false dilemma. You can look online — I've provided tons of input, with footnotes, to this board. I looked at masks as best I could.

Pulling on my cloth bandanna down, I say,

> This doesn't do anything. Maybe a cough. The studies say it may stop a few things but doesn't do much. So, why in the world do we have the Government telling us to wear these cloth things?

Leaving my bandanna down, I grab the blue medical mask underneath,

> Without telling us that this is somewhat better. 40 or 50%.

Pulling that down and showing a 3M N-95 mask with a vent,

> This protects me. It doesn't protect you because it has a vent. Some still comes through this mask.

I go on to say that it's really hard to study if you think about it.

> Here's the sad part. The death rate and the economic impact attributed to COVID is the same for the United States and Sweden. They are both bad. They both had an increase in death rate. You can name the cause whatever you want, but it is real in the United States — shown on a graph here...
>
> The trade-off between enforcement and economic impact isn't real. There are countries that have never had a lockdown and have had less economic impact.

I note Asia acted quickly in December to stop travel from Wuhan.

> They worked quickly to get widespread testing by January. Having gone through the SARS epidemic, they had built the capacity to do contact tracing and isolation ahead of time.

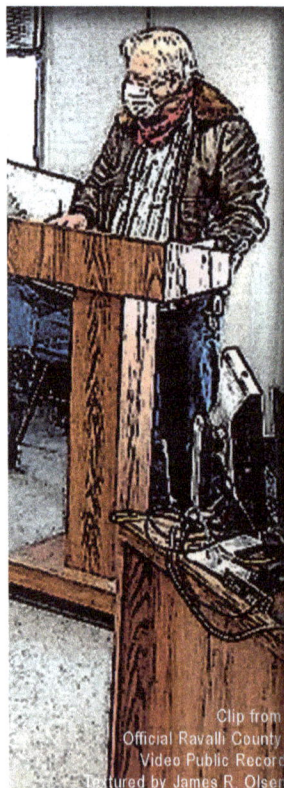

Clip from Official Ravalli County Video Public Record. Textured by James R. Olsen

*Jim gives Public Comment.*

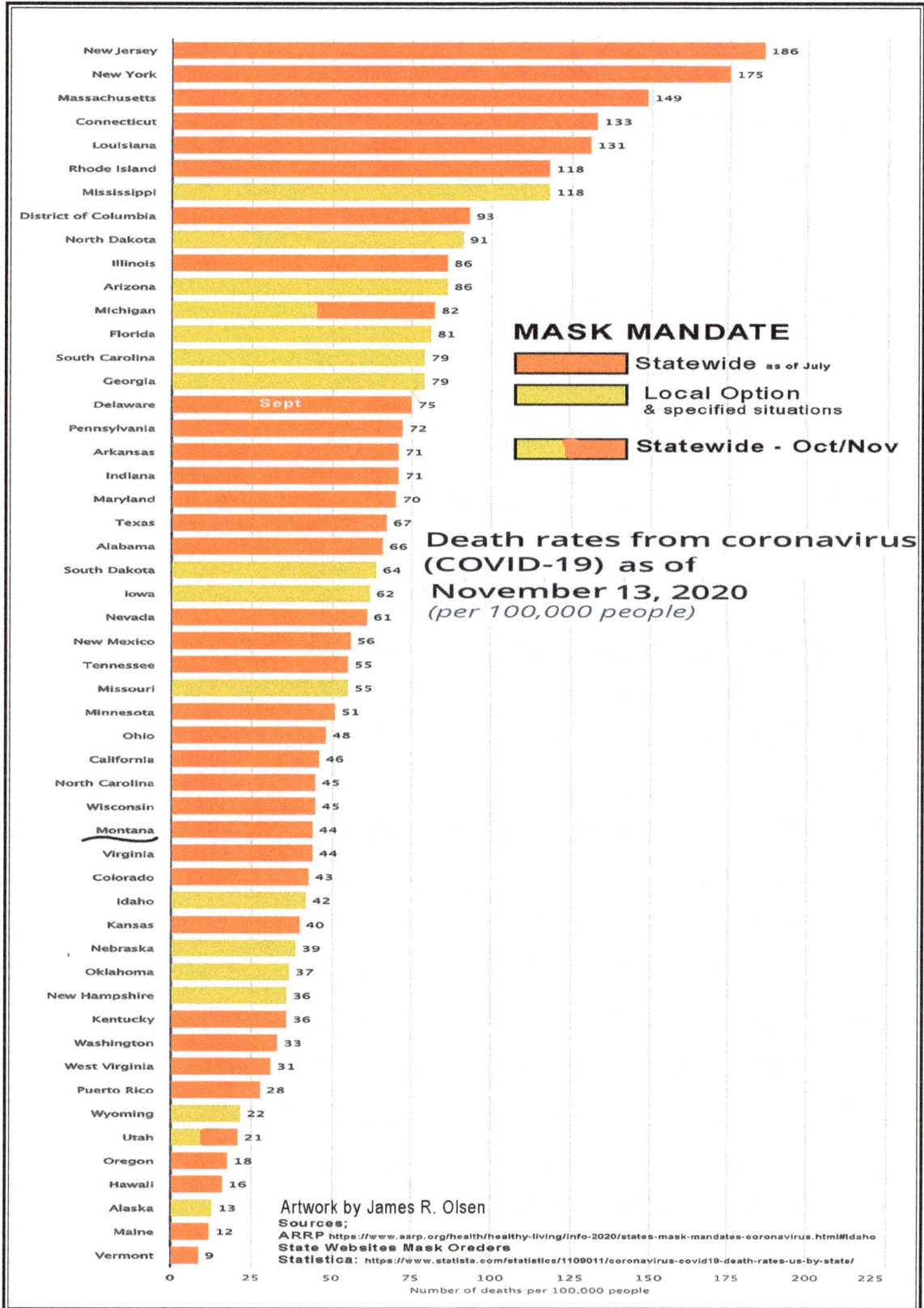

**MASK MANDATE**

Statewide as of July

Local Option & specified situations

Statewide - Oct/Nov

**Death rates from coronavirus (COVID-19) as of November 13, 2020** *(per 100,000 people)*

|  |  |
|---|---|
| New Jersey | 186 |
| New York | 175 |
| Massachusetts | 149 |
| Connecticut | 133 |
| Louisiana | 131 |
| Rhode Island | 118 |
| Mississippi | 118 |
| District of Columbia | 93 |
| North Dakota | 91 |
| Illinois | 86 |
| Arizona | 86 |
| Michigan | 82 |
| Florida | 81 |
| South Carolina | 79 |
| Georgia | 79 |
| Delaware | 75 |
| Pennsylvania | 72 |
| Arkansas | 71 |
| Indiana | 71 |
| Maryland | 70 |
| Texas | 67 |
| Alabama | 66 |
| South Dakota | 64 |
| Iowa | 62 |
| Nevada | 61 |
| New Mexico | 56 |
| Tennessee | 55 |
| Missouri | 55 |
| Minnesota | 51 |
| Ohio | 48 |
| California | 46 |
| North Carolina | 45 |
| Wisconsin | 45 |
| Montana | 44 |
| Virginia | 44 |
| Colorado | 43 |
| Idaho | 42 |
| Kansas | 40 |
| Nebraska | 39 |
| Oklahoma | 37 |
| New Hampshire | 36 |
| Kentucky | 36 |
| Washington | 33 |
| West Virginia | 31 |
| Puerto Rico | 28 |
| Wyoming | 22 |
| Utah | 21 |
| Oregon | 18 |
| Hawaii | 16 |
| Alaska | 13 |
| Maine | 12 |
| Vermont | 9 |

Number of deaths per 100,000 people

Artwork by James R. Olsen
Sources;
ARRP https://www.aarp.org/health/healthy-living/info-2020/states-mask-mandates-coronavirus.html#idaho
State Websites Mask Orders
Statistica: https://www.statista.com/statistics/1109011/coronavirus-covid19-death-rates-us-by-state/

*Handout Health Board Meeting, James R. Olsen, Mask Mandate vs Fatality 11/13/20*

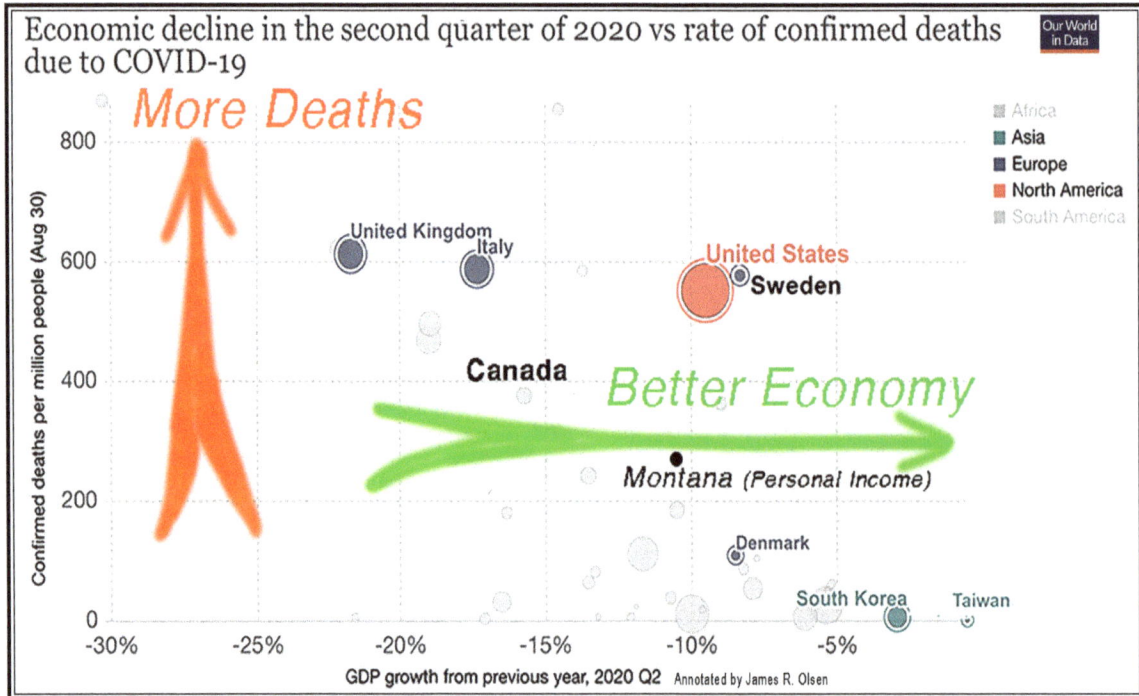

Economic decline in the second quarter of 2020 vs rate of confirmed deaths due to COVID-19

*Health Board Handout Deaths/Capita vs Economic 2nd Qtr 2020*

Find out who's sick, isolate them, is what works. And doing it quickly. The result is a death rate of one-one hundredth of the United States and an economic impact of about one-tenth.

I go on to talk about how vital medical capacity is. In Italy and New York City, when they ran out of capacity, people started dying faster.

I cursed about my own mask and apologized for my language. But the point is made.

A theme that reoccurs during the next hour and a half of public comment is the need for accurate information — particularly regarding COVID-19 cases because the case count drives fear and Government decisions. One young man notes that the Public Health Nurse does not know the false-positive rate of the PCR test,

> Claims of case numbers increasing but has no idea what the false-positive rate is... It's a lot of hype and fear. I don't know if we can get someone to get accurate numbers, but there is a lot of liberal push —as a result there is a lot of COVID cases listed when they're not.[28]

He has a point. While PCR tests often have zero false positives in a lab setting with known samples, many things create a false positive in the real world. These include contamination of the sample, contamination during the PCR test process, cross-reaction with other viruses, sample mix-ups, and software problems. But, a test is a tool for determining COVID-19 cases that need to be tracked. The terminology "false-positive," as Nurse Webber uses it, is a shorthand for a specific series of tests — an asymptomatic person testing positive and then testing negative when a second test is done to confirm the result.

Even if both tests are accurate, Public Health rightly concludes they do not have a "case." The accuracy of the test is a different matter and should be a large-scale clinical trial.

The *Ravalli Republic* reports on the meeting, "everything you need to know about where people stand":

> The sign flashes two simple messages: "Wear a mask" and "Masks save lives." It is part of a public awareness program launched by the Ravalli County Public

Health Board. It was set up in the middle of the week.

By Thursday, someone spray-painted a large red "X" across the board.[29]

We are picked up on national news, "Montana Face-Off Over Face Masks."[30]

We don't seem to be doing much right about COVID, though the Government appears to be trying. The week before Thanksgiving, Bullock doubles down with a directive limiting the size of gatherings and squeezing bars and restaurants, not closing them but limiting capacity and hours.[31]

The grim reaper is working a little overtime in Ravalli County, with 14 COVID-19-related deaths. The virus works overtime in Ravalli County — Public Health finishes the month with 785 open COVID cases.[32]

The stimulus packages in January and March provided an efficiently delivered cash infusion when most of the country was shutdown probably staved off an economic disaster. However, the impact has been uneven — some business dobetter; some go under.

While the unemployment benefit extension was needed, the added benefit have some low wage workers making more than when they worked. This and the stimulus has made job shopping a buyers-market for the worker for the first time in decades. Wages are rising and workers are slow to return to the workforce; some retire early.

Inflation, the Consumer Price Index, is starting to rise again. The injection of extra unemployment money for no work may make its way through the economy and inflationary pressures subside — but we'll never know because we a due for another massive stimulus in 2021.[33]

*30 Nov. 2020: 270,827 COVID-19 Related Deaths Reported*

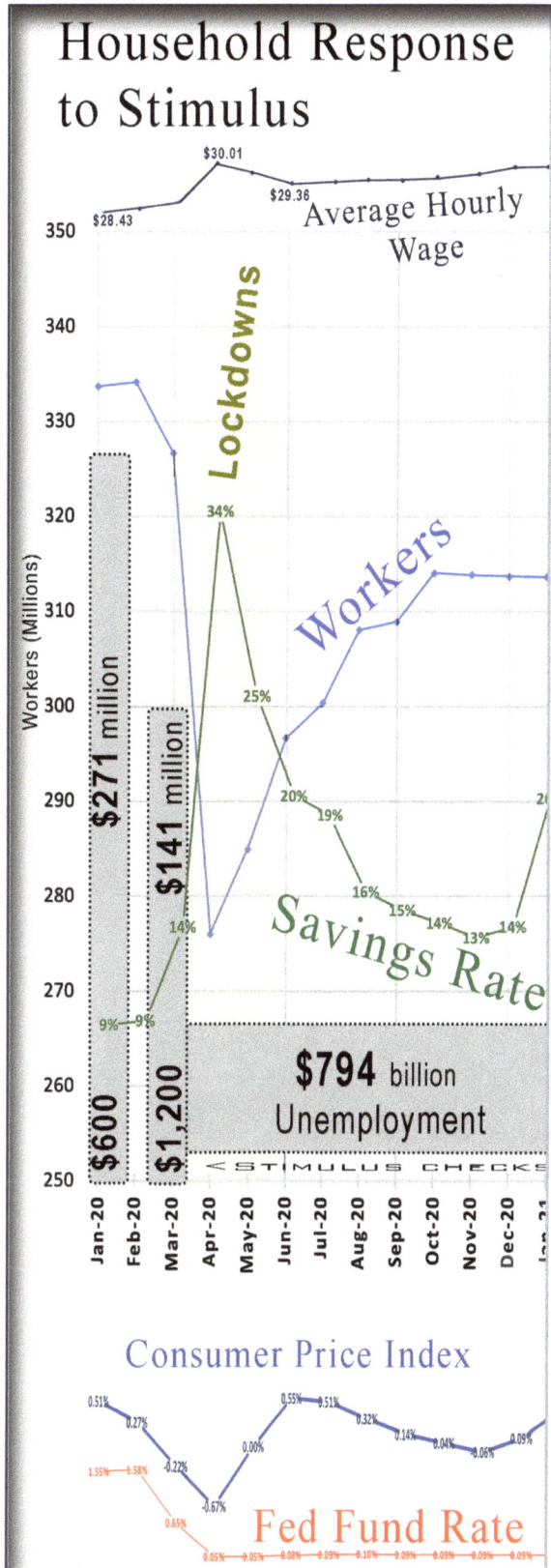

*Household Response to Economic Stimulus*

11  14  December 2020  18  21 23 24 25  29

US Death toll over 300,000
Pfizer vaccine approved by FDA
16 years and older

Congress passes 2nd relief act

Moderna vaccine approved by FDA
16 years and older

New variant in UK spreading fast

Est. 1 million vaccinated

CHRISTMAS

100 Vaccines arrive in Ravalli County

Colorado in nontraveler
Alpha, Beta, Gamma Variants
Categorized

UK Variant in

### COVID-19 Treatment Guidelines (NIAID COVID-19 Treatment Guidelines Panel)
• Recommend against using Baricitinib by itself, only in combination with Remdesivir

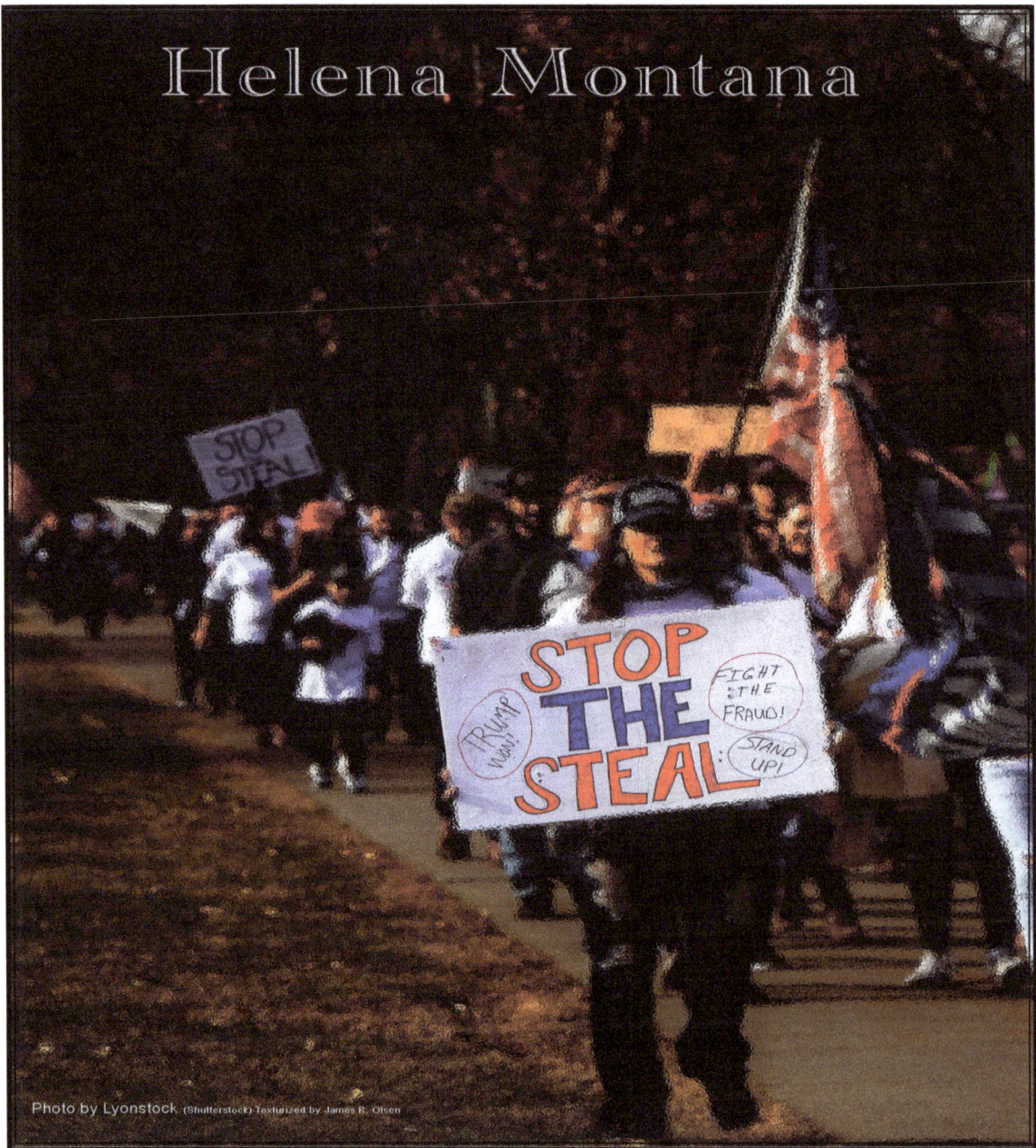

Helena Montana

Photo by Lyonstock (Shutterstock) Texturized by James R. Olsen

*Stop-the-Steal March Helena*

# 15

# THE COMING FURY

*It is statistically impossible that the person, me, that led the charge, lost.*[1]
— President Donald J. Trump

## DECEMBER 2020

The mistrust of the Government has grown. What many thought was unthinkable is now thinkable. The niggling fear that there might not be a peaceful transfer of power in the United States, as the month kicked off with a recorded speech by President Donald Trump. He's circling the abyss, appealing to the 74,216,154 people who penned in an oval on a ballot for Trump:

> All you have to do is watch the hearings and see for yourself. The evidence is overwhelming. Everyone is saying the evidence is overwhelming when they get to see it — It is too late to change the course of an election.
>
> There is still plenty of time to certify the correct winner of the election, and that is what we are fighting to do, but no matter when it happens, when they see fraud, when they see our votes, and when those votes number far more than is necessary, you cannot let another person steal that election from you.[2]

Many of the 81,268,924 who marked Biden stand aghast as the numbers cited in Trump's speech, 20,000, a hundred thousand? It seemed, well... "So what if it is 74,236,154 or 74,316,154? Its not 81 million." The pump of mistrust having been primed, the rift over COVID having increased the pressure, no longer trusted scientists adding weight, some of the 74,216,154 answered the clarion call to "Stop-the-Steal." It is like a deadly variant that arranges the right genes, the

stolen election, the bioweapon, the mandates, the violent chaos on the streets, the anticipation of vaccine mandates, the rumors of tracking micro-devices.[3]

Stop-the-Steal and COVID-19 protests merge in everyday arguments and rallies as election lawsuits multiply almost as fast as a coronavirus — the score will be 61 losses to 1 win.[4] The true believers face down the true believers, science over superstition, diversity over nationalism, safety over freedom. A pressure akin to geological compression has hardened the two narratives into two rigid granite plates, separated first by a crack, then a rift, and now the great divide.

History has been forgotten, the no-holds-barred politics, the protests in the streets that became a call to arms, that drew the battle lines, that finally set the mini-balls flying. The bloody reality of it all faded as the last American Civil

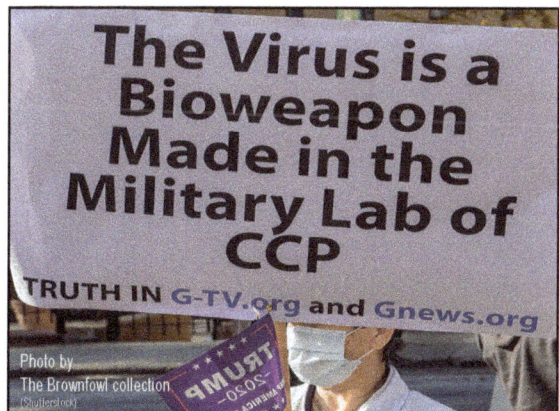

**The Virus is a Bioweapon Made in the Military Lab of CCP**

TRUTH IN G-TV.org and Gnews.org

Photo by The Brownfowl collection (Shutterstock)

*Bioweapon Sign at Stop-the-Steal Rally*

War veteran passed away the year before John F. Kennedy was elected.

Let's not kid ourselves about our history. There *is* one exception to the peaceful transfer of power in the United States. In 1860, the Democratic Party Convention had split into two conventions, each naming a presidential candidate. The newly formed Republican party named Abraham Lincoln, who had not a single vote in the Deep South because his name was not even printed on the ballot. Lincoln's election was almost universally celebrated on November 6th, 1860 — celebrated by Republicans because they won[5] — celebrated in Charleston because the cheering crowds had finally come to the banks of the Rubicon. They were ready to charge across as a local newspaper announced, "The tea has been thrown overboard and the revolution of 1860 has been initiated."[6]

I have said this at The Table when someone gets too cocky about it. I've been to places that have had civil wars, where the survivors are still alive. You don't want one, no matter what the cause célèbre. When it's over and the dust has settled, what little talk there is, is not about the cause but the suffering.

It is the holiday season in Ravalli County, and, as Health Board Member Roger DeHaan had predicted, with the days getting shorter, the air getting colder, the worries about COVID getting tiresome, more and more people risk some indoor time with families and friends. Even the advice columnist, Annie Lane, is dealing with it, responding to a woman whose family wants them to go to a big family gathering on Cape Cod — too risky, says Annie.[7] Her advice is followed by some but ignored by many.

Much to his delight, Marshall had just been inducted into the Montana Bioscience Alliance Hall of Fame.[8] However, being careful doesn't seem to make anyone exempt from getting COVID-19, though. Even Marshall Bloom, the Deputy Director of RML, who is my age, gets COVID even though he is always cautious, wearing a mask everywhere except in his office, car, and home.

But, he soon had a slight cough that didn't go away, which was noted when he filled out the required periodic survey at RML. Then he tests positive for COVID-19. On Christmas Eve, he came down with severe abdominal pain, nausea, vomiting, and diarrhea, went to the emergency room, and spent four days in ICU. They save him with an IV of electrolytes.

Doctor Bloom believes in masks, explaining that they are one of several tools. His undeniable experience makes him admit that "Anyone can get it."[9]

He suggests he is most likely to have gotten the disease at the grocery store — where there are plenty of people, most not wearing a mask.[10] When I heard this, I didn't dismiss the grocery store premise. But the engineer in me also wondered if the ventilation system in old office buildings at RML has HEPA filters. The virus could get passed from office to office, occupied by the same cohort day after day. I don't know, I haven't asked.

It will take the *New York Times* two years to report what our Health Board has heard over and over this year — *Even if masks work, mask mandates don't*.[11]

It's too bad "A Closer Look at US Deaths Due to Covid-19,"[12] became a counter-narrative source.

One can't blame the public for taking a professional's word for it, but one can hold someone with a PhD behind their name to account, just the CDC has been taken to account. Both are supposed to be trustworthy in their methods in our society. A webinar posted by a Johns Hopkins PhD[13] in economics concludes, "no evidence that COVID-19 created any excess deaths."[14] This is based looking at weekly patterns and cause of death from weekly charts using data download-

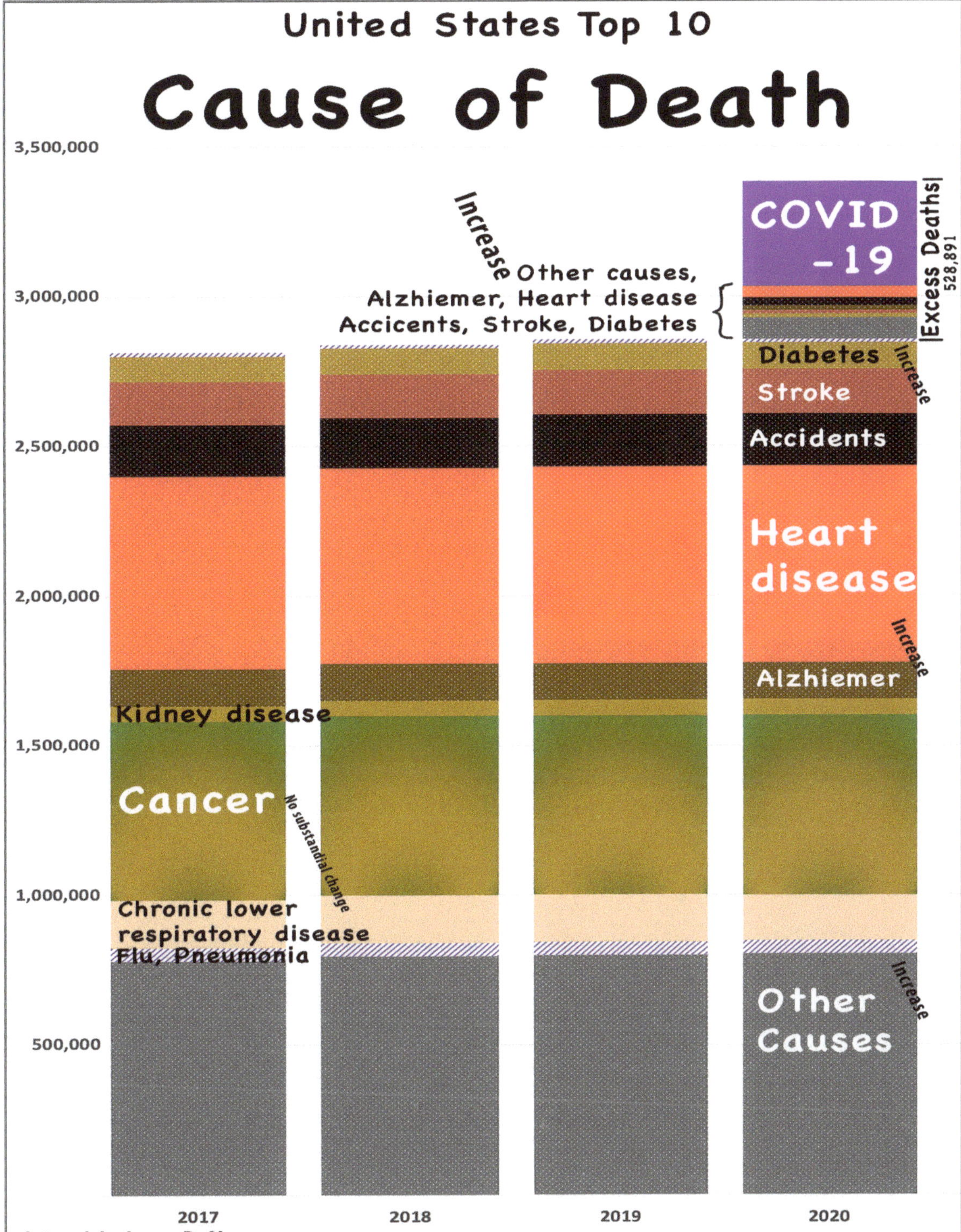

## United States Top 10
# Cause of Death

3,500,000

COVID
-19

|Excess Deaths|
528,891

*Increase*

3,000,000 — Other causes,
Alzhiemer, Heart disease
Accicents, Stroke, Diabetes }

Diabetes *Increase*

Stroke

Accidents

2,500,000

Heart
disease

2,000,000

Alzhiemer *Increase*

Kidney disease

1,500,000

Cancer

*No substantial change*

1,000,000

Chronic lower
respiratory disease
Flu, Pneumonia

Other
Causes *Increase*

500,000

2017          2018          2019          2020

**Artwork by James R. Olsen**
**Sources**
**CDC & NEJM:** Xu, "Deaths: Final Data for 2019," CDC National Vital Statistics Report, Vol. 70, No. 8, Jul. 26, 2021;
Heron, "Deaths: Leading Causes for 2017," CDC, National Vital Statistics Reports, Vol. 68, No. 6, Jun. 24, 2019
Ahmad, "The Leading Causes of Death in the US for 2020.";
Murphy, "Deaths: Final Data for 2018," Centers for Disease Control and Prevention , Jan. 12, 2021;
Murphy, "Mortality in the United States, 2020." NCHS Data Brief, no 427. Figure 4 Data;
Ahmad . Provisional Mortality Data — United States, 2020.

*Top Causes of Death and Excess Death 2020*

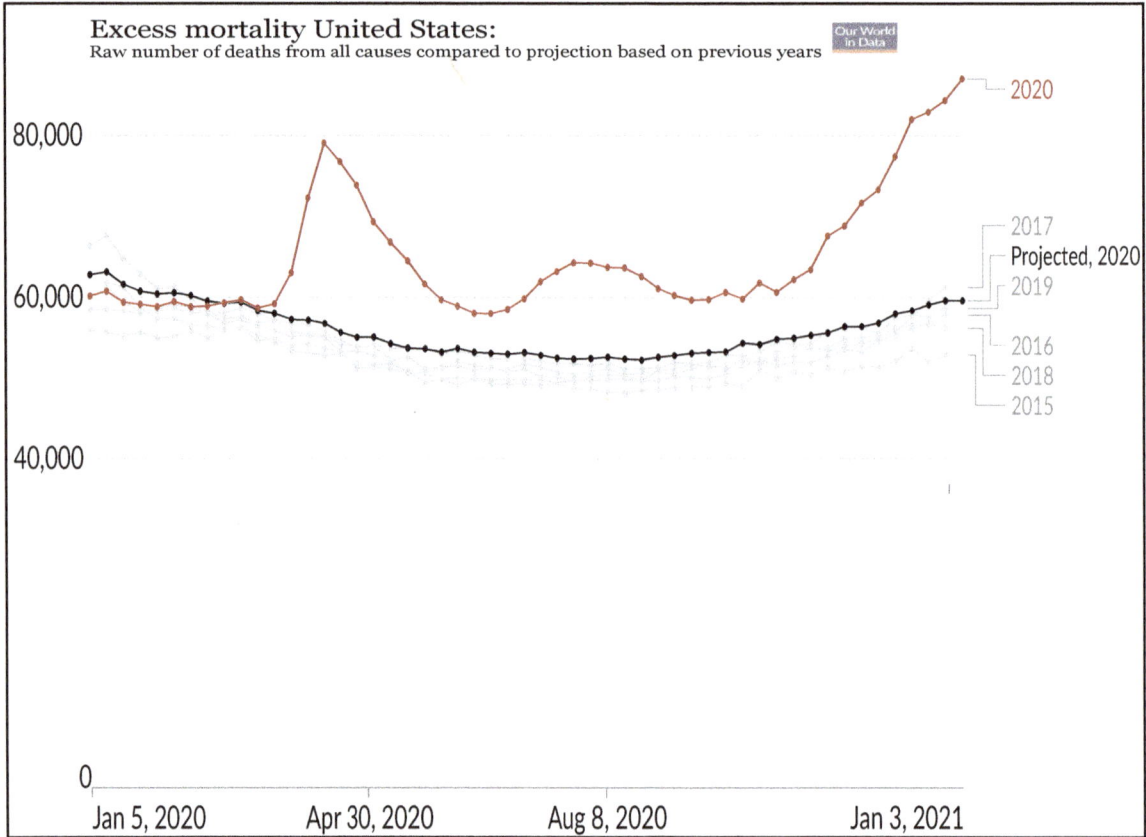

**Excess mortality United States:**
Raw number of deaths from all causes compared to projection based on previous years

*Excess Deaths 2020*

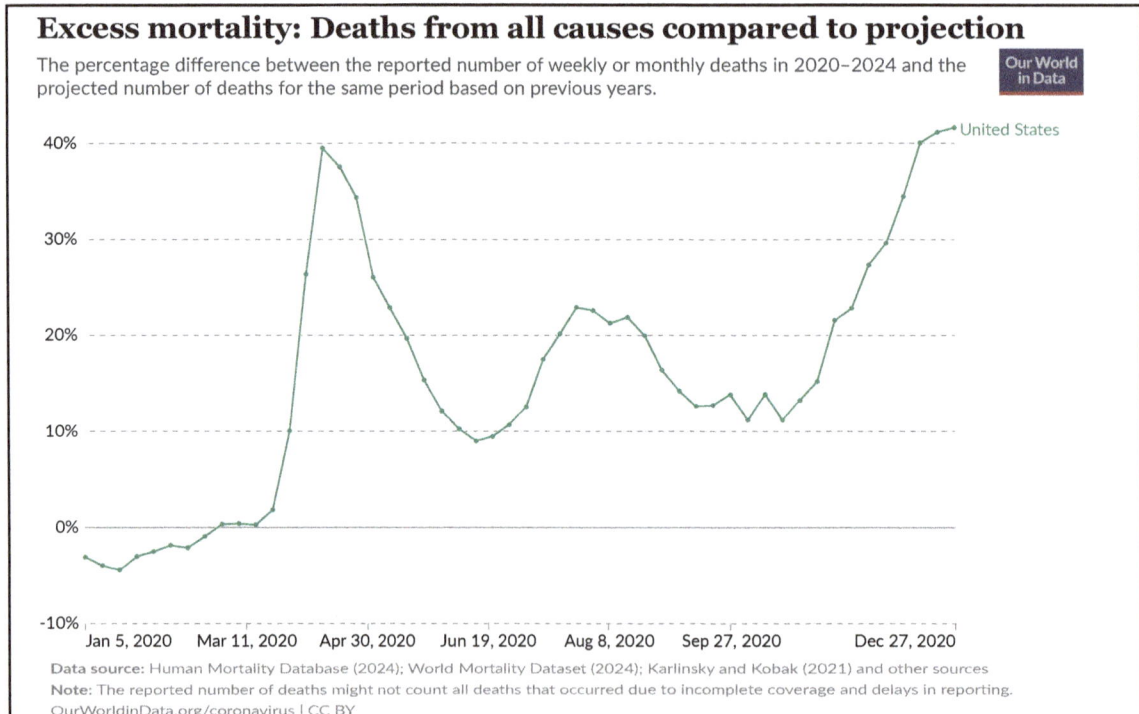

# Excess mortality: Deaths from all causes compared to projection

The percentage difference between the reported number of weekly or monthly deaths in 2020–2024 and the projected number of deaths for the same period based on previous years.

**Data source:** Human Mortality Database (2024); World Mortality Dataset (2024); Karlinsky and Kobak (2021) and other sources
**Note:** The reported number of deaths might not count all deaths that occurred due to incomplete coverage and delays in reporting.
OurWorldInData.org/coronavirus | CC BY

*Excess Deaths 2020 percent*

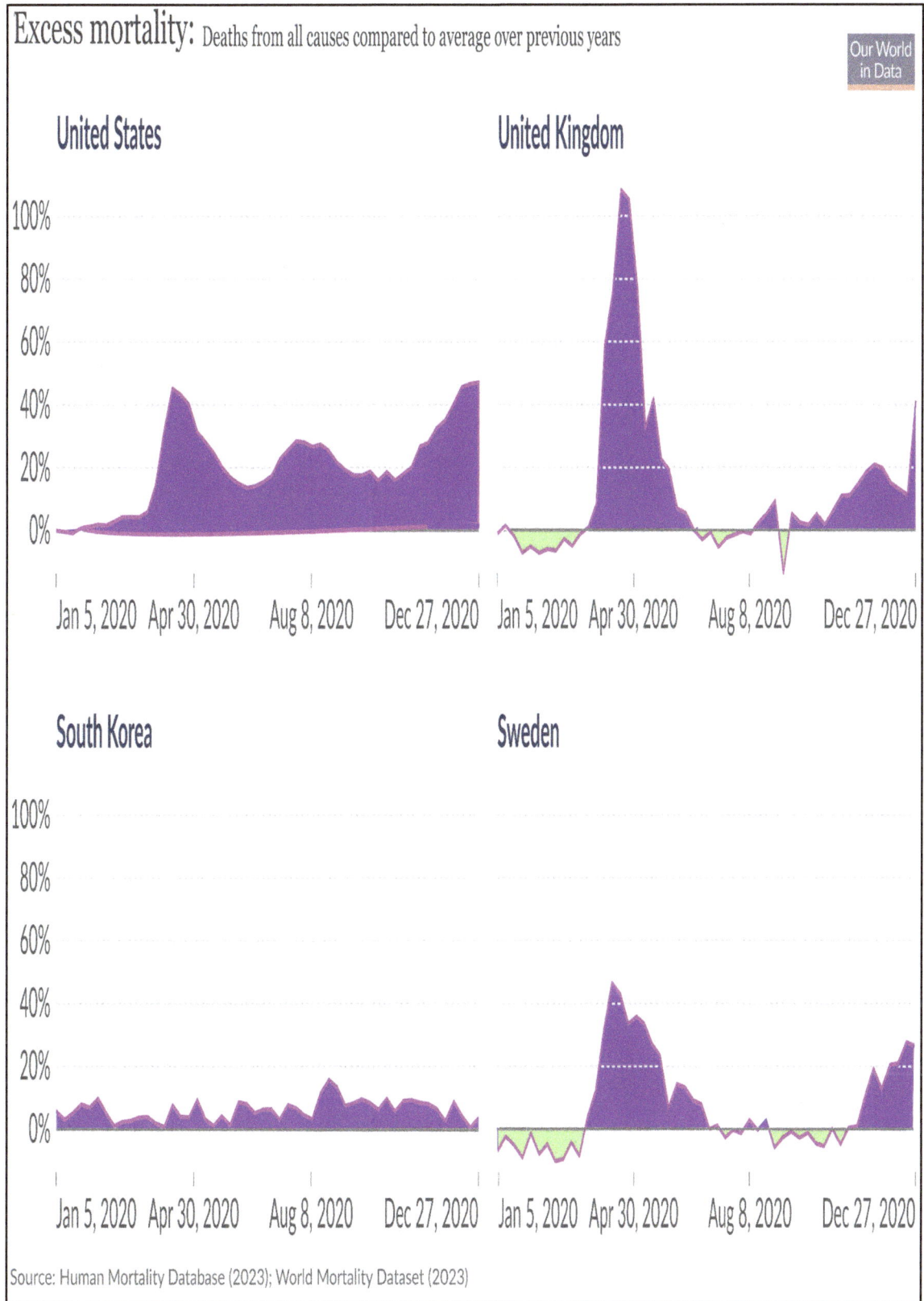

Excess mortality: Deaths from all causes compared to average over previous years

*Excess Deaths Four Countries 2020*

ed from the CDC from 2014 to 2020. Her method makes no sense to me.

The webinar is posted on YouTube, showing the Johns Hopkins PhD talking with the Johns Hopkins logo in the background.[15] An article summarizing the webinar is published on the Johns Hopkins website and then redacted, the redaction attacked, and the redaction explained. Students pointed out the obvious in a newsletter that the drop in deaths due to heart disease, etc., is because the patient also had COVID-19, and that is how it is coded as COVID-19. More importantly, the Johns Hopkins PhD does no explain why many reports and papers showing excess deaths are incorrect.[16]

The webinar's claims circle at least a quarter of the globe as a London newspaper reports on the webinar/article. The reporter has others weigh in, who note several problems.[17] All of this happens between November 13th and December 4th. By the 14th, another article pointing out the issues is posted on the Johns Hopkins website.[18]

We all make mistakes. As I said on the copyright page of this book, I have probably made mistakes in this book. Any professional can make a mistake — I certainly have in my engineering career. It is what one does when the mistake is pointed out, analyzed, and presented so I can see the error of my thinking. Do you double down? Or, admit that the point has not been proven, or even change your view? I have worked with other engineers, and I did just that.

A later article claims that the increased death rate for 2020:

> ...was not unexpected, nor alarming, but rather, is explained by the increase in population.

The Johns Hopkins PhD presents 529,894 more total deaths in 2020 than in 2019. Changes in death rates and population in the six previous years are also shown, with the death rate increase of earlier years being 13,000 to 79,000.

The author wants us to ignore the increase in deaths in 2020 compared to other years and focus on the per capita death rate, which goes from 0.9 in the previous three years to 1.0%. The author states the slight difference, only 0.1%, "is to be expected." This "slight difference" is a 10% increase, a half million people. It means something.

Now, I know some people who will say the death numbers are cooked up and not real. Given the history of the United States, I cannot deny that as a possibility; take McNamara's body counts during the Vietnam War as an example.[19] Though it is remote I cannot deny history.

This is one example out of many that made it into this book because it became part of our political dialogue: There have been too many instances of a lack of careful analysis emanating from universities. Counter-narrative purveyors find them and happily trot out the conclusions along with university credentials. I am not suggesting suppression, but a more thorough review before publishing by the institution doing the publishing — if a valid study furthers the counter-narrative, so be it — that is the very purpose academic inquiry is the purpose of a University.

The bottom line is that the United States had over a half million excess deaths in 2020. 150,000 to 200,000 were not attributed to COVID-19. Running out of medical capacity, deferred care, and less than thorough contract tracing set up conditions for these non-COVID deaths. They could have been avoided if we had prepared.

Nurse Webber starts her presentation at the December Health Board meeting by saying she usually writes down Public Health's activities beforehand. Still, she was so busy this month that she didn't have time to write. In the last month, 1,072 infectious diseases of all sorts were reported in the county. Tiffany has been working with a group from the county and DPHHS to "brain storm" what might be coming down the pike. She is unhappy that they have been reactive so far.

I don't want to get a call at 3:00 am from the road department that there's a blizzard and everyone is sick. What do we do?

The Sheriff has been beneficial in figuring out how to maintain continuity of service when many people get sick.

The Mayor of Hamilton has organized a group that helps Tiffany know what's happening in the community, schools, businesses, and long-term care facilities. She is trying to get the use of a building to do asymptomatic testing of contacts. She is also improving the process of managing isolation and release from isolation. Of course, no one checks for compliance by people told to isolate.

A person who tests positive is immediately contacted by the hospital and instructed on what to do, including self-isolation. There is a delay in being assigned to a contract tracer. There is no integrated process. The hospital does its function, and pubic health does theirs. Of course, Doctor Calderwood and Nurse Webber communicate informally frequently.

Public health officials got ten days behind in contact tracing, with the person often going through isolation and recovering before the contract tracer got to them. It's back to a three or four-day delay.

When Doctor Turner asks Nurse Webber about any problems, she starts to list them off but never gets past what she identifies as the most important one:

When she calls three or four days later, many don't know they should be isolated from their spouse.

The schools want to do "surveillance" of people like janitors who have lots of daily contacts. A person from the audience suggests testing everyone. Doctor Calderwood finds that problematic because there would be false-negatives, and false-positives who would have to go into isolation. It is obviously beyond the county's capacity.

The crowd is thinner at the Health Board Meeting. I suspect some were like me. Family and friends serve as a rationale to skip this meeting. Though, when I watch the video afterward, I am sorry I missed it.

It must have been a slap in the face when one commenter, after hearing of Webber's plight, says she can stop doing contact tracing because the whole thing is a fraud, citing a Johns Hopkins PhD.

Two weeks later, the first vaccines arrive in Ravalli County — just before Christmas, 100 doses of Moderna. The Pfizer and Moderna vaccines have not gone through the complete testing and trial process. Still, they are approved as Emergency Use Authorizations (EUA). The median time for vaccine approval in the past decade has been 7 years, a 12-month FDA review, and 7 clinical trials. While the Pfizer and Moderna trials had thousands of participants, the time has been short — too short to be sure there is not some serious but uncommon or rare side effect lurking in the shadows.[20]

Vaccines are rationed to emergency responders like the early testing rules — "emergency medical services, firefighters, police, dispatchers, dental, home health, hospice, funeral services, ski patrol and other first responder."[21] Sheriff Detective Brad Weston is pictured, in an orange t-shirt and a black mask with a Nike check-mark logo, getting the needle from retired Registered Nurse Betsy Saylor.

The *Ravalli Republic* reports on a mock drill using Commissioner Greg Chilcott as the guinea pig to simulate an prophylactic shock — noting they have epinephrine on hand. Nurse Webber talks about this new vaccine technology,

Training is super-important. As a nurse if I'm giving you something, I better know what I'm giving you.

It's not enough to be able to recover you have a reaction. I want to know what I'm giving you and explain it to you.[22]

Lab-made clones of COVID-19 antibodies were approved by the FDA last month for high-risk patients who are NOT in the hospital — antibodies need to do their work before the disease gets out of hand. It will take a while before it comes into use here. By the end of next year, the Governor will promote monoclonal antibody clinics throughout the state.[23]

A back porch BBQ is followed by my daughter hosting a Christmas dinner. True to the tradition that has grown up over the years, I cook the turkey. We don't talk much about COVID and don't worry over politics, as we share our stories of our journeys with those we love — I suspect this is the most common experience on Christmas day.

There are others, though, more and more worried about a roof over their head as more and more people move to Ravalli County, bringing cash offers, buying up rentals as leases expire, and driving up prices beyond a wage-earner's means.

We put our house on the market in 2019 with the idea of downsizing, but with downsizing to a smaller place becoming more and more expensive, we let our listing expire this year. It must be heating up even more. Our broker calls because buyers note our old listing and are willing to make an offer anyway; another buyer, family in tow, knocks on our door — we're staying put.

Laura and Peter are still in Super 8,[24] in a stasis that we know will not last forever; our search for rentals widens as it gets more half-hearted. My Christmas get-together with them has to wait until after Christmas. I sit in a chair in a hotel room, Laurel propped up in bed, exchanging gifts and best wishes.

As the year ends, some party, some worry, some march, some wait anxiously for their turn to get a vaccine. A van full make plans to go to the Capitol for President Trump's rally.

Sweden finally relents as hospitals fill during the winter surge, and many people with other issues avoid the hospital because they don't want to get COVID. Restrictions begin to be imposed, and the rules will get more stringent until Sweden's restrictions approach those of the United States by February 2021.[25]

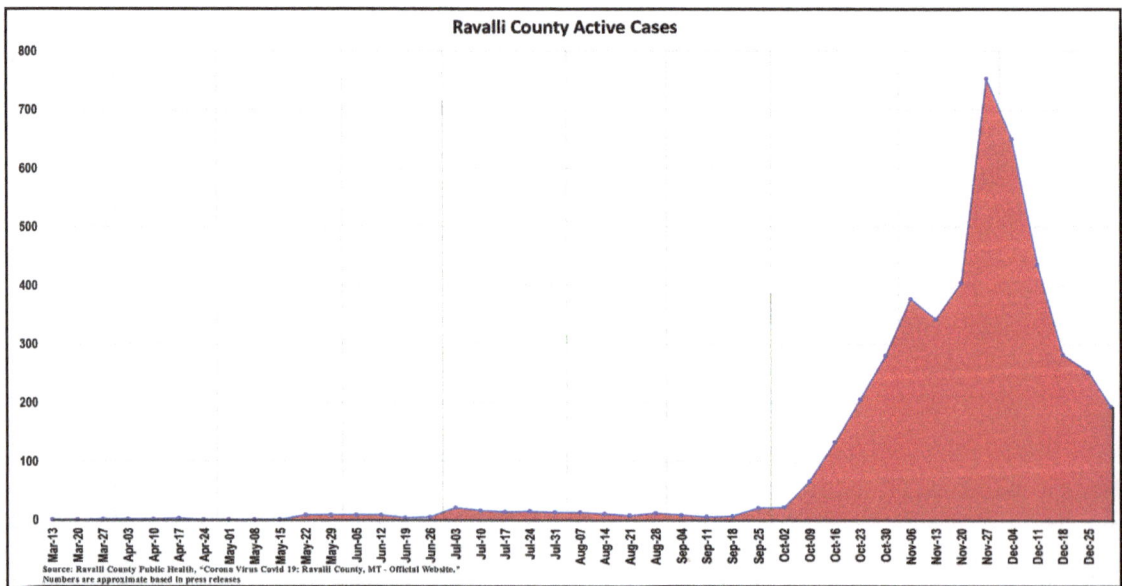

Active Cases Dec. 2020

## Excess mortality: Cumulative deaths from all causes compared to projection based on previous years, per million people

Our World in Data

The percentage difference between the cumulative number of deaths since 1 January 2020 and the cumulative projected deaths for the same period based on previous years. The reported number might not count all deaths that occurred due to incomplete coverage and delays in reporting.

**Data source:** Human Mortality Database (2024); World Mortality Dataset (2024); Karlinsky and Kobak (2021) and other sources

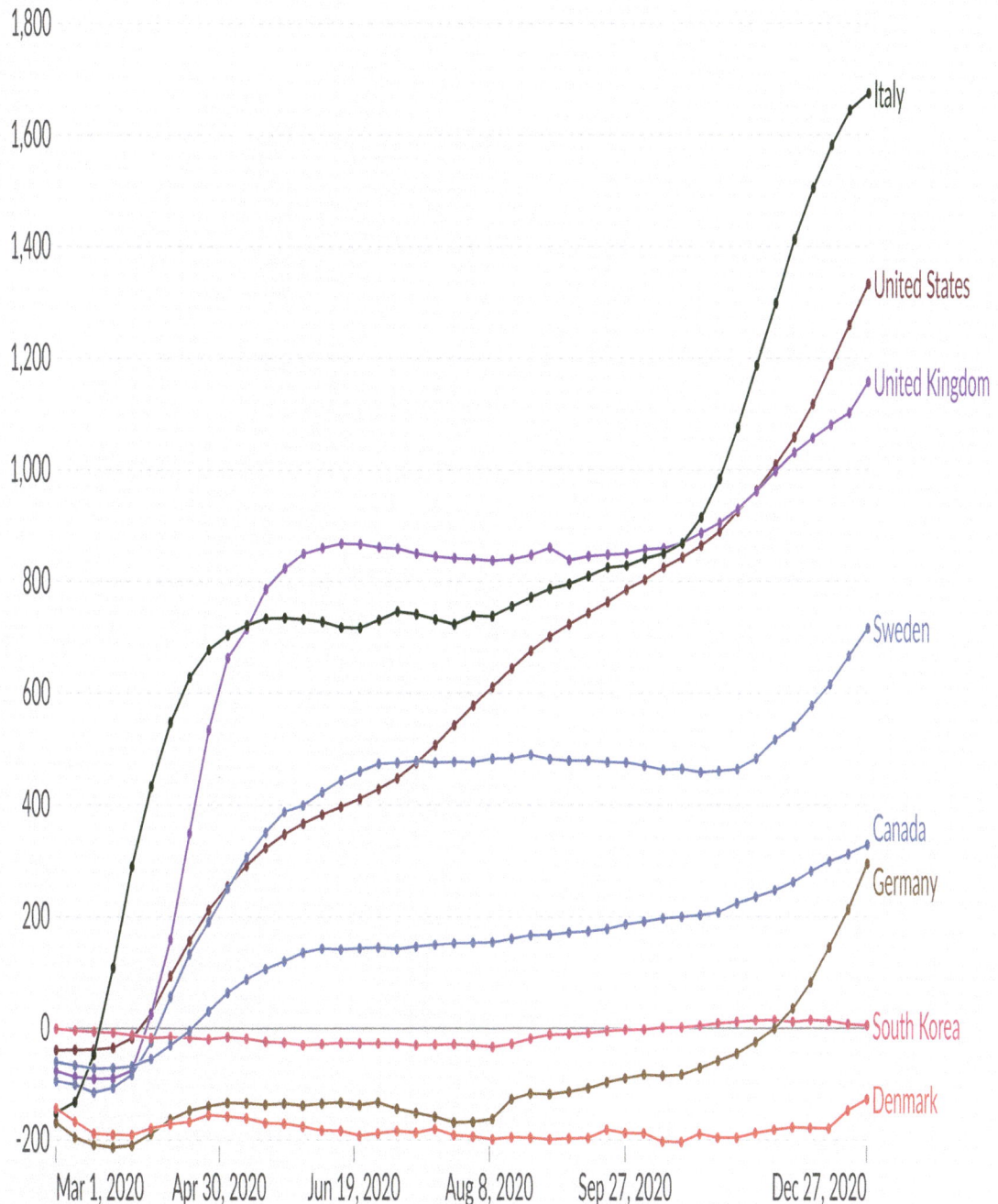

*Cumulative Deaths Per Capita 2020*

A disease is upon us, though we can't agree on what to do about it in the United States. We can't even agree on how bad it is, though everyone agrees some die. And let's face it, in a small town, you will probably know at least one of the departed.

The chances of dying in 2020 have been higher than in the previous five years. The increase is not all due to COVID-19, no matter who is counting. The numbers say people died of COVID-19. And the numbers say people died of other causes. Is it that people with other health problems didn't get help when they needed it due to lack of medical care or delayed medical care? There is some evidence of that.[26]

Is it that the stress of it all, both the fear of the disease and the fear induced by the talking heads and Twitter chatter, or the isolation, mandated and voluntary? There is evidence of that.[27] But, there is nothing definitive — except that people died. Native and Black Americans had higher death rates before the pandemic, only to get worse in 2020.[28]

Indeed, it was prescient in 2019 when Maroon 5 sang *Memories*.[29] And so, as the ballad plays again, lift a glass and salute those who we remember. For all of that, humans have dealt with worse.

Life goes on. We cook, eat, socialize, ski, ride, work, love, a back porch BBQ once in awhile — and at least one of us, on the last day of 2020, as the waning gibbous moon glows in the night sky, will indulge in a twenty dollar cigar and a bit of good bourbon.

*31 Dec. 2020: 349,938 COVID-19 Related Deaths Reported*

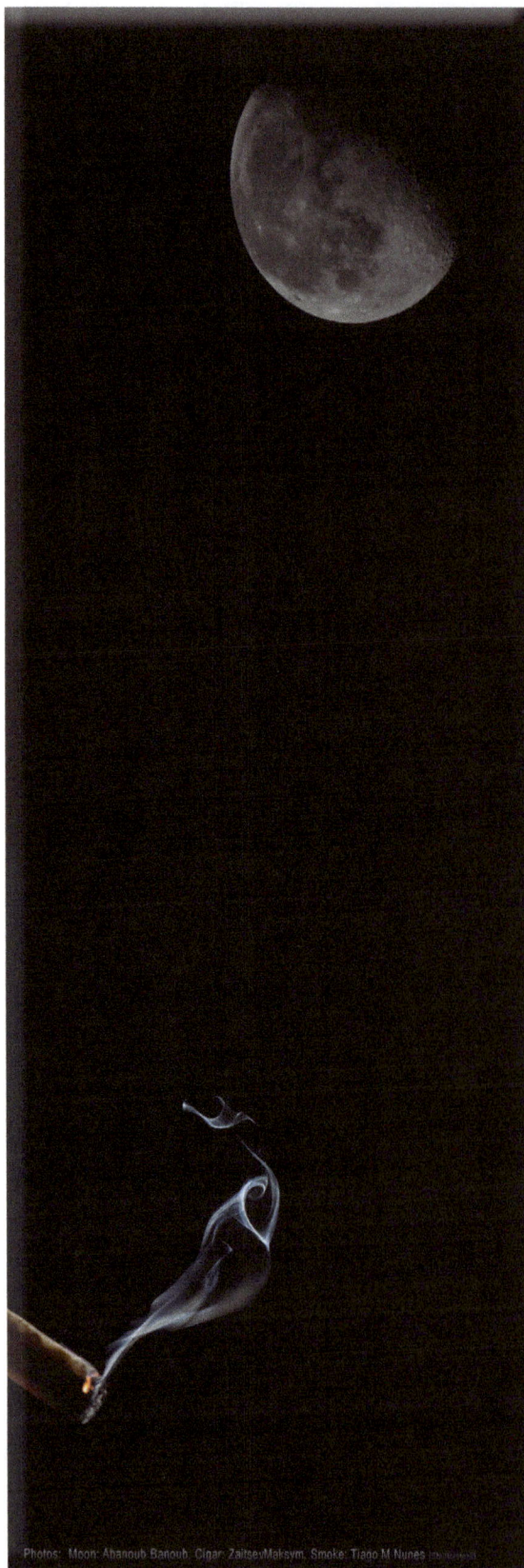

Photos: Moon: Ahanoub Banoub. Cigar: ZaitsevMaksym. Smoke: Tiago M Nunes

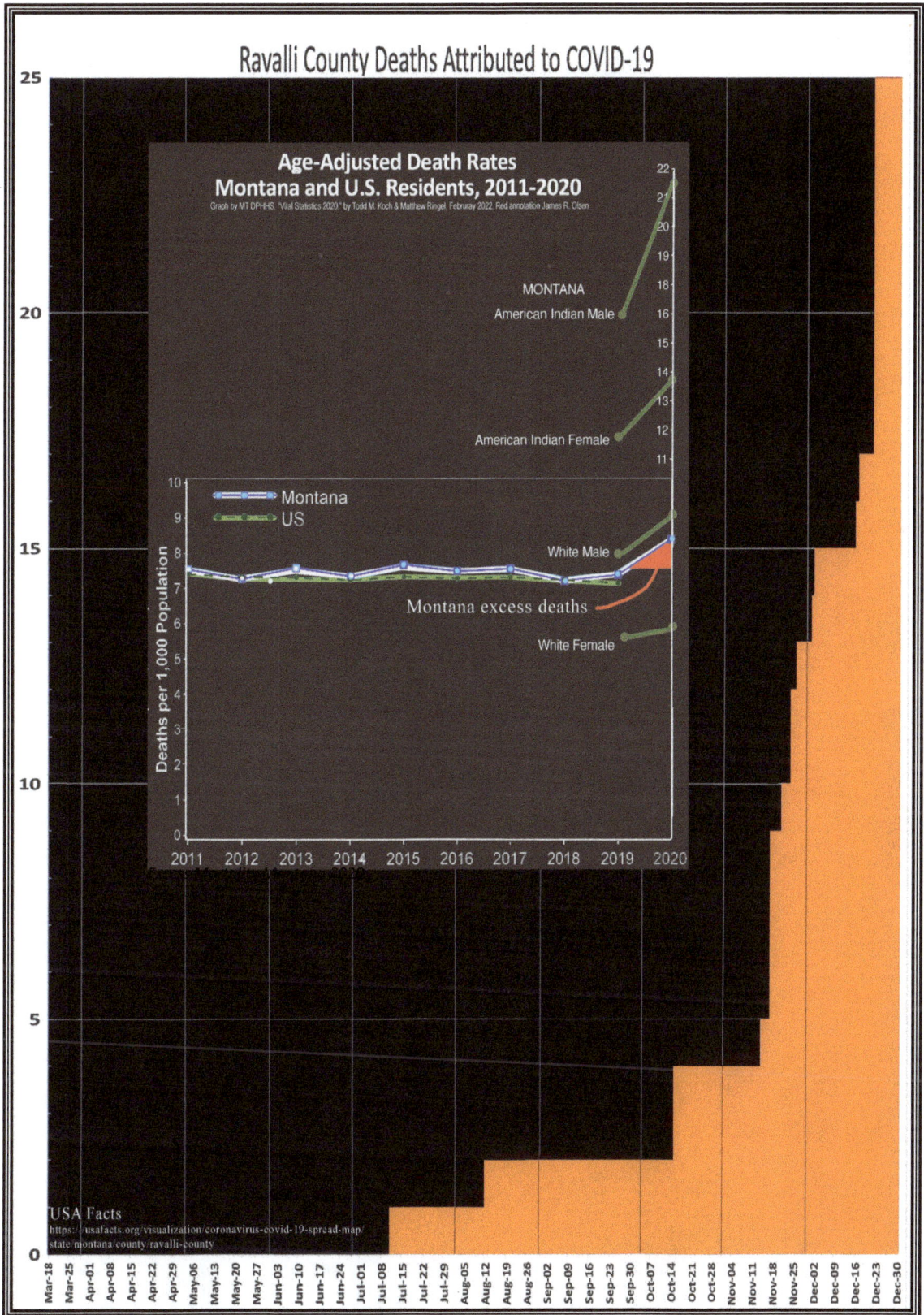

*Ravalli County COVID Fatalities 2020 with Insert of Montana and United States Fatalities 2020*

Interior of Capitol Breached     January 6, 2021[1]

**Timeline entries (by time):**

- 10:58 — Proud Boys move toward Capitol
- 11:00 — Save America Rally Begins
- 11:24 — Trump begins speech
- 12:00 — Protesters surround Sen. Young at Senate Office Building
- 12:00 — Joint Session of Congress Begins
- 12:15 — Trump urges crowd to march to Capitol "peacefully"
- 12:53 — Capitol Police Chief says National Gaurd is needed
- 12:55 — Trump ends speech
- 1:09 — Crowd confronds police
- 1:10 — Objection to Arizona causes Senators to leave for a separate session
- 1:15 — Pence Certification Process in the House Camber
- 1:49 — Crowd breachs police lines West Steps, Riot Declared
- 1:49 — House members crouch behind bullet proof benches
- 2:13 — Crowd breaks in through window
- 2:14 — Senate members leave Senate chamber
- 2:24 — Reinforced Police rush crowd out of interior of capital
- 2:41 — Trump tweets "remain peaceful. No violence!"
- 3:02 — Crowd in Senate chamber
- 3:04 — Attempt to break into House thwated, officer shoots Babbit
- 3:13 — Biden gives speech called for end of insurrection
- 4:06 — Trump releases video telling crowd to go home.
- 4:17 — Trump tweets
- 4:19 — National Guard activation approved
- 4:30 — Capital interior cleared
- 5:15 — Babbit confirmed dead
- 5:40 — Hand-to-hand combat with clubs, mace gas, and a flag.
- 9:03 — National Guard begins arriving
- Pence didn't protect the country
- Objection to Arizona fails
- House and Senate reconvenes sperately:
- 11:35 — Joint session resumes

PHOTO BY TYLER MERBLER (Flicker)
James R. Olsen: Duplicated large hat and robe to obscure identity of person taking selfie and others

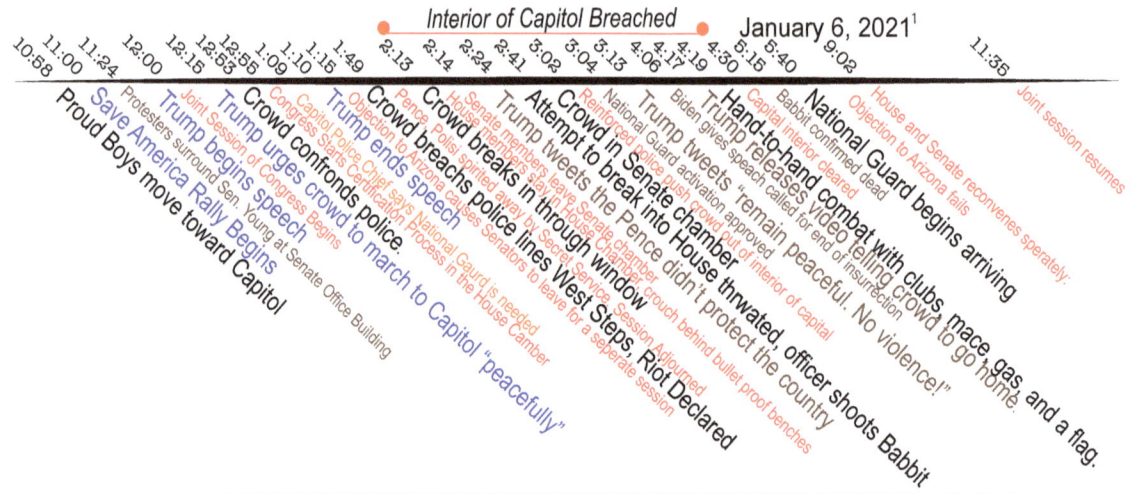

*Capitol Protest*

# 16

# TRYING TIMES IN WASHINGTON

*The guns flashed on the rim of the night like holiday fireworks.*
*The young men who fired the cannon,*
*Like the people of Charleston who watched,*
*They felt something great was beginning and were glad to be part of it.*
*The hard knowledge of war's reality would come later.*[3]
— Bruce Catton

## JANUARY 6, 2021

Where are the police? That is my question as I stare at the TV screen, having watched the speeches at Trump's rally, seeing the camera's focus shift to the Capitol, watching the crowd growing at the Capitol steps. It is indeed a thin blue line, too thin, too thin for this crowd. It is way too thin when it is obvious and expected that organized groups with violence on their mind are sure to be part of the gathering.[4]

Hoping against hope that the crowd will stay peaceful even after getting mixed messages during the Trump rally. How in the world anyone would anyone think this police presence is sufficient? It isn't.

What happens is predictable, should have been predicted, and should have been prepared for. I am still amazed no one has been held to account, but maybe some politics is involved. The presence of violent actors is no surprise. The D.C. Mayor had made a concerted effort to keep the protesters from bringing firearms, but this is not enough.[5]

When President Trump arrived at the podium, previous speakers at The Save America Rally had primed the pump. A well-known Tea Party activist reminded the crowd of why they are there, reminding the crowd that Trump is on their side, reminding the gathering of the slights they have endured. "Hi deplorables," as she calls upon them to ready themselves for the sacred mission, to fight for truth, to insist on justice, to save our democracy. Rudy Giuliani speaks, turning a phrase, a phrase some will later say should be taken literally once it is pulled out of the context of a legal battle. How many in the crowd take the phrase literally remains unknown. Guliani begins,

> In the next ten days, we get to see the machines that are crooked, the ballots that are fraudulent, and if we're wrong, we'll be made fools of. But if we're right, a lot of them will go to jail. So, *let's have trial by combat.*[6]

The facts and testimony as to what happened behind the scenes are still underway, so that will just have to play out. There is not an explicit call to violence in Trump's speech, nor a call for a violent insurrection, but a mixed message, leaving it to the listener to decide what he means and what is expected. The objective is clear: to revisit the electoral college votes in key states that Trump lost, more specifically, to get Vice President Pence and the other Republicans to send the electoral college for the swing states back to try again. The means are left to the imagination:

We gather here for one simple reason: to save our democracy..... They've used the pandemic to defraud the people...

our country will be destroyed. We are not going to stand for that...

I know that everyone here will soon be marching to the Capitol Building to peacefully and patriotically make your voices heard...

Our county has been under siege for a long time, longer than the last four years...

The [election counting] is a criminal enterprise...

We fight. We fight like hell. And if you don't fight like hell, you don't have a country anymore...

We're going to walk down Pennsylvania Avenue — I love Pennsylvania Avenue — and we're going to the Capitol. We going to try to give — the Democrats are hopeless... but we're going to try to give our Republicans some backbone — the weak ones because the strong ones don't need any of our help...

So, let's walk down Pennsylvania Avenue.[7]

Some have gathered around the Capitol as the rally crowd marches down Pennsylvania Avenue, pouring more people into the grounds around the Capitol steps like an irrigation ditch flooding a field in early spring. The sparse gathering becomes a crowd, getting increasingly crowded as people keep arriving. The air fills with chanting, yelling, and waving flags: American flags, Trump 2020 flags, "don't tread on me" flags, Confederate battle flags. Signs rise about the heads of the crowd: stop the steal, the real invisible enemy is communism, the storm is here, Trump won, Biden lost, vaccines are bioweapons.

This crowd was a long time in the making; it didn't have to be this way. It began with The Great Divide, the magnet on the left, pulling harder each time there is a death at the hands of a police officer, a death judged to be unjustified. The magnet on the right, pulling harder each time

there is a mandate, a business shuttered, a favorite treatment denied.

The crowd on January 6 is a fraction of a fraction of the people who voted for Trump. Like penguins crowded on an ice flow, they are the ones who, by will and circumstance, get pushed to the edge, to the edge of a large mass of people who slid in that direction, pressed by an eroding mistrust of government. Now, even a single slip-up gets amplified by the midway barkers, happy to be selling stories to a ready audience, edging a mass of people closer to the edge of revolt. In 2020, in a small town, a couple could hang posters around a small town that say "Freedom Meeting" and get a decent crowd.[8]

Is it a surprise that a few in that crowd have blood in their eye? Not to me. The real question is, "Does that majority still exist, those who will take time to think through the changes needed, have the final say, at the ballot box, in the legislative halls?"

What will happen at the Capitol is not left to the imagination by at least two cadres.[9] They planned for it. They hope for it. They agitate an angry crowd for it, a crowd not sure what to do next, a crowd waiting for some impetus. It often happens in such a crowd, with most people asking what we do next? The cadres answer the question; the impetus is provided. The only question is, will enough people in the crowd answer the call to deliver the bodies to overcome the police? If so, the momentum carries it forward. If not, it stall. Today there are enough willing bodies.

The cadres are ready, armed with clubs, baseball bats, hockey sticks, poles, fire extinguishers, mace, and an American Flag; they are armored in leather and helmets, equipped with walkie-talkies; they know what to do — along with some lone rangers — charge, push, hit, break, batter, swing a baseball bat.[10]

That can go either way. Are there enough people who have heard enough, believe enough, are angry enough to step over the line from peace to violence, participate in it, or tolerate it even if they don't engage in it?

Some of the crowd join in, some even pushing through the mass of bodies to get to the battle line, while many in the back don't even know what is happening. Many, many walk away, 115,000 of the 120,000 who showed up to hear President Trump; they have come to make a point, not swing a baseball bat.[11] The other 5,000 stay; half of them charge into the Capitol building. There are enough bodies to overcome the police. Some breach the doors. Others battle outside, never getting in. Others stand around the steps to watch and cheer.

The ready, willing, and adrenalin filled, the 1,200[12] break into the building. Even many of those are not prepared to swing a baseball bat. For many, finding themselves inside, act like they have crashed a Hollywood party, celebrating, taking selfies, wandering the halls. I wonder if they think about the consequences of having their picture taken.

But, among this breaking and entering crowd are a few bloody-minded people. You only have to take them at their word, "Hang Mike Pence. Hang Mike Pence. Hang Mike Pence."[13] Yelling at police, "We're gonna kill you!"; "We're gonna murder you and then them!"; "You guys are traitors and should be killed!" The only reason there isn't murder in the halls of Congress is the brave, outnumbered Capitol Police standing in the way.

Some die: from an overdose, a heart attack, wounds; A couple whose psyches are overwhelmed having citizens of their own country attack them.[14] And a bullet kills one of rioters as she tries to step through a broken door, members of Congress on the other side.

Three officers stand outside the doors of the Speaker's Lobby, about 15 steps from the door of the House of Representatives Chamber.[15] Unarmed elected representatives and staff, some shaking in fear, crouch behind bulletproof benches; the evacuation is not complete.[16] As the three officers face the crowd, the crowd chants, "Break this down, break this down."

One of the crowd says, "We don't want to hurt you," to a police officer. As a club cracks a windowpane in the door, he continues,

> We want you to go home. There are so many people. They are going to push up here. I've seen people get hurt. We'll make it past.

Outnumbered, the officers slide along the door to get out of the way; a bat, a foot, and a hand gloved by a baseball cap, pounding the glass inserts of the doors, cracking some. All the officers can do is stand along the wall and watch.

But there are officers on the other side of the doors, listening through earpieces on the net, listening to the excited chatter of hand-to-hand combat on the steps and the halls, unfolding like some 50s street gang rumble, helpless to respond to calls for help as lines are flanked, hearing the roar of the crowds still pushing in from the steps, more lines breached, and reports of members and staff still in the Chamber.

A pound at the pane on the top half of the door on the left suddenly sends it flying out of its molding.

A woman steps up on the empty window pane. A shout, "There's a gun, there's a gun." Three men trying to kick down the door keep at it. The woman leans to go through the empty pane, pushed forward by the crowd. The gun fires. She falls back.

The pounding stops. People crouch to help her. Rioters are now side-by-side with the police. Confusing shouts, "flash bang," "active shooter, get down, get her down." "We can't save her; we have to get EMS here."

Police reinforcements arrive — just in time to carry her body away.[17]

The hard knowledge of war has come.

Like any crowd of protesters, it is made up mostly of folks who come in angry peace. I've been in them, the out-of-focus bystander in the background of the telephoto news shot, going to class at Berkeley in the late '60s, picking my way around the crowds and police lines.

Here's what I saw. Sometimes, it is a planned and well-organized mass march, like in the Martin Luther King days, with nearly everybody in the crowd of the same mindset, ready to accept a specific discipline, in the case of Reverend King or the Velvet Revolution in Prague — a discipline of non-violent action, non-violent civil disobedience.

On the other hand, a cadre can be ready to accept a specific discipline *of* violence, like the October 10, 1917 uprising in Saint Petersburg, the rebellion that led to the Russian Revolution, and the foundation of a Communist-led Soviet Union.

Most of the time, the crowd is not organized, gathered because they were called by some person or group or in response to some event, such as the assassination of Martin Luther King, when neighborhoods across the country burned.

The prospect of a large crowd of angry people almost always attracts trouble-makers who have destruction and violence on their minds — 1 to 3% is always a good guess. With the adrenalin up, there becomes a tipping point — *if* the crowd sentiment doesn't hold them back, which it occasionally does.

How come the powers that be failed to provide the ready resources for the police on January 6, 2021? A question still unanswered, not even by the Senate investigation, "Examining the US Capitol Attack."[18] It is a story of miscommunication between agencies and a flawed command and control processes, it isn't that they don't have intelligence. They do.

From the Norfolk FBI office an Internet feed:

> Be Ready to Fight. Congress needs to hear glass breaking, doors being kicked in, and blood from their BLM and Pantifa slave soldiers being spilled. Get violent . . . stop calling this a march, or rally, or a protest. Go there, ready for war ...[19]

In June, the Democrat Mayor of the District of Columbia had "Black Lives Matter" painted in large yellow letters on the street in front of the White House. "The mayor criticized unidentified federal law enforcement officials for patrolling the streets of her city and operating outside 'established chains of commands'," who were inflaming protesters.[20] She was rightly concerned about a lack of command and having an assortment of armed Federal agents. Of course, this is fixable. But it wasn't fixed. Partisan politics undoubtedly got in the way.

In any case, expressing some of the same concerns, Mayor Bowser Tweeted a letter on January 5 to federal officials that she didn't want the National Guard to provide armed security. Neither the President nor Congress asked for National Guard assistance ahead of time, though Trump later claimed he suggested a large force be present and that it was rejected by House Speaker Pelosi, who denies it.

In any case, the Guard never knew about the suggestion.[21] Over a half hour after the crowd confronted the police, and 17 minutes after the Capitol is breached, a call to the President's Chief of Staff, Mark Meadows, requests the National Guard's assistance. He calls back at 2:56, saying the President approved — it will take several hours for them to arrive in full force.[22]

Inside the chambers, as the protesters gather, Vice President Mike Pence, who has held his cards close to the vest, issues a letter stating clearly that he intends to do his duty, not bending to what the crowd might want, saying he does not have the unilateral authority to decide whether an elector college vote counted. The country's founders had left that as a matter for each state.

*Noose at Capitol Building*

He does have a duty to allow Congress to consider any objections:

> I will do my duty to see to it that we open the certificates of the Electors of the several states, we hear objections raised by Senators and Representatives, and we count the votes of the Electoral College for President and Vice President in a manner consistent with our Constitution, laws, and history. So Help Me God.[23]

And so Vice President Pence does just what he said he would do after threats, evacuations, a recess, and a call from President Trump.[24] As he and his family stand in the basement, his security want him to lead Vice President Pence said, pointing his finger at the security man's chest,

> You're not hearing me, I am not leaving! I'm not giving those people the sight of a sixteen-car motorcade leaving the Capitol.[25]

Yes, Ravalli County citizens are at the Capitol. According to State Senator Theresa Manzella, "her people" are "there at the capitol steps when people start rushing in." Some of her "associates" are "detained by law enforcement for questioning and released after they had been inside the US Capitol during the insurrection."[26]

Theresa "wished the demonstration hadn't spawned an invasion. It just didn't need to happen. None of it."[27] Though, at least one Republican elected official up in Libby is unrepentant saying, "The coup is complete."[28]

Not true. Mike Pence saved our country from chaos and perdition. Within a few months, Montana will tie for the record for the most arrests of Capitol invaders per capita.[29] The FBI will continue for years combining through records, even tracking credit cards, though some field offices will refuse to go along with requests that stretch due process.[30]

Is the response to COVID-19 a factor in generating this crowd? "It is not," might be an answer. I will leave it to a woman outside the Capitol on January 6, 2021, to speak for herself.

"What's the point?" asks a CNN Reporter.

Speaking over the crowd noise with a pointed passion, she says,

> Taking our freedom, locking us down, and turning the country into a leftist socialist republic. That is not right! That's what I am doing here.[31]

*Breached Capitol 2,000 to 2,500* [32]

## Excess Deaths per 100,000
2020 to mid 2023

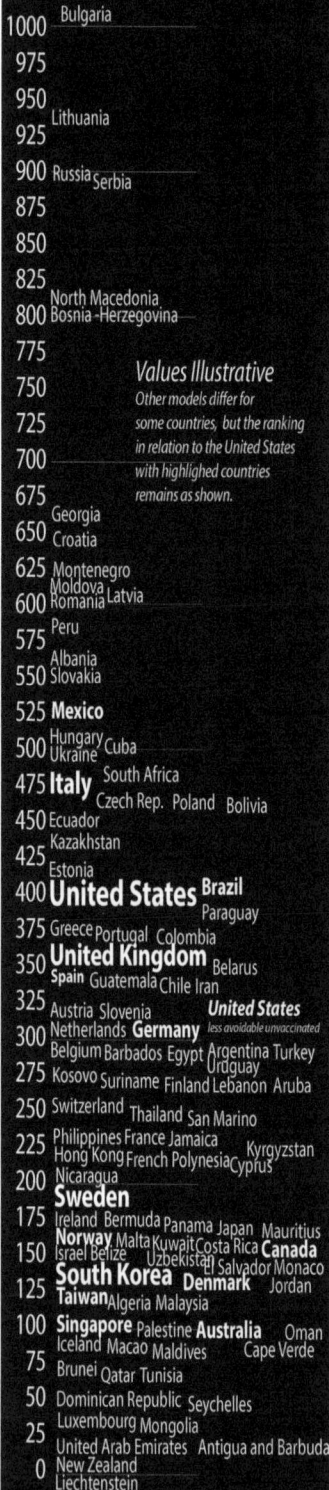

| | |
|---|---|
| 1000 | Bulgaria |
| 975 | |
| 950 | Lithuania |
| 925 | |
| 900 | Russia Serbia |
| 875 | |
| 850 | |
| 825 | North Macedonia |
| 800 | Bosnia -Herzegovina |
| 775 | |
| 750 | *Values Illustrative* |
| 725 | *Other models differ for some countries, but the ranking in relation to the United States with highlighted countries remains as shown.* |
| 700 | |
| 675 | Georgia |
| 650 | Croatia |
| 625 | Montenegro |
| 600 | Moldova Latvia / Romania |
| 575 | Peru |
| 550 | Albania Slovakia |
| 525 | **Mexico** |
| 500 | Hungary Cuba / Ukraine |
| 475 | **Italy** South Africa Czech Rep. Poland Bolivia |
| 450 | Ecuador |
| 425 | Kazakhstan |
| 400 | Estonia **United States** Brazil Paraguay |
| 375 | Greece Portugal Colombia |
| 350 | **United Kingdom** Belarus / **Spain** Guatemala Chile Iran |
| 325 | Austria Slovenia **United States** *less avoidable unvaccinated* |
| 300 | Netherlands **Germany** Belgium Barbados Egypt Argentina Turkey / Uruguay |
| 275 | Kosovo Suriname Finland Lebanon Aruba |
| 250 | Switzerland Thailand San Marino |
| 225 | Philippines France Jamaica Kyrgyzstan / Hong Kong French Polynesia Cyprus |
| 200 | Nicaragua **Sweden** |
| 175 | Ireland Bermuda Panama Japan Mauritius |
| 150 | **Norway** Malta Kuwait Costa Rica **Canada** / Israel Belize Uzbekistan El Salvador Monaco |
| 125 | **South Korea** **Denmark** Jordan / **Taiwan** Algeria Malaysia |
| 100 | **Singapore** Palestine **Australia** Oman / Iceland Macao Maldives Cape Verde |
| 75 | Brunei Qatar Tunisia |
| 50 | Dominican Republic Seychelles |
| 25 | Luxembourg Mongolia |
| 0 | United Arab Emirates Antigua and Barbuda / New Zealand Liechtenstein |

Data from: Economist, "Tracking Covid-19 Excess Deaths across Countries," *The Economist*, Jul. 10, 2023

*Excess Deaths 2020-mid 2023*

## Excess Deaths per 100,000

| | |
|---|---|
| 525 | **Mexico** |
| 500 | Hungary Cuba / Ukraine |
| 475 | **Italy** South Africa Czech Rep. Poland Bolivia |
| 450 | Ecuador |
| 425 | Kazakhstan |
| 400 | Estonia **United States** Brazil Paraguay |
| 375 | Greece Portugal Colombia |
| 350 | **United Kingdom** Belarus / **Spain** Guatemala Chile Iran |
| 325 | Austria Slovenia **United States** *less avoidable unvaccinated* |
| 300 | Netherlands **Germany** Belgium Barbados Egypt Argentina Turkey / Uruguay |
| 275 | Kosovo Suriname Finland Lebanon Aruba |
| 250 | Switzerland Thailand San Marino |
| 225 | Philippines France Jamaica Kyrgyzstan / Hong Kong French Polynesia Cyprus |
| 200 | Nicaragua **Sweden** |
| 175 | Ireland Bermuda Panama Japan Mauritius |
| 150 | **Norway** Malta Kuwait Costa Rica **Canada** / Israel Belize Uzbekistan El Salvador Monaco |
| 125 | **South Korea** **Denmark** Jordan / **Taiwan** Algeria Malaysia |
| 100 | **Singapore** Palestine **Australia** Oman / Iceland Macao Maldives Cape Verde |
| 75 | Brunei Qatar Tunisia |
| 50 | Dominican Republic Seychelles |
| 25 | Luxembourg Mongolia |
| 0 | United Arab Emirates Antigua and Barbuda / New Zealand Liechtenstein |

Data from: Economist, "Tracking Covid-19 Excess Deaths across Countries," *The Economist*, Jul. 10, 2023

# 17
# EPILOGUE - NOT OVER YET

*Ultimately, "we need vaccines" because "that is how we are going to end it."*[1]
— Doctor Anthony Fauci

Vaccines came out in record time — the primary goal of Doctor Fauci's plan to put the COVID-19 pandemic to bed. Vaccines will help, but the virus isn't done with us. The Delta variant will be more deadly, and Omicron will be less so.

The vaccines are not enough to stop COVID-19 in its tracks — it will require the virus to become less deadly and more effective treatments. Meanwhile, the anxiety, arguments, censorship, conspiracy theories, mandates, and economic impact will again enter the boxing ring for another round in 2021.

The second book in this series, *VAX WARS*, chronicles the rest of the story. The rhetoric at the health board meetings will get harsher; vaccine effectiveness and side effect risk will dominate the dialog; economic policies will set the stage for inflation; vax mandates will have neighbor fighting neighbor.

The United States will end up with one of the worst fatalities per capita in the world. The sad news is that much of it was avoidable. The third book, a short one in this series is *How To Do Pandemics Better* contains a closer examination of how the news works, the nature of the data, the nature of research, and a bold plan for how to improve health regulation and be prepared for the next pandemic —because there will be another one.

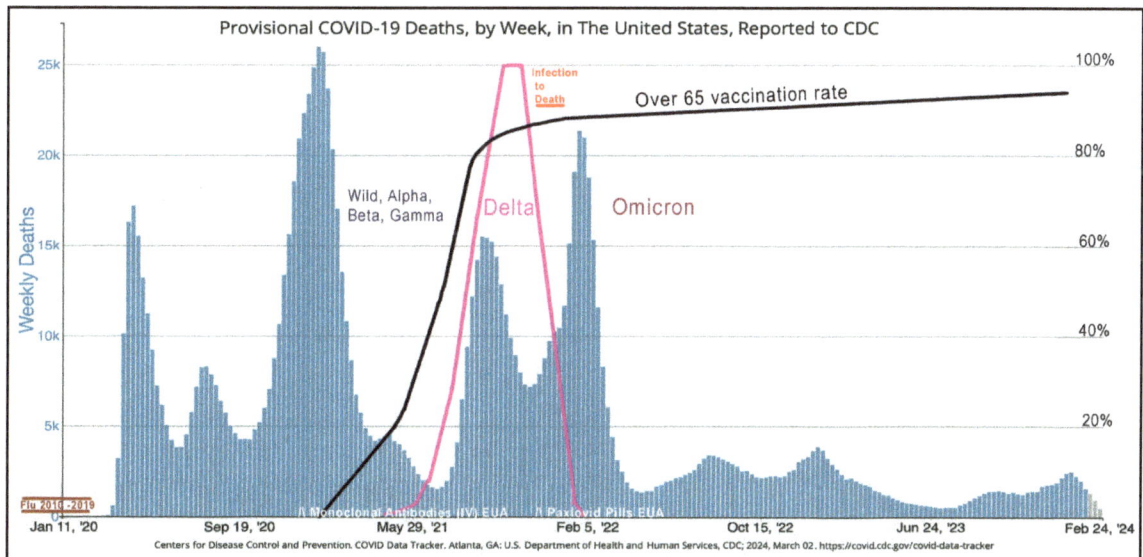

Death per week with primary variants and over 65 vax rate

# ACKNOWLEDGMENTS
## AND
# CREDITS

Thank you

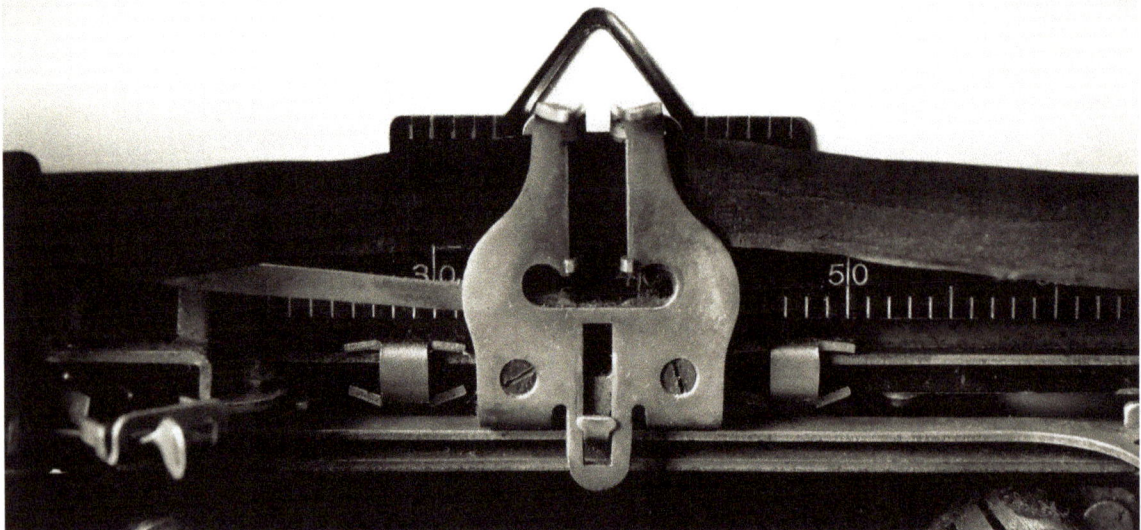

*Thank you word typed on a Vintage Typewriter.*

# Acknowledgments

The scientists and researchers who provided the documents cited by this book, people who have contributed to the our knowledge, are a critical part of the COVID-19 story. In particular, I recognize the hard and sincere work that went into the studies this book has critiqued — an unfortunate necessity to give the reader the limitations of the body of science — most were selected because they became newsworthy or were put in front of me for comment. Unfortunately there are many not cited that have much worse flaws. I was a public official in my younger years, making decisions that were not always liked. I applaud your efforts.

This book would not have been possible if not for the people who live in Ravalli County, who when through the pandemic in trying circumstances — business people who provided food, goods, and services — government officials and their staff — and those who weighed in with comments, complaints, and suggestions to our local officials.

While this book is not a memoir, it is a non-fiction narrative about people making decisions and people trying to influence decisions — including the author. It is impossible for a participant cannot stay out of a nonfiction narrative. Even the quiet journalist in the back of the room with a note pad affects what happens in the room. I didn't even try. I, like other people in this book made decisions, made public comment, and attempted to influence government decisions. So, I'm in it along with everyone else.

I believe everyone from Ravalli County in this book was being honest, stating what they believed to be true. There are always some that seemed to be making something up for some self serving purpose; for Ravalli County, for private persons, I left them out — they didn't seem to have much influence on outcomes. For public officials, the reader can be the judge.

I wish to acknowledge and thank the people of Ravalli County who made public comment, the officials who listened to them, the business owners who made decisions, whose statements and correspondence are reported in this narrative. The fact that we don't always agree takes nothing away from the heartfelt comments and decisions. As Skip Chisholm said while giving public input at nearly every Health Board meeting,speaking of the public process in this county as "a wonderful process."[1] As of 2023, he is now on the other side of the table as a member of the Board of Health.

## ART CREDITS

Art means all artwork: photographs, graphics, images, illustrations, and unretyped copies of public records.
Art is copyright James R. Olsen and others as stated in the following notes, exceptions, and additions:
• Graphics which have been licensed for this book. The restrictions for the license is carried forward to this book. This includes Shutterstock . (https://www.shutterstock.com) and Adobe Stock (https://stock.adobe.com) and other licensors. Their rights are reserved. Some licenses are restricted for use in this book.
• Graphics from Creative Commons or Public Domain have been included. Please read the relevant license or release before copying. Any additions or modifications by James R. Olsen made to a creative commons artwork are provided under the same terms. In most cases credit is required by the Creative Commons License.
• Artwork is copyrighted, all rights reserved, as stated on the copyright page, unless otherwise stated herein, in which case they are subject to copyrights and limitations of the copyright holder — the limitations stated herein are for guidance and are controlled by the rights and limitations of the source of the ma-

---

terial, the copyright holder, or creator, which **the reader is hereby notified that they need to check the source before any use.**

- Items marked Our World in Data are credited to a long list, so it is provide here:
  Edouard Mathieu, Hannah Ritchie, Lucas Rodés-Guirao, Cameron Appel, Charlie Giattino, Joe Hasell, Bobbie Macdonald, Saloni Dattani, Diana Beltekian, Esteban Ortiz-Ospina and Max Roser (2020) - "Coronavirus Pandemic (COVID-19)". Fatality rates: Carl Bergstrom, Bernadeta Dadonaite, Natalie Dean, Joel Hellewell, Jason Hendry, Adam Kucharski, Moritz Kraemer, and Eric Topol Published online at Our World In Data.org. Retrieved from: . https://ourworldindata.org/coronavirus [Online Resource] and https://ourworldindata.org/covid-deaths

- Art marked *Creative Commons* have been provided and are provided subject to the following licenses.

  CC-BY-4.0 Creative Commons Attribution, CC-BY-3.0 Creative Commons Attribution-Share Alike, CC-BY-2.0 Creative Commons Attribution-Share Alike, Released to Public Domain:
  https://creativecommons.org/licenses/by/4.0/legalcode • https://creativecommons.org/licenses/by/3.0/legalcode
  https://creativecommons.org/licenses/by/4.0/ • https://creativecommons.org/licenses/by/2.0/legalcode

- The graphics in the book usually show the Create Commons Icon, but this is for guidance only. See the link to the original art for the actual license.
- For art marked, "Work of the Government," see 17 US Code § 105: copyright protection is not available for works of the Federal Government with some exceptions.
- Art shown as the work of the State of Montana or its political subdivisions is provided as a copy the work under the provisions of Montana Code Annotated (MCA), Title 2, Chapter 6, including MCA § 2-6-1001, which implements the The Constitution of the State of Montana, Article II, Part II, Section 9. Right to Know. The copy in this book of a document that is property of the State of Montana.
- Citations, without a URL can be found in References Cited.
- Artwork without an embedded credit was modified to included an embedded citation and, if created commons, a created commons logo.

†† These were posted on Flickr by NIH or NIAID and marked as licensed under CC-BY-2.0, but based on the credit under which is listed, it *may* be a work of the government and subject to 17 US Code § 105.

## EFFECTS ADDED TO VARIOUS ART
Paper Effect Background, Lazy Dazie, Licensed through Adobe Stock. Glass Effect - Text, Marcel, Licensed through Adobe Stock.

## COVER
Flying Nurse. Illustration by Chritos Georhiou. Licensed through Shutterstock..
Protest Revolution Silhouette by Nosyrevy. Modified and clipped by James R. Olsen. Licensed through Shutterstock.
Background: Brown Linen Texture by Phatthanit. Modified by James R. Olsen. Llicensed through Shutterstock.

## FRONTMATTER
"Alternate Doorway: Neuroinvason by SARS-Co V-2." Illustration Doctor Disha Chauhan. Stretched and clipped by James R. Olsen. Cell Space 2020 for Artists, The Scrippps Research Institute,Creative Commons. https://ccsb.scripps.edu/new-ways-of-living/cellspace-2020-artists/.
"Human Made logo," Author's Guild.
"Abandoned," Fritz von Uhde, National Museum In Warsaw, James R,Olsen, Clipped Greyscale. Out of Copyright.
"Ms Corona," by Mary Byers. Photo by James R. Olsen. Work of the Author.
Untitled, Canadian Goose, Photo by James R. Olsen. Author's own work.

## PREFACE
"Ravalli Location Map." Illustration by David Benbennick. James R. Olsen art modification: Cropped. Public Domain. **Facing**
"Ravalli Map." Mapped by US Geological Survey. James R. Olsen art modification: Annotated. Work of the Federal Government. **Facing**.
"Bitterroot River. "Photo by James R. Olsen. Work of the Author. **Facing**.
"Illustration of Corona Virus." Illustration by Alissa Eckert, MSMI; Dan Higgins, MAMS Centers for Disease Control and Prevention (CDC). Work of the Federal Government. **Page 3**

## CHAPTER 1 - Nov 2019
"SARS-Co V-2 Genome Sequence." Illustration by James R. Olsen, Source of information Genome NIH Public Domain, Genome:NIH, Inset: Nucleotides: Nucleotides Bruas. James R. Olsen art modification: Nucleotides: Creative Commons 3.0. **Page 4**

## CHAPTER 2 - Dec 2019
"Dec. 2019 Timeline," by James R. Olsen. Work of the Author. **Page 5**
"COVID Air." Illustration by Anton Balazh, Corona Virus: Alissa Eckert, MSMI; Dan Higgins, MAMS Centers for Disease Control and Prevention (CDC). James R. Olsen art modification: Added covid icons (Illustration by Alissa Eckert, MSMI; Dan Higgins, MAMS Centers for Disease Control and Prevention (CDC). Work of the Federal Government.). Earth Flights: Licensed through Sutterstock. **Page 6**

## CHAPTER 3 - Jan 2020
"Jan. 2020 Timeline," by James R. Olsen. Work of the Author. **Page 8**
" Aerial view of Wuhan skyline and Yangtze river." Photo by sleepingpanda. James R. Olsen art modification: Stretched. Licensed through Sutterstock. **Page 8**
"The Table." Photo by James R. Olsen. Work of the Author. **Page 10**
"Bat Captured in Net." Photo by Sudarat89. Licensed through Sutterstock. **Page 12**
"Big Whites, Street Scene Shutdown in Wuhan." Photo by Robert Way. James R. Olsen art modification: Stretched. Licensed through Sutterstock. **Page 21**
"Diamond Princess Cases and Deaths." Graphic by Timothy W Russell, Joel Hellewell, Christopher I Jarvis, Kevin Van Zandvoort, Sam Abbott, Ruwan Ratnayake, CMMIDCOVID-19 working group,Stefan Flasche, Rosalind M Eggo, W John Edmunds, Adam J Kucharski • Rearranged by James R. Olsen. Creative Commons. https://www.eurosurveillance.org/content/10.2807/1560-7917. ES.2020.25.12.2000256 **Page 23**
""Hamilton City Map,RML Location." Graphic by authority of Dominic Farrenkopf, Mayor of City of Hamilton. James R. Olsen art modification: Annotated. Work of the City of Hamilton, MT. **Page 24**
"BSL-4 Laboratory at Hamilton,Montana." Photo by NIH/ Bryan Kercher. James R. Olsen art modification: Inset: Screen clip of NIHYou Tube BSL-4. Work of the Federal Government. **Page 25**
"World Map COVID-19 Cases January 2020." Map by Andrie Minsk. Colored and annotated by James R. Olsen. Licensed through Sutterstock. **Page 27**

## CHAPTER 4 - Feb 2020
"Feb. 2020 Timeline," by James R. Olsen. Work of the Author. **Page 28**
"Storm Clouds." Photo by STILL FX. Corona Virus: Alissa Eckert, MSMI; Dan Higgins, MAMS Centers for Disease Control and Prevention (CDC). James R. Olsen art modification: Stretch down. Clouds Licensed through Sutterstock. **Page 28**
"South Korea Travel Restrictions and Cases," Graphic by Gwee, Sylvia Xiao Wei, Pearleen Ee Yong Chua, Min Xian Wang, and Junxiong Pang. Expanded and moved annotations by James R. Olsen. BMC Infectious Diseases, "Impact of travel ban implementation on COVID-19 spread in Singapore, Taiwan, Hong Kong and South Korea during the early phase of the pandemic: a comparative study." Creative Commons. https://bmcinfectdis.biomedcentral.com/articles/10.1186/s12879-021-06449-1/figures/1 **Page 29**
"International Travel Restrictions Feb. 2020." Graphic by Our World in

DATA. James R. Olsen art modification: Modified graphics. Creative Commons.. https://ourworldindata.org/grapher/international-travel-covid?time=latest **Page 30**

"International Travel Restrictions Jan. 2020." Graphic by OUR WORLD IN DATA. James R. Olsen art modification: Modified graphics. Creative Commons. . https://ourworldindata.org/grapher/international-travel-covid?time=latest **Page 30**

"South Korea's Response to COVID-19." Graphic by Eunsun Jeong, Munire Hagose , Hyungul Jung, Moran Ki, Antoine Flahault. Creative Commons.. https://www.mdpi.com/1660-4601/17/24/9571 **Page 31**

"Vials:SARS-Co V-2CDCPCR Test Kit." Photo by CDC. James R. Olsen art modification: Clipped, cross-through, annotated. Work of the Federal Government. **Page 32**

"COVID Tests and Cases Feb 2020" " Graphic by OUR WORLD IN DATA. James R. Olsen art modification: Modified graphics. Creative Commons. https://ourworldindata.org/grapher/international-travel-covid?time=latest **Page 33**

"Colorized Electron Microscope Image of SARS-Cov-2." Electron Microscope Colorized by NIAID RML. Creative Commons[TT]. James R.Olsen modification: clipped https://www.flickr.com/photos/niaid/49531042877 **Page 35**

"SARS-Co V-2 Virus Illustration." Illustration by Coronavirus Structural Task Force - Thomas Splettstößer. James R. Olsen art modification: Combined 3 art piece and annotated. Creative Commons. https://insidecorona.net/media-resources/beautiful-illustrations/ **Page 36**

"Antibody Illustration." Illustration by National Human Genome Project colorized by Fvasconcellos. Annotated by James R. Olsen. Released to public domain. https://commons.wikimedia.org/w/index.php?title=File:Antibody.svg **Page 38**

"Immune Response to SARS-Co V-2," Credit Funk CD, Laferrière C and Ardakani A. Front. "A Snapshot of the Global Race for Vaccines Targeting SARS-Co V-2 and the COVID-19 Pandemic." Pharmacol. 11:937. doi: 10.3389/fphar.2020.00937. Art by Ian Dennis. Creative Commons. https://www.frontiersin.org/articles/10.3389/fphar.2020.00937/full **Page 38**

"SARS-Cov-2 Virus on Human Cell," Fort Detrick electron micrograph, colorized (Omigron) 7 Feb 2023. Image by NIAID,Creative Commons.[TT] https://www.flickr.com/photos/niaid/52675842325/ **Page 39**

"Immune Response to Virus Time Graph." Graphic by James R. Olsen. Work of the Author. **Pages 40-41**

"Flu Dose Human Challenge." Graphic by James R. Olsen. Work of the Author. **Page 42**

"Asymptomatic Infections with PCR Tests." Graphic by James R. Olsen. Work of the Author. **Page 42**

"Figure 4.1 - Fatalities Including COVID and other causes." Graphic by James R. Olsen. Work of the Author. **Page 44-45**

"COVID Risk Factors." Graphic by CDC. Work of the Federal Government. **Page 45**

Doctor Anthony Fauci." Photo by Courtesy: National Institute of Allergy and Infectious Diseases. James R. Olsen art modification: Changed Color Balance. Work of the Federal Government. **Page 46**

"Genetic manipulation concept." Illustration by natali_mis. Licensed through Adobe Stock. Stretched and Modified by James R. Olsen. **Pages 48-49**

"Wuhan Institute of Virology entrance." Photo by UReem2805. James R. Olsen art: Clipped. Creative Commons.. https://commons.wikimedia.org/wiki/File:Wuhan_Institute_of_Virology_main_entrance.jpg **Page 51**

"Fauci Briefing to President Trump." Video frame by James R. Olsen, NIH. James R. Olsen art modification: Clips and redraw of clip from video. Insert with Fauci and Trump at table. Made Public Domain by C-Span license. **Pages 54-55**

"COVID-19 Infection and Severe Outcomes by Bloodtype." Graphic by James R. Olsen. Work of the Author. **Page 57**

## CHAPTER 5 - Epidemiology

"Mapping a Case." Illustration by pandpstockk001. Licensed through Adobe Stock. **Page 58**

"Buruli Ulcer." Photo by Johnson PDR, Stinear T, Small PLC, Pluschke G,Merritt RW, Portaels F, et al. (2005) Buruli Ulcer (M. ulcerans Infection): New Insights, New Hope for Disease Control. PLoS Med 2(4): e108.. Creative Commons. https://journals.plos.org/plosmedicine/article?id=10.1371/journal.pmed.0020108 **Page 59**

"Buruli Ulcer Distribution." Graphic by Johnson PDR, Stinear T, Small PLC, Pluschke G,Merritt RW, Portaels F, et al. (2005) Buruli Ulcer (M. ulcerans Infection): New Insights, New Hope for Disease Control. PLoS Med 2(4): e108.. Creative Commons. https://journals.plos.org/

plosmedicine/article?id=10.1371/journal.pmed.0020108 **Page 60**

## CHAPTER 6 - Mar 2020

"Mar. 2020 Timeline," by James R. Olsen. Work of the Author. **Page 62**

"Fatalities & State by State Lockdowns, March 2020." Graphic and Illustration by OUR WORLD IN DATA. James R. Olsen art modification and combined art pieces: Stretched, Added Insert graph from OUR WORLD IN DATA. Creative Commons. Insert 4 Horsemen TPYXA and Lock Icons svyatoslavius: Licensed through Shutterstock. https://ourworldindata.org/covid-deaths **Page 62**

"Mary Byers Sewing Mask." Photo by James R. Olsen. Hand model Mary Byers. Work of the Author. **Page 64**

"Doctor Bloom." Photo by NIAID. Work of the Federal Government. **Page 65**

"Hydroxychloroquine on Twitter." Graphic by Alessandro R. Marcon and Timothy Caulfield. Creative Commons.. https://journals.uic.edu/ojs/index.php/fm/article/view/11707/10223 **Page 68**

"HCQ Reduce Risk of Symptoms." Schilling et al. "Evaluation of hydroxychloroquine or chloroquine for the prevention of COVID-19 (COPCOV): A double-blind, randomised, placebo-controlled trial" Creative Commons. https://journals.plos.org/plosmedicine/article?id=10.1371/journal.pmed.1004428 **Page 69**

"RT-PRC Chemical Process Illustration." Illustration by Lokesh Thimmana. James R. Olsen art modification: Removed paragraph numbers, added annotation. Creative Commons.. https://commons.wikimedia.org/wiki/File:RT_PCR_Model.jpg **Page 72**

"Figure 5.1 — COVID Tests." Credit Funk CD, Laferrière C and Ardakani A. Front. "A Snapshot of the Global Race for Vaccines Targeting SARS-Co V-2 and the COVID-19 Pandemic." Pharmacol. 11:937. doi: 10.3389/fphar.2020.00937. Art by Ian Dennis. James R. Olsen annotated for tests. Creative Commons.. https://www.frontiersin.org/articles/10.3389/fphar.2020.00937/full **Page 73**

"Figure 5.2 — COVID Feedback Delay. Illustrated by James R. Olsen. Work of the Author. **Page 75**

"Figure 5.3 —United States,South Korea, Cases and Tests Per Capita." Graphic by OUR WORLD IN DATA. James R. Olsen art modification: Combined Two OUR WORLD IN DATA Graphics and scaled to match. Creative Commons.. https://ourworldindata.org/covid-deaths **Page 75**

"Figure 5.4 — Confirmed COVID-19 Cases through March 16, 2020." Graphic by OUR WORLD IN DATA. James R. Olsen art modification: Added Annotation from OUR WORLD IN DATA, Added Smoothed Line. Creative Commons.. https://ourworldindata.org/covid-deaths **Page 76**

"Health Officer Authority 2020." Copy of Laws by James R. Olsen. Original documents owned by the State of Montana. Work of the Author. **Page 79**

"Fatality Rate, Cases, and Responses." Graphic by OUR WORLD IN DATA. James R. Olsen art modification: Combined two OUR WORLD IN DATA Graphics added highlights for shutdown periods, annotated. Creative Commons https://ourworldindata.org/covid-deaths **Page 83**

"Stock Market Indexes Jan. to Mar. 2020," . " Illustration by James R. Olsen, Work of the Author. **Page 84**

"Montana Botanicals Tinctures." Photo by James R. Olsen. Label by Hillery Daily. Photograph with Permission. Work of the Author. **Page 85**

## CHAPTER 7 - Apr 2020

"Apr. 2020 Timeline," by James R. Olsen. Work of the Author. **Page 86**

"Ms Corona," by Mary Byers. Photo by James R. Olsen, . Work of the Author. **Page 86**

"Exhausted." Photo by DC Studio + Model Release. James R. Olsen art modification: Clipped. Licensed through Sutterstock. **Page 86**

"Intubation." Photo by Tyler Olsen + Model and Property Release. James R. Olsen art modification: Clipped and Textured. Licensed through Sutterstock. **Page 88**

"Aware Prone Position." Photo by Andrew Williams. James R. Olsen art modification: Clipped Textured. Work of the Federal Government. **Page 88**

"Survival Rate Aware Prone Position." Graphic by James R. Olsen. Work of the Author. **Page 89**

"Improved ARDS Treatment Over Time." Graphic by Eric O. Yeates, Jeffry Nahmias, Justine Chinn, Brittany Sullivan, Stephen Stopenski, Alpesh N., Ninh T. Nguyen. James R. Olsen art modification: Added % Intubated line, Annotated. Creative Commons.. https://journals.plos.org/plosone/article?id=10.1371/journal.pone.0253767 **Page 89**

"Cumulative Fatalities Through Apr 8." Graphic by OUR WORLD IN DATA. James R. Olsen art modification: Added Wuhan and New York City, background color. Creative Commons. https://ourworldindata.org/

nio Guilem, Mnt House: Vacclav, Scale: SciPro, Licensed through Shutterstock. Text and Layout Work of the Author. **Pages 175-176**

**CHAPTER 13 - Oct 2020**
"Oct. 2020 Timeline," by James R. Olsen. Work of the Author. **Page 178**
"Tip of the Iceberg and Titanic." Photo of Iceberg by Alone. Drawing of Titanic by Aman Artist. James R. Olsen Combined and distorted Titanic. Licensed through Sutterstock. **Page 178**
"Active Cases mid-Oct," Graphic by James R. Olsen. Work of the Author. **Page 181**
"Commissioners Statement on COVID." Copy Montana Official Letter by Ravalli County Board of Commissioners. Work of Ravalli County, original owned by the State of Montana. **Page 184**
"Cumulative Deaths per Capita Jan-Oct2020." Graphic by OUR WORLD IN DATA. James R. Olsen art modification: Modified graphics. Creative Commons.. https://ourworldindata.org/covid-deaths **Page 187**

**CHAPTER 14 - Nov 2020**
"Nov. 2020 Timeline," by James R. Olsen. Work of the Author **Page 188**
"Moment of truth on a vintage typewriter.." Photo by Nokuro. Licensed through Sutterstock. **Page 188**
"Electoral Collage 2020." Graphic by Star Boy X. James R. Olsen art modification: Added Visual Key Icons, Annotated. Creative Commons. **Page 191**
"Dough Boy Memorial. Photo by James R. Olsen. Work of the Author. **Page 192**
"Active Cases Nov," Graphic by James R. Olsen. Work of the Author. **Page 193**
"Jim Give Public Comment," Video Clip from Public Meeting, Public Record,Ravalli County, MT. Copy of Original created and owned by the State of Montana. James R. Olsen clip and texturized. **Page 194**
"Handout Health Board Meeting, James R. Olsen, Mask Mandate vs Fatality 11/13/20." Graphic by James R. Olsen. Work of the Author. Submitted to Public Record. **Page 197**
"Health Board Handout Deaths/Capita vs Economic 2nd Qtr 2020." Graphic by OUR WORLD IN DATA. James R. Olsen art modification: Added graphics and data. Creative Commons. Submitted for the public record. https://ourworldindata.org/grapher/q2-gdp-growth-vs-confirmed-deaths-due-to-covid-19-per-million-people **Page 198**
"Household Response to Economic Stimulus." Illustration by James R. Olsen. Work of the Author. **Page 199**

**CHAPTER 15 - Dec 2020**
"Dec. 2020 Timeline," by James R. Olsen. Work of the Author **Page 200**
"Stop-the-Steal March Helena." Photo by Lyonstock. James R. Olsen art modification: Clipped and Textured. Licensed through Sutterstock. **Page 200**
"Bioweapon Sign at Stop-the-Steal Rally." Photo by The Brownfowl Collection. Licensed through Sutterstock. **Page 201**
Top Causes of Death and Excess Causes 2020." Graphic by James R. Olsen. Work of the Author. **Page 203**
"Excess Deaths 2020[13]. " Graphic by OUR WORLD IN DATA. James R. Olsen art modification: Stretched, Added excess deaths from CDC data. Creative Commons. **Page 204**
"Excess Deaths Four Countries." Graphic by OUR WORLD IN DATA. James R. Olsen art modification: Modified graphics. Creative Commons. https://ourworldindata.org/covid-deaths **Page 205**
"Active Cases Dec," Graphic by James R. Olsen. Work of the Author.
Cumulative Fatalities per capita 2020" Graphic by OUR WORLD IN DATA. Creative Commons. James R. Olsen added excess death line. . https://ourworldindata.org/covid-deaths **Page 207**
"Ravalli County Cases." Graphic by James R. Olsen. Work of the Author. **Page 208**
"Waning Gibbous Moon." Combined art. Moon: Abanoub Banough. Clipped cigar: Zaitsev Maksym, Smoke: Tigao M Nunes. All Licensed through Sutterstock. **Page 210**
"Ravalli County COVID Fatalities." Graphic by James R. Olsen. Work of the Author. Insert: "Age-Adjusted Death Rates Montana and US Residents 2011-2020."Graph by MT DPHHS. "Vital Statistics 2020." by Todd M. Koch & Matthew Ringel, Februray 2022, Fig. 7, p. 13. Modified layout, Red annotation, and added American Indian and gender lines by James R. Olsen. Copy of Original created by the State of Montana. **Page 211**

**CHAPTER 16 - Jan. 6, 2020**
"Jan. 2020 Timeline. Jan 6, Attack on the Capitol." Work of the Author **Page 212**

"Capitol Riot." Photo by Tyler Merbler. James R. Olsen art modification: Duplicated part of picture "hat and cape" to obscure identity. Creative Commons.. https://flickr.com/photos/37527185@N05/50821579347 **Page 212**
"Noose at Capitol Building." Photo by Tyler Merbler. Creative Commons https://www.flickr.com/photos/37527185@N05/50826223403/in/photostream/ **Page 217**

**CHAPTER 17 - Epilogue**
""Excess Deaths 2020 to mid-2023." Work of the Author, Data from the Economist Magazine, **Page 218**
"Death per week with primary variants and over 65 vax rate " Graph, CDC: https://covid.cdc.gov/covid-data-tracker#trends_weeklydeaths_select_00 Art Work of Federal Government. James R. Olsen modifications from data from OneWorldData, USAFacts., "What's the nations progress on vaccines?" May 2023, Dukee, Alison, **"Stunning Vaccine Stat: 98.5% Of U.S. Seniors Have Had Shot."** *Forbes*, November 1, 2021. **Page 219**

## ACKNOWLEDGMENTS
"Thank you typed on vintage typewriter" Photo by JOKE_PHA-TRAPONG. Annotated by James R. Olsen. Licensed through Sutterstock. **Page 220**

### END NOTES
"Notes on Wall," Photo by OliverFoerstener. Licensed throiugh Adobe Stock. Clipped, modified, change to Sepia by James R. Olsen **Page 226**

### Bibliography
"Paper Stack, Africa Studio. Textured by James R. Olsen. Licensed through Sutterstock. **Page 256**

## INDEX
"Index." Photo by Lane V. Erickson. Clipped and Stretched by James R. Olsen. Licensed through Sutterstock. **Page 258**
"Ms Corona," by Mary Byers. Photo by James R. Olsen, . Work of the Author. **Page 270**

## ABOUT THE AUTHOR
"Photo of Author." Photo by Mary Byers. Used with Permission. **Page 271**

Photo by OliverFoerstner Clipped, modified, sepia James R. Olsen

*Notes on Wall*

# ENDNOTES

Endnotes that are ~ Arial Narrow ~ cite an entry in **References Cited**. News and social media items are dated in these endnotes. Medical journal citation dates can be found in References Cited. Added commentary and casual references are in ~ Times New Roman ~. Note British English spelling is used in some citations.

➤ The Arial Narrow citiations are in *References Cited* (2,500 of them) in Book 3 or, more conveniently, **https://www.jamesrolsen.com/covid-wars-references** (hyperlinks included).

## PREFACE
**Pages 1 - 5**

1 Catton, *History of the Civil War*, Vol, 1, *The Coming Fury*, p. 12.
2 Gorbalenya, "Severe Acute Respiratory Syndrome-Related Coronavirus: The Species and Its Viruses – a Statement of the Coronavirus Study Group."
Maurier, "A Complete Protocol for Whole-Genome Sequencing of Virus from Clinical Samples: Application to Coronavirus OC43."
3 Murray, *Medical Microbiology*, p. 306. The most common cause of the common cold is a rhinovirus. A minority of are caused by coronaviruses. Estimates range from 10% to 30% of common colds being cause the a coronavirus, of which there are several types that infect humans: CDC, "Common Human Coronaviruses," Centers for Disease Control and Prevention, Feb. 13, 2020.
4 Middle Eastern Syndrome Corona Virus, with most infections from camels, through close contact between humans happens. Identified in 2012. WHO, "Middle East Respiratory Syndrome Coronavirus (MERS-COV)," World Health Organization, Aug. 5, 2022.
5 Severe acute respiratory syndrome, or SARS-CoV-1, is the first of the novel Coronavirus outbreaks in 2003. There were 8,422 cases in South China, of which 11% died. Wikipedia, "SARS," https://en.wikipedia.org/wiki/SARS.
6 Vazquez, "Trump Again Defends Use of the Term 'China Virus'," *CNN*;
Little, "Trump's 'Chinese' Virus Is Part of a Long History of Blaming Other Countries for Disease," *Time Magazine*, Mar. 20. 2020.

*Epoch Times* Editors, "Statement on Special Edition on Coronavirus Pandemic," *Epoch Times*, May 31, 2020.
Few governments are immune to political delays and misinformation.
7 The People's Republic of China (PRC) is a one party state ruled by the Chinese Communist Party (CCP), founded by Mao Zedong and his cohorts after prevailing in a long civil war that started before World War II and ended shortly after the world war ended. The Nationalist Chinese Party, whose most well known leader was Chiang Kai-shek, lost the civil war. He and his followers fled to the island of Taiwan (known to Europeans up to that time as Formosa — a designation given to the island by Portuguese explorers). Chiang Kai-shek's Republic of China became synonymous with Taiwan (and small close-by islands). The island was ruled under martial law until 1992. While it is a de-facto nation-state, it's status is a source of tension. The CCP claims sovereignty over Taiwan, though Taiwan is now a self governing Constitutional Republic.
8 Wllace-Wells, "Dr. Fauci Looks Back: 'Something Clearly Went Wrong'." *New York Times*, April 24, 2023.
9 Ibid.
10 American Battlefield Trust, Civil War Casualties: https://www.battlefields.org/learn/articles/civil-war-casualties
11 Thorning, "No Guns Allowed," *Ravalli Republic*, Mar. 3, 1995.
Bethel, Martha A. "Terror in Montana," *New York Times*, Jul. 20, 1995, Page A-2.
12 Schwatz, "Embattled CDC Rethinks Response to the Pandemic," *Wall Street Journal*, p. A3, Feb. 9, 2022.

## 1. Darwin's Dream - Nov. 2019

1 Megginson, "Lessons from Europe for American Business."
2 Xu, "The Analysis of Six Patients With Severe Pneumonia Caused By Unknown Viruses.";
Zhou,"Identification of novel bat coronaviruses sheds light on the evolutionary origins of SARS-CoV-2 and related viruses."
Huang,"Novel Virus Discovery in Bat and the Exploration of Receptor of Bat Coronavirus HKU9.";
Huang , "A Bat-Derived Putative Cross-Family Recombinant Coronavirus with a Reovirus Gene."
3 IMDb, "Run Silent, Run Deep," IMDb, 1958.
4 Daszak, "Understanding the Risk fo Bat Coronavirus Emergence," NIAID Grant 5R01AI110964-05, Interim Report. May 3, 2021. pp. 5.
5 Worobey, "Dissecting the early COVID-19 cases in Wuhan.";
Nsoesi., "Analysis of Hospital Traffic ...in Wuhan... Indicates Early Disease... Fall of 2019." Though this has methodology issues: See Chen, "Comment on 'Analysis of... ";

Allam, "The First 50 days of COVID-19: A Detailed Chronological Timeline & Extensive Review of Literature Documenting the Pandemic."
6 Fongaro, "The Presence of SARS-COV-2 RNA in Human Sewage in Santa Catarina, Brazil, November 2019.";
Xu, "The Analysis of Six Patients With Severe Pneumonia Caused By Unknown Viruses [from 2012].";
Montomoli, "Timeline of SARS-COV-2 Spread in Italy: Results from an Independent Serological Retesting."; The test used has been questioned and the results may not be valid: Parodi, "Dispute over Italian coronavirus study shows challenges of probing origins," *Reuters*, July.23, 2021
Chan, Viral: "The Search for the Origin of Covid-19," p. 14. It is dismissed as "only six were sick and exposed to high dose," but if was as benign for most people as COVID-19, and others were exposed but to less of a dose, it is possible other miners were spreading it around without knowing it;
Quammen, *Breathless the Scientific Race to Defeat a Deadly Virus*, pp. 230-290 for a more detailed look;

Xu, "The Analysis of Six Patients With Severe Pneumonia Caused By Unknown Viruses.";

## 2. Flying Covid Air - Dec. 2019

1 Dwight D. Eisenhower, speech at Northwestern University, address to the Second Assembly of the World Council of Churches, 1954. He attributed the quote to the prior president of the university.
2 International Trade Association, "International Visitation to and from the United States," Used 2019 entries 79 million/365 (Accessed October 19, 2023) https://www.trade.gov/sites/default/files/2021-03/Fact%20Sheet%20Internation-al%20Visitation%20FINAL.pdf
3 Huang, "Clinical Features of Patients Infected with 2019 Novel Coronavirus in Wuhan, China.";
Page, "How It All Started: China's Early Coronavirus Missteps," Wall Street Journal, Mar. 6, 2020.
4 Daley, "The original Sars virus disappeared – here's why coronavirus won't do the same," The Conversation, June 5, 2020.
5 Gomez-Barroso, "Spread of Ebola Virus Disease Based on the Density of Roads in West Africa.";

## 3. Trying Times in Wuhan - Jan. 2020

1 Wang, In the same Breath, HBO, movie, minute 40.
2 Willoughby, The Copper Kings;
Malone, The Battle for Butte: Mining and Politics on the Northern Frontier, 1864-1906, 180-181.;
3 Baumler, "Tammany, Marcus Daly's Famous Horse," Great Falls Tribune, Oct. 7, 2015.
4 Montana Human Rights Network, https://mhrn.org
5 Baucus, "First Graders Celebrate Martin Luther King Day," Ravalli Republic, Jan. 22, 2020.
6 Bitterroot College, Spring 2020, College, Continuing Ed, FabLab & Workforce Training (Pamphlet), p. 19;
Olsen, "Effective Grass Roots Activism."
7 Gandhi, Non-Violent Resistance (Satyagraha).
8 Young, An Easy Burden: a Memoir of the Civil Rights Movement, p. 58.
9 Matthew 5:44 KJV.
10 King, "'I Have a Dream' (Video)."
11 Dixon, "The Morality of Anti-Abortion Civil Disobedience" makes a case, using the criteria of non-violent civil rights movement.
Gandhi, The Story of My Experiments with Truth, p. 66;
Andrew Young noting, "Reading Gandhi, I understood that the Gospel could be applied to a political situation" (Young, An Easy Burden: a Memoir of the Civil Rights Movement, p. 58);
Julia Butterfly Hill, The Legacy of Luna: the Story of a Tree, a Woman, and the Struggle to Save the Redwoods, p. 133.
12 Zuo, "Hong Kong Takes Emergency Steps as Mystery 'Pneumonia' Infects 27 in Wuhan," South China Morning Post, Dec. 31, 2019;
Huang, "WHO in Touch with Beijing after Mystery Viral Pneumonia Outbreak," South China Morning Post, Jan.1, 2020;
Rimmer, "The Race to Patent the SARS Virus ..."
13 Zuo, "Hong Kong Takes Emergency Steps as Mystery 'Pneumonia' Infects 27 in Wuhan," South China Morning Post, Dec. 31, 2019.
14 Qin, "China Reports First Death From New Virus," New York Times. Jan. 10, 2020
15 Wee, "China Grapples With Mystery Pneumonia-Like Illness," New York Times, Jan. 6, 2020;
Nectar Gan, "China Pneumonia Outbreak: It's Not SARS,

Zhou, "Addendum: A pneumonia outbreak associated with a new coronavirus of probable bat origin."
7 Berche, "History of Measles."

Sprecher, "Understanding the Key to Outbreak Control — Sudan Virus Disease in Uganda."
Day, "Covid-19: Four Fifths of Cases Are Asymptomatic, China Figures Indicate."
6 Lerer, "Debate Night: The 'on Politics' Breakdown," New York Times, Dec. 20, 2019;
Cochrane, "2nd Senate Republican Questions Impartiality of Impeachment Trial," New York Times, Dec. 31, 2019.
7 Basavaraju, "Serologic testing of U.S. blood donations to identify SARS-CoV-2-reactive antibodies: December 2019-January 2020"; The method used for the assay. ELISA, can have false positvies reported here as 10% for a differnt disease: Ruitenberg, "Reliability of the enzyme-linked immunosorbent assay (ELISA) for the serodiagnosis of Trichinella spiralis infections in conventionally raised pigs."
Althoff, "Antibodies to Severe Acute Respiratory Syndrome Coronavirus 2 (SARS-COV-2) in All of US Research Program Participants, p. 2.

so What Is It?," CNN, Jan. 7, 2020 - 2 AM.
16 Su, "How Taiwan, a Non-WHO Member, Takes Actions in Response to COVID-19."
17 WHO, "Novel Coronavirus – China | Disease Outbreak News," WHO Website, Jan. 12, 2020;
Tan, "Singapore's Pandemic Preparedness: An Overview of the First Wave of Covid-19.";
Guardian Staff, "Coronavirus: What Airport Measures Are in Place to Detect Sick Passengers?," The Guardian, Jan. 18, 2020;
Culver, "The Wuhan Market at the Center of the Outbreak Remains Tightly Sealed," CNN, Jan. 22, 2020;
Li, Facts and Analysis: Canvassing Covid-19 Responses;
WHO, "Updated WHO Advice for International Traffic in Relation to the Outbreak of the Novel Coronavirus 2019-NCoV," WHO Website, Jan. 27, 2020.
18 Though more Chinese companies are using Amazon as a store front, with the blessing of China's government: Fu, "China Courts US Consumers Through a Familiar Storefront in Amazon," Epoch Times, Jul. 4, 2021.
19 Zhou, "A Pneumonia Outbreak Associated with a New Coronavirus of Probable Bat Origin.";
Huang, "Clinical Features of Patients Infected with 2019 Novel Coronavirus in Wuhan, China," p. 500;
Farrar, Spike: The virus vs. the people - the inside story, p. 6.
20 Wang, "Serological Evidence of Bat SARS-Related Coronavirus Infection in Humans, China."
Lu, "Genomic Characterisation and Epidemiology of 2019 Novel Coronavirus: Implications ....";
Cohen, "Mining Coronavirus Genomes for Clues to the Outbreak's Origins," Science Magazine, Jan. 31, 2020;
Racaniello, Vincent (Interviewer), Edward Holmes (Interviewee). "TWiV 1019: Eddie Holmes on SARS-CoV-2 origins," (Vide) MicrobeTV, Jun. 24, 2023;
Wang,. "Bats, civets and the emergence of SARS.";
Gostin, "The Origins of Covid-19 — Why It Matters (and Why It Doesn't)".
21 Al Jazeera, "3 Days that Stopped the World," (Video). Jan. 18, 2020. Minute 6.
22 Farrar, Spike: The virus vs. the people - the inside story, pp. 16-22;

Quammen, *Breathless*, pp. 18-19.

23 Holmes, Edward (Posted by) and On behalf of: Yongzhen Yongzhen, "Novel 2019 Coronavirus Genome," Virological (Virological (posted on GenBank, accession MN908947), Jan. 10, 2020. (May have been the 11ᵗʰ in Australia)
   Cohan, "Chinese Researchers Reveal Draft Genome of Virus Implicated in Wuhan ... Outbreak," *Science Insider*, Jan. 11, 2020;
   Campbell, "Zhang Yongzhen Speaks out about Controversies around His Work," *Time*, Aug. 25, 2020;
   AP, "China Suspends Social Media Accounts of Covid Policy Critics," *AP NEWS*, Jan. 7, 2023;
   Schnirring, "China Releases Genetic Data on New Coronavirus, Now Deadly.";
   AP, "China Delayed Releasing Coronavirus Info, Frustrating Who," *AP News*, Jun. 11, 2021.

24 Page, "How It All Started: China's Early Coronavirus Missteps," *Wall Street Journal*, Mar. 6, 2020.

25 World Health Organization (WHO), "Preliminary Investigations Conducted..." Twitter, Jan. 14, 2020;
   Gavis, "WHO Haunted by Old Tweet Saying China Found No Human Transmission...," *New York Post*, March 20, 2020

26 Page, "How It All Started: China's Early Coronavirus Missteps," *Wall Street Journal*, Mar. 6, 2020;
   Lee, "Chinese Citizen Journalist Jailed for Reporting Virus Outbreak Continues Hunger Strike," *Epoch Times*, Aug. 5, 2021;
   Victims of Communism Editorial Staff, "The Coronavirus Coverup," Victims of Communism, Apr. 10, 2021. This is an organization was founded by Congress in 1993. Very conservative outlook, but generally accurate.

27 Calvert, "China, the Who and the Power Grab That Fuelled a Pandemic," *The Sunday London Times*, Aug. 14, 2021.

28 Wang, "Human-to-Human Transmission Confirmed in China Coronavirus," *AP News*, Jan. 19, 2020.
   Al Jazeera, "3 Days that Stopped the World," (Video) YouTube. Jan. 18, 2021. Min. 6:50.

29 Rabin, "First Patient With Wuhan Coronavirus Is Identified in the US," *New York Times*, Jan. 21, 2020;
   Stone, "1st US Case of Coronavirus Confirmed in Washington State," *National Public Radio (NPR)*, Jan. 22, 2020;
   CDC Media Relations, "First Travel-Related Case of 2019 Novel Coronavirus Detected in United States," CDC, Jan. 21, 2020;
   Holshue, "First Case of 2019 Novel Coronavirus in the United States."

30 Gibney, *Totally Under Control*, Minute 16.
   Frontline PBS. "Coronavirus Pandemic (full documentary)." PBS. April 21, 2020;
   Gibney, *Totally Under Control*, Minute 16.

31 Han, "Digestive Symptoms in COVID-19 Patients with Mild Disease...: Clinical Presentation, Stool Viral RNA Testing, ....";

32 Tian, "Review Article: Gastrointestinal Features in COVID-19 and the Possibility of Faecal Transmission."

33 Wu, "Coronavirus Disease 19 with Gastrointestinal Symptoms as Initial Manifestations: A Case Report."

34 Cohan, "1st novel coronavirus case detected in US; Expert says China outbreak is 'tip of the iceberg'." *Boston Herald*, Jan. 21, 2020

35 Kernen, Joe and President Donald Trump. "Interview: Joe Kernen Interviews Donald Trump on CNBC From Davos," *CNBC*, Jan. 22, 2020.

36 Ghinai, "First Known Person-to-Person Transmission of ... (SARS-CoV-2) in the USA."

37 Gostic, "Estimated Effectiveness of Symptom and Risk Screening to Prevent the Spread of COVID-19."

38 Abutaleb, "Trump Administration Announces Mandatory Quarantines ...," *Washington Post*, Jan. 31, 2020;
   Corkery, "Trump Administration Restricts Entry Into US From China," *New York Times*, Jan. 31, 2020.

39 Brockell, "Yes, there really was a 'Typhoid Mary,' an asymptomatic carrier who infected her patrons," *Washington Post*, Mar.17, 2020.

40 Azar, "Coronavirus News Conference HHS," (Video). C-SPAN. January 28, 2020.

41 Trump, President. "Proclamation on Suspension of Entry as Immigrants and Nonimmigrants of Persons who Pose a Risk of Transmitting 2019 Novel Coronavirus," White House Archives. Jan 31, 2020.

42 Kadlec, "U.S. Department of Health and Human Services (HHS) Assistant Secretary for Preparedness and Response (ASPR) Indicent Response Framework," ASPR HHS, Apr. 3, 2019;
   Duke, *National Incident Management System*, 3rd Edition. Oct. 2017;
   FEMA. 2019 *National Threat and Hazard Identification and Risk Assessment (THIRA) Overview and Methodology.*

43 Executive Office of the President of the United States. *Playbook for Early Response to High-Consequence Emerging Infectious Disease Threats and Biological Incidents;*
   Knight, "Obama team left pandemic playbook for Trump administration, officials confirm," *PBS News Hour*. May 15, 2020;
   Gibney, *Totally Under Control*, Minute 33.

44 Schuchat, "Community Mitigation Guidelines to Prevent Pandemic Influenza — United States, 2017." CDC. Apr. 17, 2017.

45 Kadlec, "Crimson Contagion 2019 Functional Excercise After Action Report," ASPR HHS, (by way of FOIA Response) Dec.9, 2019.
   Kadlec, Robert. DRAFT (FOUO) "Crimson Contagion 2019 Functional Excercise After Action Report," ASPR HHS, (by way of New York Times). 2019.

46 MAGA —Make America Great Again, Trump's campaign slogan.

47 Abutaleb, "The U.S. was beset by denial and dysfunction as the coronavirus raged," *Washington Post*, Apr. 4, 2020;
   Gibney, *Totally Under Control.*
   Bright, "Retaliation for Whistleblowing" Office of Special Counsel: OSC Form-14, May 5, 2020;

48 Bright, "Retaliation for Whistleblowing" Office of Special Counsel: OSC Form-14, May 5, 2020. pp. 14-16.
   White House (President Obama), "FACT SHEET: Progress in Our Ebola Response at Home and Abroad," Press Release. Feb. 11, 2015.

49 Kadlec, "Crimson Contagion 2019 Functional Excercise After Action Report," ASPR (Administration of Perparedness and Response) HHS, Dec.9, 2019, p. iv. para. 3.c;
   Gibney, *Totally Under Control*, Minute 19;
   Bright, "Retaliation for Whistleblowing" Office of Special Counsel: OSC Form-14, May 5, 2020, p. 47.

50 HHS Press Office, "Secretary Azar Declares Public Health Emergency for United States for 2019 Novel Coronavirus," US HHS, Jan. 31, 2020.

51 Kernen, Joe and President Donald Trump. "Interview: Joe Kernen Interviews Donald Trump on CNBC From Davos," *CNBC*, Jan. 22, 2020.

52 Wang, "Xi Jinping Warns That China's Coronavirus Outbreak Must Be Taken Seriously," *The Diplomat*. Jan. 21 2020. The cases being reported are being signficantly under-reported from all offical Chinese sources including Chairman Xi.

53 Andersen, *Fairy Tales of Hans Christian Andersen*. Project Gutenenberg. Nov. 8, 2008. "The Emperor's New Suit."

54 Fauci. "HIH, Dr. Fauci Emails, Nov 2019 through Jan 2020 Informed Consent Action Network in response to FOIA Request # 54106." NIH FOIA Library, pp. NIH 002820, NIH-002839.

55 IMDb "Red Dawn," IMDb.,1984.

56 Gibney, *Totally Under Control*, Minute 17.

57 Gibney, *Totally Under Control*, Minute 18.

58 Haberman, "Trade Adviser Warned White House in January of Risks of a Pandemic," *New York Times*, Apr. 6, 2020;

Peter Navarro is a prolific writer with several books on China including: *The Coming China Wars*, 2006. *Death by China: Confronting the Dragon - A Global Call to Action*. 2011

59 CDC, Global Disease Detection (GDD) Operations Center. (Accessed 10/13/2023) https://www.cdc.gov/globalhealth/healthprotection/gddopscenter/

Gill, "What is Open-Source Intelligence," SAS, Feb. 23, 2023.

60 Hegarty, "The Chinese Doctor Who Tried to Warn Others About Coronavirus," BBC News. Feb.ry 6, 2020

Wang, *In the same breath*, HBO, movie, Minute 4.

61 Hegarty, "The Chinese Doctor Who Tried to Warn Others About Coronavirus," *BBC News*. Feb. 6, 2020;

Wikipedia, Li Wenliang, https://en.wikipedia.org/wiki/Li_Wenliang.

62 Zuo, "Hong Kong Takes Emergency Steps as Mystery 'Pneumonia' Infects 27 in Wuhan," *South China Morning Post*. Dec. 31, 2019.

63 Wang, *In the same Breath*, HBO, movie.

64 Reed, Christopher. "A Conversation With Nanfu Wang (In The Same Breath)," *Hammer to Nail*. August 19, 2021;

Wikipedia, "Nanfu Wang," https://en.wikipedia.org/wiki/Nanfu_Wang

65 Al Jazeera, "3 Days that Stopped the World" (Video), Jan. 18, 2021.

66 Al Jazeera. "China expands coronavirus outbreak lockdown to 56 million people," Imbeded video of Hospital. Jan. 25, 2020.

67 Page, "How It All Started: China's Early Coronavirus Missteps," *Wall Street Journal*. Mar. 6, 2020

68 Fang, *Wuhan Diary*;

Gao "In Depth: How Wuhan lost the fight to contain the coronavirus," CAIXIN. Feb. 6, 2020.

69 Wikipedia: Sina Weibo. https://en.wikipedia.org/wiki/Sina_Weibo

70 Su, "Two Months into Coronavirus Lockdown, Her Online Diary Is a Window into Life and Death in Wuhan," *Los Angeles Times*, Mar. 21, 2020;

Most Chinese speak Mandarin, the official language. Cantonese, is spoken in Hong Kong and surrounding area. The same written language can be read by all literate Chinese.

71 Fang, *Wuhan Diary*, Translator Michael Berry's Afterword, p. 362.

72 Su, "Two Months into Coronavirus Lockdown, Her Online Diary ..," *Los Angeles Times*, Mar. 21, 2020.

73 Fang, *Wuhan Diary*, p. 3–4.

74 Gong, "The One Who Whistled." People (*Renwu*) Magazine.

75 Ibid.

76 Fang, *Wuhan Diary*, p. 3–4.;

Griffiths, "Weibo's Free-Speech Failure," *The Atlantic*, Apr. 10, 2019;

Economist Editors, "China Invents the Digital Totalitarian State," *The Economist*, Dec. 17, 2017.

China is an autocracy with one party rule. The party is involved in every level of government: Congressional Research Service, "China's Political System in Charts: A Snapshot Before the 20th Party Congress," Report R46977, Nov. 24, 2021.

Kuo, "China's Communist Party Isn't Really Afraid of the Internet," *The Atlantic*, Jul. 17, 2013.;

Knockel, *We Chat, They Watch*.

77 Heman, "Fang Fang: The 'Conscience of Wuhan' amid Coronavirus Quarantine," The Diplomat, Mar. 23, 2020

78 Gabler, *An Empire of Their Own: How the Jews Invented Hollywood.*.

79 Bruinius, "Why Free Speech Is under Attack from Right and Left," *Christian Science Monitor*, "Daily" for Jul. 1, 2021;

Matthew, "US Library Defunded after Refusing to Censor LGBTQ Authors: 'We Will Not Ban ...'," *The Guardian*, Aug.5, 2022.

80 Fang, *Wuhan Diary*, p. xi.

81 Zheng. "'Big White' Army on China's Covid Front Lines Stands Ready to Take the Fall," *South China Morning Post*, Dec. 5, 2022

82 Fang, *Wuhan Diary*, p. 21.

83 Netizen: A Net Citizen, an active and frequent commentator on political and social issues in the Internet.

84 Fang, *Wuhan Diary*, p. 187.

85 Wang, *In the same breath*, Minute 41.

86 Kemps, *COVID Reference*, p. 505;

Huang, "Clinical features of patients infected with 2019 novel coronavirus in Wuhan, China," p. 449, Figure 1.

Schnirring, "China Releases Genetic Data on New Coronavirus, Now Deadly."

87 ProMed. A moderated web service for emails regarding infectious diseases.

88 McArdle and Bloom. "Dr. Marshall Bloom on Investigating Infectious Diseases. (Video), minutes 3 to 5.

89 Nakazawa, "Chronology of Covid-19 Cases on the Diamond Princess Cruise Ship and Ethical Considerations: A Report from Japan.";

Yeung, "Here's the Latest on the Coronavirus Outbreak," *CNN*, Feb. 7, 2020;

Mallapaty, "What the Cruise-Ship Outbreaks Reveal about COVID-19," *Nature News*, Mar. 26, 2020;

Farrar, *Spike: The virus vs. the people - the inside story*, p. 77.

Russel, "Estimating the Infection and Case Fatality Ratio for Coronavirus Disease (Covid-19) Using Age-Adjusted Data from the Outbreak on the Diamond Princess Cruise Ship, Feb. 2020.";

Pezzullo, "Age-Stratified Infection Fatality Rate of COVID-19 in the Non-Elderly Population.";

COVID-19 Forecasting Team, "Variation in the COVID-19 Infection–Fatality Ratio by Age, Time, and Geography during the Pre-Vaccine Era: a Systematic Analysis," *The Lancet*, Feb. 24, 2022.

Moriarty, "Public Health Responses to COVID-19 Outbreaks on Cruise Ships — Worldwide, February–March 2020." MMWR Morb Mortal Wkly Rep 2020;69:347-352.;

Inui, "Chest CT Findings in Cases from the Cruise Ship Diamond Princess with Coronavirus Disease (COVID-19).";

Zhang, "Estimation of the Reproductive Number of Novel Coronavirus (COVID-19) and the Probable Outbreak Size on the Diamond Princess Cruise Ship: A Data-Driven Analysis."

90 White House, "President George W. Bush Signs S.15-Project Bioshield Act," National Archives and Records Administration. National Archives and Records Administration, Jul. 21, 2004.

91 Olsen, James R, (For edit and signature by James Miller, President of Friends of the Bitterroot). Letter to Dr. Anthony S. Fauci. "Scoping Comments on Rocky Mountain Laboratories (RML) BSL-4 Facility" 2002.

Armstrong, *Germ Wars*.

Leonard, Acting Director, Officer of Research Facilities Development and Operations, NIAID. "Final Environmental Impact Statement, Rocky Mountain Laboratories, Integrated Research Facility," p. 5-231 Appendix C. Title page of Presentation, Contents not included in EIS.

92 Olsen, "RML Final EIS," Hamilton City Council Meeting, Jun. 1, 2004. Similar to RML Advisory Council Presentation.

93 *Ravalli Republic,* "Laboratory Visiting Hours Cut Off Indefinitely Because of Q Fever," *Ravalli Republic,* May 4, 1960.

94 It was known as the Spanish Flu because that is where wide-read reports that this where disease was first occurred, but it is probably not the origin. Wikipedia, "Spanish Flu," https://en.wikipedia.org/wiki/Spanish_flu;
In a way the COVID-19 pandemic has some of the same earmarks. The disease had been noted by doctors earlier, in reported on the occurrence of a mystery disease seen as early as 1916: Hammond, "Purulent Bronchitis.," *The Lancet,* Jul. 14, 1917.

95 USSR = Union of Soviet Republics, also known as the Soviet Union, was ruled by Russians with its capital in Moscow and covered a territory that reach from the Atlantic to the Pacific. It broke up in 1991: Wikipedia, "Dissolution of the Soviet_Union," https://en.wikipedia.org/wiki/Dissolution_of_the_Soviet_Union

96 Kendal, "Antigenic Similarity of Influenza A(H1N1) Viruses from Epidemics in 1977–1978 to 'Scandinavian' Strains Isolated in Epidemics of 1950–1951.";
Wikipedia, "1977 Russian Flu," https://en.wikipedia.org/wiki/1977_Russian_flu

97 Mecklin, "Threatened Pandemics and Laboratory Escapes: Self-Fulfilling Prophecies," *Bulletin of the Atomic Scientists,* Mar. 31, 2014.;
See also: Califano. *The Swine Flu Affair: Decision-Making on a Slippery Disease,* App.C.; Kennedy, *The Real Anthony Fauci,* pp. 746–50.

98 Parker, "National Science Advisory Board for Biosecurity (NSABB) Meeting Minutes," Jan. 24, 2020.
Howell, "Rosendale aims to defund RML research," *Bitterroot Star.* Dec. 6, 2023.

99 Race, "Evaluation of the Public Review Process and Risk Communication at High-Level Biocontainment Laboratories."

100 We certainly weren't the only ones — a number of groups in areas around the country, dealing with the same issues, most of whom we contracted. See Armstrong, *Germ Wars: The Politics of Microbes and America's Landscape of Fear.* Our campaign is in it along with others, most of whom we had contacted.

101 Harmon, "Nearly 400 Accidents with Dangerous Pathogens & Biotoxins Reported in US Labs over 7 Years," *Scientific American,* Oct. 3, 2011.

102 Code of Federal Regulations, Title 42 § 73.12 Biosafety. "Biosafety in Microbiological and Biomedical Laboratories (BMBL)," CDC and NIH (6th Edition), Jun. 2020;
Mining: Code of Federal Regulations, Title 30.
Flight Safety: Code of Federal Regulations, Title 14.
In comparing these, the regulations for Biosafety amount to guidance. I can find no required interaction with the government agency once a lab is licensed. This is quite different for the two examples cited, Flight and Mines.

103 GAO — Government Accounting Office, and office of the US. Congress, sometimes referred to as the "Congressional Watch Dog,"https://www.gao.gov;
See also Board on Life Sciences, *Potential Risks and Benefits of Gain-of-Function Research: Summary of a Workshop,* Chapter 5.

104 Kingsbury, "Gao-14-785T, High-Containment Laboratories: Recent Incidences of Biosafety Lapses," GAO, July 16, 2014, pp. 1-2.

105 Associated Press, "Documents Show Lapses at Hamilton Federal Lab," Independent-Record (Helena), p. A5., Nov. 24, 2014.

106 WION, "Why Scientists Are Concerned about Virus Leaks at Biolabs," WION, no date available.

107 The Freedom of Information Act, (FOIA) 5 USC. § 552 provides the right for citizens to get a copy of information and records from the Federal Government. The term is also used for equivalent state laws in Montana, which has the right to inspect any government document that is not subject to privacy in the state constitution (Montana Constitution Art. II, § 9. Right to know). https://www.foia.gov.

108 Young, "Newly Disclosed CDC Biolab Failures 'like a Screenplay for a Disaster Movie'." *USA Today.* June 2, 2016

109 Wolf, Julie, and Marshall Bloom. "Powassan virus and tick biology with Marshall Bloom (Audio Podcast).";
NIAID, "The Integrated Research Facility at NIAID's Rocky Mountain Laboratories," (Video), Sep. 11, 2012.
As part of the court approved mediation agreement for RML, an emergency treatment plan was required for exposures at the RML campus, leading ended up including an Isolation treatment room at Saint Patrick's Hospital in Missoula. Risi, Bloom, et al. "Preparing a Community Hospital to Manage Work-Related Exposures ... in Biosafety Level 3 and 4 Laboratories."

110 Kessler, "Analysis | Timeline: How the Wuhan Lab-Leak Theory Suddenly Became Credible," *Washington Post,* May 27, 2021;
Select Subcommittee on the Coronavirus Pandemic, " COVID Origins Hearing Wrap Up: Facts, Science, Evidence Point to a Wuhan Lab Leak," Mar. 8, 2023.

111 GreatGarmeIndia, *Journal of Geopolitics and International Relations,* Registered as a quarterly magazine in India, online articles updated frequently. https://greatgameindia.com.

112 ZeroHedge: https://www.zerohedge.com. Archives require subscription.

113 Menachery, "Correction: ...A Sars-like Cluster of Circulating Bat Coronaviruses Shows Potential for Human Emergence.";
Roy, et al..Members of Congress, "Letter to Dr. Fauci," Feb. 28, 2021;..

114 Knowledge Fight, founded by Dan Friensen, with Jordan Holmes, https://knowledgefight.libsyn.com;

115 Sonny and Cher, "The Beat Goes On," (Video) YouTube (Sonny and Cher (Video), 1967.

116 Williamson, "Alex Jones's Podcasting Hecklers Face Their Foil's Downward Slide," *New York Times,* Apr. 18, 2021.

117 Rahhal, "Wuhan, China's Coronavirus Epicentre, Has SARS and Ebola Lab," *Daily Mail,* Jan.24, 2020;
Gertz, "Coronavirus Link to China Biowarfare Program Possible, Analyst Says," *Washington Times,* Jan. 26, 2020. The Washington Times would walk this claim back in March, but the genie was out of the bottle.

118 Marshall, Roger, Senator, Bob Kadlec et al. "Muddy Waters, The Origins of COVID-19 Report," US Congress, April 17, 2023;
Asher, David (Hudson Institute) Host, Robert Kadlec, (Asst. Sec. [former] HHS Preparedness and Response), "COVID-19 Origins," C-Span (Video). June 1, 2023. Dr. Kadlec summarized the report well — though he adds a claim at min. 43 that seems very far off from other sources. Taiwan had their first case was in April 2022. See: Taiwan Centers for Disease Control, "Taiwan timely identifies first imported case of 2019 novel coronavirus infection returning from Wuhan..." Taiwan CDC, January 21, 2020.

ODNI Office of Strategic Communications, "Updated Assessment of COVID-19 Origins,"Office of Director of National Intelligence (United States) Undated);

119 Cotton, "Senator Cotton Urges China Travel Ban in Wake of Coronavirus," (Video) US Congress, Armed Services Committee, Jan. 30, 2020.

120 Kinetz, "Anatomy of a conspiracy: With COVID, China took leading role," AP, Feb. 14, 2021

Yang, "One-fourth of coronavirus misinformation in Taiwan comes from Chinese trolls: CIB, *Taiwan News*, March 25, 2020;

Reed, "China clamping down on coronavirus research, deleted pages suggest," *The Guardian*, April 11, 2020;

Molter, "Virality Project (China): Coronavirus Conspiracy Claims," Stanford, Blog. March 31 2020;

Fang, *Wuhan Diary*, p. 14;

Amnesty International. "Write for Rights 2021: China - Zhang Zhan. Oct. 28, 2021.

121 It is unclear if the campaign has any impact on COVID-19 fatalities since the Philipinnes fatalty rate as one quarter of that of the United States.

122 Bing, "Pentagon ran secret anti-vax campaign to undermine China during the pandemic," R*euters*, Jun. 14, 2021;

Democracy Now. "Pentagon Ran a Secret Anti-Vax Campaign to Undermine China at the Height of the Pandemic: Reuters," Democracy Now, Jun. 20, 2024.

123 Hegarty, "The Chinese Doctor Who Tried to Warn Others About Coronavirus," *BBC News*. Feb. 6, 2020.

Public apologies are different than in the US, a more frequent ritual in China, to save the face of the institution and restore a sense of an orderly society: Kadar, "Public Ritual Apology – a Case Study of Chinese," Discourse, Context & Media;

124 Fang, *Wuhan Diary*, p. 26.

# 4. Storm Clouds Coming Fast - Feb. 2020 ────────────

1 Whoriskey, "How US Coronavirus Testing Stalled: Flawed Tests, Red Tape and Resistance to Using the Millions of Tests Produced by the Who," *Washington Post*, Mar. 20, 2020

2 Shoichet, "The US Coronavirus Travel Ban Could Backfire. Here's How," *CNN*, Feb. 7, 2020;

Bouie, "The Racism at the Heart of Trump's 'Travel Ban'," *New York Times*, Feb. 4, 2020,.

3 Gwee, **"Impact of Travel Ban Implementation on Covid-19 Spread in Singapore, Taiwan, Hong Kong and South Korea during the Early Phase of the Pandemic: A Comparative Study."**

4 Neidler, "TechNote: What Are the Differences between PCR, RT-PCR, QPCR, and RT-Qpcr?," Enzo Life Sciences, May 31, 2021

5 Watson, "How This South Korean Company Created Coronavirus Test Kits in Three Weeks," *CNN*, Mar. 13, 2020;

6 Wikipedia, COVID-19 pandemic in South Korea: https://en.wikipedia.org/wiki/COVID-19_pandemic_in_South_Korea

7 Terhune, "Special Report: How Korea trounced U.S. in race to test people for coronavirus," Reuters. Mar. 18, 2020;

Gibney, *Totally Under Control*, Minute 23.

8 Terhune, "Special Report: How Korea trounced U.S. in race to test people for coronavirus," Reuters. Mar. 18, 2020

9 Scott, "South Korea's Covid-19 Success Story Started with Failure," Vox, Apr. 19, 2021.

10 Watson, "How This South Korean Company Created Coronavirus Test Kits in Three Weeks," *CNN*, Mar. 13, 2020;

11 Ibid;

Republic of Korea, "All about Korea's Response to COVID-19 View: Key Strategies," Republic of Korea, Ministry of Foreign Affairs, Oct. 2020.

12 Whoriskey, "How US Coronavirus Testing Stalled: Flawed Tests, Red Tape and Resistance to Using the Millions of Tests Produced by the Who," *Washington Post*, Mar. 20, 2020.

13 Scott, "South Korea's Covid-19 Success Story Started with Failure," Vox, Apr. 19, 2021;

You, "Lessons From South Korea's Covid-19 Policy Response."

14 Armour, "What Derailed America's Covid Testing: Three Lost Weeks," *Wall Street Journal*, Aug. 18, 2020;

15 Apparently because the controversy over test contamination, Lindstrom is no longer at the Respiratory Lab at CDC.

16 Bandler, "Inside the Fall of the CDC," *ProPublica*, Oct.15, 2020.;

Temple-Raston, "CDC Report: Officials Knew Coronavirus Test Was Flawed but Released It Anyway," *NPR Morning Edition*, Nov. 6, 2020.

17 Bandler, "Inside the Fall of the CDC," *ProPublica*, Oct. 15, 2020.

18 Temple-Raston, "CDC Report: Officials Knew Coronavirus Test Was Flawed but Released It Anyway," *NPR Morning Edition*, Nov. 6, 2020.

19 Lindstrom, Interview at CDC OGC Office, by [Redacted], "Dr. Steve Lindstrom Interview," Document Cloud. March 4, 2020 4pm

20 Temple-Raston, "CDC Report: Officials Knew Coronavirus Test Was Flawed but Released It Anyway," *NPR Morning Edition*, Nov. 6, 2020.

21 Yeung, "Here's the Latest on the Coronavirus Outbreak," *CNN*, Feb. 7, 2020;

Mallapaty, "What the Cruise-Ship Outbreaks Reveal about COVID-19," *Nature News*, Mar. 26, 2020;

Russel, "Estimating the Infection and Case Fatality Ratio for Coronavirus Disease (Covid-19) Using Age-Adjusted Data from the Outbreak on the Diamond Princess Cruise Ship, Feb. 2020.";

Moriarty, "Public Health Responses to COVID-19 Outbreaks on Cruise Ships — Worldwide, February–March 2020." MMWR Morb Mortal Wkly Rep 2020;69:347-352.

Zhang, "Estimation of the Reproductive Number of Novel Coronavirus (COVID-19) and the Probable Outbreak Size on the Diamond Princess Cruise Ship: A Data-Driven Analysis."

22 Gibney, *Totally Under Control*, Hour 1, Minute 1.

23 CDC, "CDC 2019-Novel Coronavirus (2019-nCoV) Real-Time RT-PCR Diagnostic Panel. For Emergency Use Only. Instructions for Use," CDC-006-00019, Revision: 08, CDC, March 7, 2023

24 Gibney, *Totally Under Control*, Minute 44;

Lee, "Analysis of the initial lot of the CDC 2019-Novel Coronavirus (2019-nCoV) real-time RT-PCR diagnostic panel."

25 Kusher, "Press Release: LabCorp Launches Test for Coronavirus Disease 2019." Mar. 5, 2020

26 Sharfstein, "Diagnostic Testing for the Novel Coronavirus."

27 Zhu, "A Novel Coronavirus... Patients with Pneumonia in China, 2019."

28 Capital Chemist. "How contagious are colds?" (Acc. 10/15/2023) https://www.capitalchemist.com.au/articles.aspx?title=How-contagious-are-colds?

29 American Hospital Association, "CDC Confirms First US Coronavirus Case Not Linked to Travel, Exposure," American Hospital Association | AHA News, Feb. 27, 2020.

30 Brownstone, "King County Patient Is First in US to Die of Covid-19 as Officials Scramble to Stem Spread of Novel Coronavirus," *The Seattle Times*, Feb. 29, 2020.

31 CDC, "CDC, Washington State Report First Covid-19 Death," Centers for Disease Control and Prevention, News Room, Feb. 29, 2020.

32 Cull, "Daughter of Patricia Dowd, the First US Covid-19 Victim, Breaks Her Silence," *NBC Bay Area*, Feb. 12, 2021;

Santa Clara Public Health, "County of Santa Clara Identifies Three Additional Early COVID-19 Deaths," County of Santa Clara Emergency Operations Center, Apr. 21, 2020.;

33 Rowan, "Mysterious COVID data: Was San Jose woman really first U.S. death?." *Mercury News*, April 8, 2021

Rowan, "Exclusive: First U.S. COVID deaths came earlier — and in different places — than previously thought," M*ercury News*, Aug. 22, 2020

Rowan, "Exclusive: How did a Kansas grandmother just become the first U.S. COVID death? Not even her family knew until this week," M*ercury News*, Sep.2, 2020

34 Taylor, *Final Environmental Impact Statement RML Integrated Research Facility*, 4.1.1 Proposed Action, p. 4-14.

35 Ibid., p. 5-205.

36 Monninger, "Preparation of Viral Samples within Biocontainment for Ultrastructural Analysis";

Franken, "A Technical Introduction to Transmission Electron Microscopy for Soft-Matter."

37 van Doremalen, "Aerosol and Surface Stability of SARS-CoV-2 as Compared with SARS-CoV-1."

38 Cuffari, "What Are Spike Proteins?," *News Medical Life Sciences*, Feb. 24, 2021.

39 Pappas, "Scientists Figure out How New Coronavirus Breaks into Human Cells," *LiveScience*, Mar. 11, 2020;

Wrapp, "Cryo-EM Structure of the 2019-Ncov Spike in the Prefusion Conformation," BioRxiv (Preprint), Feb. 15, 2020;

Wrapp, "Cryo-EM Structure of the 2019-Ncov Spike in the Prefusion Conformation.";

Yan, "Structural Basis for the Recognition of SARS-COV-2 by Full-Length Human ACE2,"

Whittaker, "SARS-COV-2 Spike and Its Adaptable Furin Cleavage Site.";

Xia, "The Role of Furin Cleavage Site in SARS-COV-2 Spike Protein-Mediated Membrane Fusion in the Presence or Absence of Trypsin."

40 Doctor Klioze gives video of how the ACE-2 and various organs interact, which suggests and explanation for a variety of symptoms: Klioze, "Covid and the ACE-2 Surface Protein," (Video), May 3, 2020. Some of Kloize's later videos do not dismiss vaccines entirely, but depart from the CDC narrative of vaccinating all;

See also paper related to ACE-2: Sahin, "The Role of Renin Angiotensin Aldosterone System.. Pathogenesis &Pathophysiology of COVID-19."

41 Coronavirus structural taskforce: https://thorn-lab.com/coronavirus-structural-taskforce/

42 Louten, *Essential Human Virology*. pp.287-288

TMPRSS2: Transmembrane protease, serine 2, Wikipedia, "TMPRSS2," https://en.wikipedia.org/wiki/TMPRSS2;

Furin: Wikipedia, "Furin," https://en.wikipedia.org/wiki/Furin;

43 Janeway, *Immunobiology: The Immune System in Health and Disease*, Chapter 1, pp. 1-34.

44 Zou, "SARS-COV-2 Viral Load in Upper Respiratory Specimens of Infected Patients";

Puhach,, "SARS-COV-2 Viral Load and Shedding Kinetics";

45 Bolze, "SARS-COV-2 Variant Delta Rapidly Displaced Variant Alpha in the United States and Led to Higher Viral Loads";

Ke, "Daily Longitudinal Sampling of SARS-COV-2 Infection Reveals Substantial Heterogeneity in Infectiousness."

46 Louten,. *Essential Human Virology*. pp. 284-291

47 Specifically, a viron is the virus intact and infectious as opposed to its form while replicating and a cell. Burrell, "Virion Structure and Composition," Fenner and White's Medical Virology, Nov. 11, 2017, pp. 27-37,.

48 Janeway, *Immunobiology: The Immune System in Health and Disease*, Section 2-1, pp 36-37.

49 Ibid;

NIH, "Marvels of Mucus and Phlegm," National Institutes of Health, Sep. 1, 2020.

50 Loscalzo, *Harrison's principles of Internal Medicine*, pp. 6576-79;

Janeway, *Immunobiology: The Immune System in Health and Disease*, Section 2-25 to 2-27, pp 81-84;

Murray, *Medical Microbiology*, p. 88;

Flagg "Low level of tonic interferon signalling is associated with enhanced susceptibility to SARS-CoV-2 variants of concern in human lung organoids.";

Banerjee, "Experimental and Natural Evidence of SARS-COV-2 Infection-Induced Activation of Type I Interferon Responses.";

Schiuma, "Innate Immune Response in SARS-COV-2 Infection.";

Bastard, "Autoantibodies against Type I IFNs in Patients with Life-Threatening COVID-19.";

Hadjadj, "Impaired type I interferon activity and inflammatory responses in severe COVID-19 patients."

51 Cheemarla, "Dynamic Innate Immune Response Deters Susceptibility to SARS-COV-2 Infection and Early Replication Kinetics."

52 Liu, "Cd8+ T Cells Predicted the Conversion of Common Covid-19 to Severe";

Wang, "IGG against Human Betacoronavirus Spike Proteins Correlates with SARS-COV-2 Anti-Spike IGG Responses and Covid-19 Disease Severity."

53 Murray, *Medical Microbiology*, pp. 89-91;

Janeway, *Immunobiology: The Immune System in Health and Disease*, Part II, pp. 93-153;

Loscalzo, *Harrison's principles of Internal Medicine*, pp. 11343-11453;

Thomas, "SARS-COV-2 Neucleocapside (N) Protein Is Heavily Glycosylated," *News Medical Life Sciences*, Aug. 31, 2020;

Ruigrok, "Nucleoproteins and nucleocapsids of negative-strand RNA viruses."

54 Wikipedia, Antibody, https://en.wikipedia.org/wiki/Antibody

55 Stankiewicz. "Trajectory of Viral RNA Load Among Persons With Incident SARS-CoV-2 G614 Infection (Wuhan Strain) in Association With COVID-19 Symptom Onset and Severity.";

Xia, "The Role of Furin Cleavage Site in SARS-COV-2 Spike Protein-Mediated Membrane Fusion in the Presence or Absence of Trypsin."

56 Kresge, "Coronavirus Tally May Be Tip of Iceberg as Sick Go Untested," B*loomberg*, Mar. 14, 2020;

Community Developed Resource, "What Is Viral Load and Why Are so Many Health Workers Getting Sick? - SCCM.";

Hogan, "How Much of the Coronavirus Does It Take to Make You Sick? The Science, Explained."

Lindsley, "Quantity and Size Distribution of Cough-Generated Aerosol Particles Produced by Influenza Patients during and after Illness.";

Papineni, "The Size Distribution of Droplets in the Exhaled Breath of Healthy Human Subjects."

57 Memoli, "Validation of the Wild-Type Influenza A Human Challenge Model H1N1PDMIST: An A(H1N1)PDM09 Dose-Finding Investigational New Drug Study.".

58 Murchu "Immune Response Following Infection with SARS-CoV-2 and Other Coronaviruses: A Rapid Review."

59 Zhou, "Viral Emissions into the Air and Environment after SARS-COV-2 Human Challenge: A Phase 1, Open Label, First-in-Human Study.";

Goldberg, . "Viral Determinants of Acute COVID-19 Symptoms in a Nonhospitalized Adult Population in the Pre-Omicron Era."

60 Hurst, "Severe Acute Respiratory Syndrome Coronavirus 2 Infections among Children in the Biospecimens from Respiratory Virus-Exposed Kids (Brave Kids) Study," Figure 3c, p. e2879;

Emery, "The Contribution of Asymptomatic SARS-COV-2 Infections to Transmission - a Model-Based Analysis of the Diamond Princess Outbreak.";

Loscalzo, *Harrison's Principles of Internal Medicine, Twenty-First Edition*, p. 6582.

61 Sutton, "Universal Screening for SARS-COV-2 in Women Admitted for Delivery," Obstetric Anesthesia Digest 40, no. 4 (April 20, 2020): pp. 199-199, https://doi.org/10.1097/01.aoa.0000719556.78821.15;

Chest CT Findings in Cases from the Cruise Ship Diamond Princess with Coronavirus Disease (COVID-19).";

Graham, "SARS-COV-2 Infection, Clinical Features and Outcome of COVID-19 in United Kingdom Nursing Homes.";

Arons, "Presymptomatic SARS-COV-2 Infections and Transmission in a Skilled Nursing Facility.";

Ma, "Global Percentage of Asymptomatic SARS-COV-2 Infections among the Tested Population and Individuals with Confirmed Covid-19 Diagnosis.";

Abdelmoniem, "SARS-COV-2 Infection among Asymptomatic Healthcare Workers of the Emergency Department in a Tertiary Care Facility.";

Buitrago-Garcia, "Occurrence and Transmission Potential of Asymptomatic and Presymptomatic SARS-COV-2 Infections: A Living Systematic Review and Meta-Analysis.";

Emery, "The Contribution of Asymptomatic SARS-COV-2 Infections to Transmission - a Model-Based Analysis of the Diamond Princess Outbreak.";

Petersen, "Three Quarters of People with SARS-COV-2 Infection Are Asymptomatic: Analysis of English Household Survey Data,";

Hurst, "Severe Acute Respiratory Syndrome Coronavirus 2 Infections among Children in the Biospecimens from Respiratory Virus-Exposed Kids (Brave Kids) Study."

62 Felman, "Signs and Symptoms: Definition, Importance, and Uses."

63 Meltzer, "Association of Vitamin D Status and Other Clinical Characteristics with COVID-19 Test Results.";

Mahmoudii, *Nutrition and Immunity*;

Kumar, "Role of Vitamins and Minerals as Immunity Boosters in Covid-19";

Beck, "Selenium and vitamin E status: impact in viral pathology.";

Meisel, "Folate Levels in Patients Hospitalized with Coronavirus Disease 2019.";

Nimavat, "Vitamin D Deficiency and COVID-19: A Case-Control Study at a Tertiary Care Hospital in India.";

Pal, "Zinc and Covid-19: Basis of Current Clinical Trials.";

Shakeri, "Evaluation of the Relationship between Serum Levels of Zinc, Vitamin B12, Vitamin D, & Clinical Outcomes in Patients with COVID-19.";

Tan, "Cohort Study to Evaluate the Effect of Vitamin D, Magnesium, and Vitamin B12 in Combination on Progression to Severe Outcomes in Older Patients with Coronavirus (COVID-19).";

Zhang, "Association between Regional Selenium Status and Reported Outcome of Covid-19 Cases in China."

64 ZOE Editorial Staff, "What Are the New Top 10 Covid Symptoms?," ZOE, Dec. 13, 2022.

65 Finlay, "Persistent Post–COVID-19 Smell Loss Is Associated with Immune Cell Infiltration and Altered Gene Expression in Olfactory Epithelium.";

Wellford, S.A., Moseman, E.A. "Olfactory immune response to SARS-CoV-2."

Marcos-Pérez, "Association of inflammatory mediators with frailty status in older adults: results from a systematic review and meta-analysis."

66 Cao "Covid-19 and Its Effects on the Digestive System."; Shahbaz "Erythroid Precursors and Progenitors Suppress Adaptive Immunity and Get Invaded by SARS-COV-2.";

Huizen, "Gastrointestinal Symptoms of COVID-19: What We Know so Far," Medical News Today, Feb. 5, 2022;

67 Galiatsato, "What Does Covid Do to Your Blood?," Johns Hopkins Medicine, Mar.3, 2022;

Sperati, "Coronavirus: Kidney Damage Caused by Covid-19," Johns Hopkins Medicine, Nov.17, 2022;

Post, "Heart Problems after Covid-19," Johns Hopkins Medicine, Apr. 28, 2022.

68 CDC, "Underlying Medical Conditions Associated with Higher Risk for Severe COVID-19: Information for Healthcare Professionals," Centers for Disease Control and Prevention, Dec. 5, 2022;

Mayo Clinic Staff, "Covid-19: Long-Term Effects," Mayo Clinic, Jun. 28, 2022),

Blauer, "Collecting Comorbidity Data To Identify COVID-19 Correlations,"Johns Hopkins Coronavirus Resource Center, Oct. 4, 2021;

Chae,, "Diabetes and Covid-19: Johns Hopkins Diabetes Guide," Diabetes and COVID-19 | Johns Hopkins Diabetes Guide, Nov. 16, 2020;

69 Maragakis, "Who Is at High Risk for Severe Coronavirus Disease?," Who Is at High Risk for Severe Coronavirus Disease? | Johns Hopkins Medicine, Feb. 15, 2022;

Mayo Clinic Staff, "Covid-19: Who's at Higher Risk of Serious Symptoms?," Mayo Clinic, Sep. 27, 2022;

Loscalzo, *Harrison's principles of Internal Medicine. 21st ed.*, p. 6579.

70 Gupta, "The Mystery of Why the Coronavirus Kills Some Young People," *CNN*, Apr. 6, 2020.

71 Chen, "People with blood type A may be more vulnerable to coronavirus, China study finds)." *South China Morning Post*, March 17 2020;

Ray, "Association Between ABO and Rh Blood Groups and SARS-CoV-2 Infection or Severe COVID-19 Illness : A Population-Based Cohort Study."

72 CDC, "Risk for COVID-19 Infection, Hospitalization, and Death by Age Group," CDC, Dec. 28, 2022.

73 Torrey, "ICD 10 Codes and How to Look Them Up," Verywell Health, Jul. 22, 2022;

CMS, "ICD Code Lists," Centers for Medicare & Medicaid Services, Sep. 28, 2022.

74 Howdon, "Death Certificate Data: Covid-19 as the Underlying Cause of Death," Oxford University/The Centre for Evidence-Based Medicine, Sep. 17, 2020.

75 Clancy, "A First Draft of the History of Treating Coronavirus Disease 2019: Use of Repurposed Medications in United States Hospitals."

76 Fauci, "Glucocorticosteroid therapy: mechanisms of action and clinical considerations."

77 HIV, Human Immunodeficiency Disease which can lead to AIDS, Acquired Immunodeficiency Syndrome.

78 Lock, "A Generation Plowed Under," *San Francisco Examiner*, Image Section, Jun 28, 1988. pp. 17-22;

Fauci. *On Call, A Doctors' Journey in Publlic Health*, p. 99.

79 Kramer, "I Call You Murderers, Image Essay," *San Francisco Examiner*, Image Section, Jun 28, 1988. pp. 23-24.

80 Mullens, "Virus Mutations Suspected as Possible AIDS Origin," *Vancouver Sun*, June 11, 1987.

81 Boghardt, "Soviet Bloc Intelligence and Its AIDS Disinformation Campaign."

82 Clinical Info HIV.gov, "Guidelines for the Prevention and Treatment of Opportunistic Infections in Adults and Adolescents with HIV, Pneumocystis Pneumonia.";
Kennedy, *The Real Anthony Fauci*, p. 342.

83 Fauci, "Dr. Anthony Fauci on Larry Kramer: I'll Miss His Warmth Most," *Time*, May 29, 2020;

84 Lamb, "Q&A With Dr. Anthony Fauci," C-Span (Video), Jan. 8, 2015;

85 Anthony Fauci, Wikipedia, https://en.wikipedia.org/wiki/Anthony_Fauci

86 Lamb, "Q&A With Dr. Anthony Fauci," C-Span (Video), Jan. 8, 2015.
Ibid, Repeat of quote from *Science Magazine* 2003.

87 Fauci, "NIH, Dr. Fauci Emails, Feb through April 2020 in Response to FOIA Request from *Washington Post* and Buzz Feed News," p. NIH-002434.

88 Ibid. pp. NIH-001992 - 001993.

89 Fauci, "NIH, Dr. Fauci Emails, ..." pp. NIH-002433 - 002445.

90 Cohen, "Mining Coronavirus Genomes for Clues to the Outbreak's Origins," *Science Magazine*, Jan. 31, 2020;

91 Fauci, "NIH, Dr. Fauci Emails, Feb through April 2020 in Response to FOIA Request from *Washington Post* and Buzz Feed News," p. NIH-002396, Email. Kristian Andersen to Fauci, CC Farrar. Subject: FW Science: Mining Coronavirus genomes for clues to the outbreaks origins, Jan. 31. 2020. 10:30 AM. The "evolution" referred to is the furin cleavage.

92 Commer, James , Committee on Oversight and Reform and Jim Jordan, Committee on Judiciary (Ranking members), letter to: Honoral Xavier Becerra, Secretary US HHS. No Subject (request for transcribed interview and delivery of Fauci emails). January 11, 2022;
. Young,."'I remember it very well': Dr. Fauci describes a secret 2020 meeting to talk about COVID origins." *USA Today*. Jun. 17, 2021

93 Andersen, "The Proximal Origin of SARS-COV-2."

94 Johnson, "Identifying rare variants inconsistent with identity-by-descent in population-scale whole-genome sequencing data.";

95 Segreto, "The Genetic Structure of SARS-COV-2 Does Not Rule out a Laboratory Origin.";
Tobias, "Evolution of a Theory/ Unredacted NIH Emails Show Efforts to Rule Out Lab Origin of Covid," *The Intercept,* Jan. 13, 2023;
Altermann, "Grand Challenges in Microbiotechnology: Through the Prism of Microbiotechnology."
Gorman, "Scientist Opens Up About His Early Email to Fauci on Virus Origins," *New York Times*, Jun. 14, 2021;
Fauci, "NIH, Dr. Fauci Emails, Feb 2, 2020. Response to FOIA Request: Except posted by Jimmy Tobia" Document Cloud;
_____. "NIH, Dr. Fauci Emails, Nov 2019 through Jan 2020 Informed Consent Action Network in response to FOIA Request # 54106." NIH FOIA Library.
_____. "NIH, Dr. Fauci Emails, Feb through April 2020 in Response to FOIA Request #53768 from Washington Post and Buzz Feed News," Document Cloud, June 1, 2020 and NIH FOIA Library.
Zhang, "Probable Pangolin Origin of SARS-COV-2 Associated with the COVID-19 Outbreak."

96 Wikipedia, Robert_Kadlec. https://en.wikipedia.org/wiki/Robert_Kadlec

97 P3 = Potential pandemic pathogens

98 Zhengli-Li Shi's name is 石正丽. These Chinese characters are in the traditional Asian name order, surname, given name, the opposite of the American/European tradition. 石, Shi, is her "last name," The "-" is usually left off and rendered as Zhengli. Thus, the name in the Chinese order as Shi Zhengli. Her given name 正丽 seems to translate as upright and beautiful (https://chinese.yabla.com/chinese-english-pinyin-dictionary.php)

99 UNC Directory, "Ralph Baric, Phd," UNC School of Medicine, Department of Microbiology and Immunology.

100 P3 = Potential pandemic pathogens: NIH, "Framework for guiding funding decisions about proposed research involving enhanced potential pandemic pathogens," US DHHS, Administrative for Strategic Preparedness and Response § (2017). See also: National Science Advisory Board, "Recommendations for the Evaluation and Oversight of Proposed Gain-of-Function Research," National Science Advisory Board, May 2016
Xu, "Less Is More, Natural Loss-of-Function Mutation Is a Strategy for Adaptation."

101 NIH, "Framework for guiding funding decisions about proposed research involving enhanced potential pandemic pathogens." Wikipedia, Gain of Function: Research: https://en.wikipdia.org/wiki/Gain-of-function_research
Yong, "Mutant-Flu Paper Published," *Nature News*, May 2, 2012;
Selgelid, "Gain-of-Function Research: Ethical Analysis."

102 US. HHS, "Framework for Guiding Funding Decisions about Proposed Research Involving Enhanced Potential Pandemic Pathogens.";
18 USC § 175 - Prohibitions with respect to biological weapons.
CDC, "Select Agents and Toxins List," Select Agents and Toxins List";
CDC, "Guidance on the Regulation of Select Agent and Toxin Nucleic Acids."

103 Bozue, *Medical Aspects of Biological Warfare*, p. 5.

104 Reardon, "US Government Lifts Ban on Risky Pathogen Research," Nature 553, no. 7686 (2017): pp. 11. XXX-Menachery, "Correction: Corrigendum: A Sars-like Cluster of Circulating Bat Coronaviruses Shows Potential for Human Emergence."

105 Wenstrup. "An Evaluation of the Evidence Surrounding EcoHealth Allicane, Inc.'s Research Activities," Interim Staff Report Select Subcommittee on the Coronavirus Pandemic Committee on Oversight and Accountability of the U.S. House of Representatives, May 1, 2024.

106 Hu, "Discovery of a rich gene pool of bat SARS- related coronaviruses provides new insights into the origin of SARS coronavirus.

107 Dance, "The Shifting Sands of 'Gain-of-Function' Research," *Nature News*, Oct. 27, 2021;
Paul, "Dr. Rand Paul Stops Gain-of-Function Research Funding in China," US Senate, May 25, 2021;
Board on Life Sciences, *Potential Risks and Benefits of Gain-of-Function Research: Summary of a Workshop.*
Jacobsen, "'We Never Created a Supervirus," Ralph Baric Explains Gain-of-Function Research," MIT Technology Review, Jul. 26, 2021;
Kingston and Posted by Rod of Iron, "Fauci Got Millions to Dr Ralph Baric for Bioweapon (video)," Broadcast Freedom, Sep.1, 2021;
Kingston Biotech Analyst: https://karenkingston.net;

108 Wikipedia, "Wuhan Institute of Virology," https://en.wikipedia.org/wiki/Wuhan_Institute_of_Virology.
Pone, Laura (NAIAD Grants Management Officer), Letter to Peter Daszak (Ecohealth Allicance). Project Title, "Understanding the Risk of Bat Coronavirus Emergence," Notice of Award. May 27, 2014, p. 12.

109 Chen, "Pioneering Bat Scientist Wins Both Praise and Online Vitriol (Print Edition)," *South China Morning Post*, Feb. 6, 2020.

110 Hu, "Discovery of a rich gene pool of bat SARS- related coronaviruses provides new insights into the origin of SARS coronavirus."

111 Chen, "Pioneering Bat Scientist Wins Both Praise and Online Vitriol (Print Edition)," *South China Morning Post*, Feb. 6, 2020

112 Hu, "Discovery of a rich gene pool of bat SARS- related coronaviruses provides new insights ."into the origin of SARS coronavirus.

113 Mallapaty, "Covid-Origins Study Links Raccoon Dogs to Wuhan Market: What Scientists Think," *Nature News*, Mar. 21, 2023;

114 Racaniello, Vincent (Interviewer), Edward Holmes (Interviewee). "TWiV 1019: Eddie Holmes on SARS-CoV-2 origins," (Video) *MicrobeTV*. Jun. 24, 2023.

115 Menachery, et al. Shi, Baric, "A Sars-like Cluster of Circulating Bat Coronaviruses Shows Potential for Human Emergence," pp. 1508-1514;
Wordvice. "How to Order Authors in Scientific Papers. Wordvice. Aug. 18, 2022

116 Menachery, et al. Shi, Baric, "A Sars-like Cluster of Circulating Bat Coronaviruses Shows Potential for Human Emergence." p. 1512.

117 Ibid, p. 1508.

118 Fauci, "Doctor Anthony Fauci's Calendar. Nov. 1, 2019 to Mary 31, 2020.," Open the Books, FOIA Production (Judicial Watch FOIA), Feb 11, p. 103.

119 Menachery, "A Sars-like Cluster of Circulating Bat Coronaviruses Shows Potential for Human Emergence," p. 1508.

120 Menachery, "Author Correction: A Sars-like Cluster of Circulating Bat Coronaviruses Shows Potential for Human Emergence."

121 Pone, Laura (NAIAD Grants Management Officer), Letter to Peter Daszak (Ecohealth Allicance). Project Title, "Understanding the Risk of Bat Coronavirus Emergence," Notice of Award. May 27, 2014. "C3: Specific Aim 3: Testing Predictions on CoV inter-species transmission": pp. 117-119.
Daszak, "Understanding the Risk fo Bat Coronavirus Emergence," NIAID Grant 5R01AI110964-05, Interim Report. May 3, 2021. pp. 4, 25.

122 NIH, "Framework for guiding funding decision about proposed research involving enhanced potential pandemic pathogens § 2017."

123 Hu, "Development of cell-based pseudovirus entry assay to identify potential viral entry inhibitors and neutralizing antibodies against SARS-CoV-2.";
Dadonaite, "A pseudovirus system enables deep mutational scanning of the full SARS-CoV-2 spike.";
Integral Molecular. "Establishing your pseudovirus neutralization assay," (Accessed 12/24/2023);
Bilska, Miroslawa, Haili Tang, and David C. Montefiori. "Short Communication: Potential Risk of Replication-Competent Virus in HIV-1 Env-Pseudotyped Virus Preparations."

124 Wenstrup. "An Evaluation of the Evidence Surrounding EcoHealth Allicane, Inc.'s Research Activities," Interim Staff Report Select Subcommittee on the Coronavirus Pandemic Committee on Oversight and Accountability of the U.S. House of Representatives, May 1, 2024;
Select Subcommittee. *After Action Review of the COVID-19 Pandemic*, Dec. 4, 2024. pp.463- 481.

125 Kaiser, "NIH says grantee failed to report experiment in Wuhan that created a bat virus that made mice sicker," *Science Insider*. Oct. 21, 2021;

Lerner, "NIH Officials Worked With EcoHealth Alliance To Evade Restrictions on Coronavirus Experiments," *The Intercept*. November 3, 2021;
Lenharo, Mariana, Lauren Wolf. "Controversial virus-hunting scientist skewered at US COVID-origins hearing," *Nature*. May 1, 2024
Tabak, Lawrence (Principle Deputy Director NIH) letter to Representative James Comer, Implied subject: "Grant R01AI110964 Award to EcoHealth Alliance," Oct. 20, 2021.
EchoHealth Alliance would end up being Debarred from Federal Contracts in 2024:
Brisbon, "Action Referral Memorandum for EcoHealth Alliance, Inc. (EHA)." Effective May 15, 2024;
GAO. "NIH Could Take Additional Actions to Manage Risks of Involving Foreign Subrecipients," GAO-23-106119. Government Accounting Office. June 2023;
Lenharo, "US halts funding to controversial virus-hunting group: what researchers think" Nature. May 16, 2024.

126 NIH, Framework for guiding funding decision about proposed research involving enhanced potential pandemic pathogens § 2017;
Whittaker, "SARS-COV-2 Spike and Its Adaptable Furin Cleavage Site.";
Xia, "The Role of Furin Cleavage Site in SARS-COV-2 Spike Protein-Mediated Membrane Fusion in the Presence or Absence of Trypsin.;
Lendman, "Coronavirus Is a Biowarfare Weapon Developed by the US - Law Professor Claims," Al Bawaba, Feb. 5, 2020;
Phillips, Joshua (Interviewer) and Casey Fleming (Interviewee), "Did the Chinese Coronavirus Start as a Bioweapon Program?: The China Report," *Epoch Times* (Video), Feb. 3, 2020.

127 Robertson, "The Wuhan Lab and the Gain-of-Function Disagreement," FactCheck.Org. May 21, 2021 (Update Ju;y 1, 2021);
Wenstrup. "An Evaluation of the Evidence Surrounding EcoHealth Allicane, Inc.'s Research Activities," Interim Staff Report Select Subcommittee on the Coronavirus Pandemic Committee on Oversight and Accountability of the U.S. House of Representatives, May 1, 2024.

128 Fauci. *On Call, A Doctors' Journey in Public Health*, p. 417.

129 Paul, Rand and Anthony Fauci, "Senator Rand Paul Accuses Dr. Fauci of Lying about Gain-of-Function Research at Wuhan Lab," US Senate Hearings. C-Span (Video), Jul. 20, 2021.
Paul, *Deception, The Great COVID Cover-Up*. There are a number of diffences in interpretation in this book and other sources. For example, the first paragrpaph of Chapter 2 in *Deception* differs signficantly from Farrar, *Spike: The virus vs. the people - the inside story*, pp. 63-68.

130 Paul, Rand and Anthony Fauci, "Senator Rand Paul Accuses Dr. Fauci of Lying about Gain-of-Function Research at Wuhan Lab," US Senate Hearings. *C-Span* (Video), Jul. 20, 2021

131 Chen, "Spike and nsp6 are key determinants of SARS-CoV-2 Omicron BA.1 attenuation."

132 Salzberg, "Gain-of-Function Experiments at Boston University Create a Deadly New Covid-19 Virus. Who Thought This Was a Good Idea?," *Forbes Magazine*, Oct. 26, 2022

133 Last, John M. *A Dictionary of Epidemiology*, p. 132.

134 Harrison, "Mechanisms of SARS-COV-2 Transmission and Pathogenesis.";
Wang., "SARS-COV-2: Structure, Biology, and Structure-Based Therapeutics Development.";

Saito, "Enhanced Fusogenicity and Pathogenicity of SARS-COV-2 Delta P681R Mutation."; .

135 Sharples, "Potential Risks and Benefits of Gain-of-Function Research: Summary of a Workshop."

Gryphon Scientific, "Risk and Benefit Analysis of Gain of Function Research," NIH, Dec. 15, 2015;

NIH, "Gain of Function Research," NIH, Jan. 10, 2023 (Accessed 3/13/23).

136 Salzberg, "Gain-of-Function Experiments at Boston University Create a Deadly New Covid-19 Virus. Who Thought This Was a Good Idea?," *Forbes Magazine*, Oct. 26, 2022.

137 Rogin, "Opinion State Department cables warned of safety issues at Wuhan lab studying bat coronaviruses," *Washington Post*, Apr. 14, 2020;

An updated policy from the White House in May 2024, improvements of oversight of Enhanced PPP, but only for Federally Funded work. There is still no law or regulation: Executive Office of the President, "United States Government Policy for Oversight of Dual Use Research of Concern and Pathogens with Enhanced Pandemic Potential," May 2024.

138 Select Subcommittee. *After Action Review of the COVID-19 Pandemic*, Dec. 4, 2024. p. 4 footnote (Alina Chan, Why the Pandemic Probably Started in a Lab, in 5 Key Points, THE N.Y. TIMES June 3, 2024).

139 Science Safety Security, "Science Safety Security – Finding the Balance Together," US HHS, May 2018.

140 O'Day, "Top disease official: Risk of coronavirus in USA is 'minuscule'; skip mask and wash hands."(Embedded Video) *USA Today*. Feb. 17, 2020.

Smerconish, (Inverviewer), Fauci (Interviewee). "Immunologist: We are clearly at the brink of a pandemic," (Video) *CNN*. Feb. 28, 2020.

140 Gibney, *Totally Under Control*, Minute 54.

141 CNN, "Obama administration resignations and firings." *CNN Politics*. Oct. 2, 2015

142 Trump, "President Trump addresses coronavirus outbreak at news conference – 2/26/2020." (Video) *CNBC*. Feb. 26, 2020.

143 Vazquez, "Trump puts Pence in charge of US coronavirus response," CNN. February 27, 2020.

Fauci. *On Call, A Doctors' Journey in Public Health*, p. 356.

144 C-SPAN. "President Trump with Coronavirus Task Force Briefing," (Video) *C-SPAN*, Feb. 29, 2020;

Kuznia, "The Timetable for a Coronavirus Vaccine Is 18 Months. Experts Say That's Risky," *CNN*, Apr. 1, 2020;

Zimmer, Carol (Interviewer), Anthony Fuaci (Interviewee), "A conversation with Dr. Anthony Fauci," *Now What?* Jan. 9, 2024

145 Messonnier, "Transcript for the CDC Telebriefing Update On Covid-19," CDC, Feb. 28, 2020.

146 Langley, "US Stocks Slide into a Correction as Virus Fears Show No Sign of Easing," *Wall Street Journal*, Feb. 28, 2020

147 Yang, "Clinical Course and Outcomes of Critically Ill Patients with SARS-COV-2 Pneumonia in Wuhan, China: A Single-Centered, Retrospective, Observational Study," Table 2, p. 478.

# 5. Epidemiology

1 Johnson PDR, Stinear, Small PLC, et al. "Buruli Ulcer (M. ulcerans Infection): New Insights, New Hope for Disease Control."

# 6. The Fifth Horseman Is Fear - Mar. 2020

1 Czech Movie with the theme that fear underlies what seems normal during the WWII German occupation. Wikipedia, The_Fifth_Horseman_Is_Fear: https://en.wikipedia.org/wiki/The_Fifth_Horseman_Is_Fear

2 "Pestilence" is added to expound upon "pale," which is also sometimes translated as a sickly green, often associated with disease and plague in commentaries.

3 Wikipedia, (Ghost) Riders in the Sky: A Cowboy Legend. https://en.wikipedia.org/wiki/(Ghost)_Riders_in_the_Sky:_A_Cowboy_Legend; The Highwaymen (Video) https://www.youtube.com/watch?v=nOWjX4BpC24

4 Revelation 6:8, KJV.

5 Di Donato,, "All of Italy Is in Lockdown as Coronavirus Cases Rise," CNN, Mar.13, 2020.;

Farrar, *Spike: The virus vs. the people - the inside story*, p. 95.

Silver, "Italy's Nightmare Offers a Chilling Preview of What's Coming," *Bloomberg*, Mar. 13, 2020.

6 Self identified by people who spend a lot of time in a Laboratory. Wiktionary, "Lab Rat," https://en.wiktionary.org/wiki/lab_rat;

IMDb, "Lab Rats," IMDb.

7 PPE = Personal Protective Equipment, N-95 is a standard for filtration of a mask.

8 Neergaard, "Age Not the Only Risk for Severe Coronavirus Disease," *Missoulian* (AP), Mar. 30, 2020)

9 Fox and Friends, "US Surgeon General Urges Americans to Stop Buying, Wearing Masks amid Coronavirus," *Fox News*, Mar.2, 2020

10 Ibid.

11 Cramerh, "Surgeon General Urges the Public to Stop Buying Face Masks," *The New York Times*, Feb. 29, 2020.

12 Stahl, "Dr. Anthony Fauci Talks about COVID-19," 60 Minutes Overtime, Mar. 8, 2020;

13 Cowling, "Face Masks to Prevent Transmission of Influenza Virus: A Systematic Review," 2010.

One of several studies, this one a literature review, that indicate masks don't impact population infection rates for Flu. There are few exceptions, not cited, that show some difference in reduced infection rates but they are small diffferences. Several studies show that the quality of the mask, esp. if it is N95, increases protection, but no population study has everyone, or most people, wearing N-95 masks.

14 Stahl, "Dr. Anthony Fauci Talks about COVID-19," 60 Minutes Overtime, Mar. 8, 2020.

15 Backus, "Bitterroot Sees Mask, Sanitizer Shortages," *Ravalli Republic*, Mar. 4, 2020, pp. A-1 and A-3.

16 Hamilton Christian Academy. https://www.hamiltonchristianacademy.org/about-us/

17 Backus, "Bitterroot Sees Mask, Sanitizer Shortages," *Ravalli Republic*, Mar. 4, 2020, pp. A-1 and A-3.

18 The "Associate Director" (Associate Director for Scientific Management) Division of Intramural Research is the local "face of RML," Dr. Fauci is the official director. Staff members in Bethesda, MD, make policy decisions such as adding a BSL-4 complex to the RML Campus.

19 NIAID, "Marshall E. Bloom, M.D."

20 Cowboy Channel, "Highest Scoring Saddle Bronc Horses of 2020," The Cowboy Channel, Oct. 20, 2020.

Lily Stomper, https://ur-pk.facebook.com/BurchRodeoCompany/videos/324-lilly-stomper/604238017154395/ (Accessed 7/30/2024)

21 The video appears to be produced by xPBx32xEntertainment. There are several copies on the Internet. https://www.youtube.com/watch?v=MHOW3h5mDcQ https://www.youtube.com/watch?v=y2QrPZNiBVo (Accessed 7/30/2024)

22 Ferguson, "Report 9: Impact of Non-Pharmaceutical Interventions (NPIs) to Reduce COVID-19 Mortality and Healthcare Demand.";
White, "Bad Data and Flawed Models? Fact-Checking a Case against Lockdowns."

23 Keeling, "Fitting to the UK Covid-19 Outbreak, Short-Term Forecasts and Estimating the Reproductive Number," Sectin 10, p. 1732.

24 Marcon, "The Hydroxychloroquine Twitter War: A Case Study Examining Polarization in Science Communication."

25 Liu, "Hydroxychloroquine, a Less Toxic Derivative of Chloroquine, Is Effective in Inhibiting SARS-COV-2 Infection in Vitro."
Liu, "Is hydroxychloroquine beneficial for COVID-19 Patients?"

26 The principle, supervising, or team leader is usually listed last in the list of authors on a paper, The person who does the most work and leads the writing is listed first. Wordvice, "How to Order Authors in Scientific Papers,"Wordvice, Aug. 18, 2022;
There has been some pre-COVID controversy around Didier Raoult, Wikipedia, "Dinier Raoult," https://en.wikipedia.org/wiki/Didier_Raoult

27 Gautret, "Hydroxychloroquine and Azithromycin as a Treatment of COVID-19," Mar. 17, 2020

28 Gautret, "Hydroxychloroquine and Azithromycin as a Treatment of COVID-19: Results of an Open-Label Non-Randomized Clinical Trial," p. 13.

29 Raoult, "Coronavirus : Diagnostiquons Et Traitons ! Premiers Résultats Pour La Chloroquine (Coronavirus: Let's Diagnose and Treat! First Results for Chloroquine)." (Video In French), Mar. 16, 2020;
Raoult, "Interview Prof. Didier Raoult with Dr Oz," Dr. Oz (Video), Mar. 28, 2020..

30 Niburski, "Impact of Trump's Promotion of Unproven COVID-19 Treatments on Social Media and Subsequent Internet Trends: Observational Study."

31 A twitter bot is software that controls the actions of an account. (Wikipedia, Twitter Bot, https://en.wikipedia.org/wiki/Twitter_bot).

32 Fauci. *On Call, A Doctors' Journey in Public Health*, p. 374-75.

33 Cohen, "Yet Another Study Shows Hydroxychloroquine Doesn't Work against Covid-19," CNN, May 11, 2020.

34 Rosenberg, "Association of Treatment with Hydroxychloroquine or Azithromycin with in-Hospital Mortality in Patients with COVID-19 in New York State."

35 Gautret, "Hydroxychloroquine and Azithromycin as a Treatment of COVID-19: Results of an Open Label Non-Randomized Clinical Trial Revisited.";
Mehra, "Retracted: Hydroxychloroquine or Chloroquine with or without a Macrolide for Treatment of COVID-19: A Multinational Registry Analysis," (RETRACTED).

36 Meeus, "Efficacy and safety of in-hospital treatment of Covid-19 infection with low-dose hydroxychloroquine and azithromycin in hospitalized patients: A retrospective controlled cohort study.";
Catteau, "Low-dose hydroxychloroquine therapy and mortality in hospitalised patients with COVID-19: a nationwide observational study of 8075 participants."
Chief Investigators of the Randomised Evaluation of COVid-19 therapy (RECOVERY) Trial, Press Release: "No Clinical Benefit from Use of Hydroxychloroquine in Hospitalised Patients with COVID-19," Jun. 5, 2020.;

Barnabas, "Hydroxychloroquine as postexposure prophylaxis to prevent severe acute respiratory syndrome coronavirus 2 infection: a randomized trial.";
Schilling, "Evaluation of hydroxychloroquine or chloroquine for the prevention of COVID-19 (COPCOV): A double-blind, randomised, placebo-controlled trial."

37 Hinton. "Request for Emergency Use Authorization For Use of Chloroquine Phosphate or Hydroxychloroquine Sulfate Supplied From the Strategic National Stockpile for Treatment of 2019 Coronavirus Disease," Mar. 28, 2020.;
Taylor, , "Exclusive: Pressed by Trump, U.S. pushed unproven coronavirus treatment guidance" Reuters. Apr. 4, 2020

38 Waldrop, "Fearing Coronavirus, Arizona Man Dies after Taking a Form of Chloroquine Used in Aquariums," CNN, Mar. 25, 2020)

39 House Oversight and Reform Committee Hearing on Coronavirus Response, Day 1 (Video), House of Rep. Committee Room, Mar. 11, 2020.

40 Ibid.
41 Ibid.
42 Ibid.
43 Ibid.

44 Achenbach, "Coronavirus Forecasts Are Grim: 'It's Going to Get Worse'," The Washington Post, Mar. 11, 2020.

45 Bermas, "The Event 201 Exercise Watch along! 'Countermeasures' Part 1," (Video) YouTube (@JasonBermas, March 21, 2020);
Kennedy, *The Real Anthony Fauci*, p. 876.

46 Johns Hopkins Center for Health Security, "Event 201, a Pandemic Exercise to Illustrate Preparedness Efforts," Event 201. Website and Videos of Table Top Exercise, October 18, 2019.

47 Kadlec, "Crimson Contagion 2019," Government Attic (US Dept. of Health and Human Services, December 9, 2019), https://www.governmentattic.org/38docs/HHSaarCrimsonContAAR_2020.pdf.

48 Kertscher, "Politifact - No, Empty Hospital Beds Do Not Indicate Covid-19 Is 'Fake Crisis'," @politifact, April 3, 2020

49 Roads, ed., "Coronavirus: What Real Science Would Look like, If It Existed," Nexus Newsfeed, Mar. 10, 2020

50 Ai, "Correlation of Chest CT and RT-PCR Testing for Coronavirus Disease 2019 (Covid-19) in China: A Report of 1014 Cases.";
Rousan et al., "Chest X-Ray Findings and Temporal Lung Changes in Patients with Covid-19 Pneumonia."

51 CT is Computed Tomography using a rotating x-ray beam to do a 3-D image (See CT Scan wikipedia).

52 Ai, "Correlation of Chest CT and RT-PCR Testing in Coronavirus Disease 2019 (COVID-19) in China: A Report of 1014 Cases."

53 Dramé, "Should RT-PCR Be Considered a Gold Standard in the Diagnosis of COVID-19?."

54 Axell-House, "The Estimation of Diagnostic Accuracy of Tests for Covid-19: A Scoping Review,";
Caliendo, "COVID-19: Diagnosis," Up To Date, Wolters Kluwer, Dec. 24, 2021.

55 Chen, "A COVID-19 Patient with Multiple Negative Results for PCR Assays Outside Wuhan, China: A Case Report."

56 Funk, "A Snapshot of the Global Race for Vaccines Targeting SARS-COV-2 and the COVID-19 Pandemic."

57 Tsao, "Accuracy of Rapid Antigen vs Reverse Transcriptase–Polymerase Chain Reaction Testing for SARS-CoV-2 Infection in College Athletes During Prevalence of the Omicron Variant."

58 Gans, "False-Positive Results in Rapid Antigen Tests for SARS-COV-2."

59 Kent, "Asymptomatic Lateral Flow Testing May Be Doing More Harm than Good," Medical device Network, Jul. 15, 2021.

60 Bichell,, "Who Has COVID-19? One Colorado County Is Offering Blood Tests To All Its Residents To Find Out," Mar. 21, 2020. *All Things Considered*, Wyoming Public Media.

61 Public Health in 406, "Interim Analysis of COVID-19 Cases in Montana as of 7/30/2021."

The date of the first cases in Montana are reported in other media a few days later, but this may have been a reporting delay to the CDC and the press.

62 Fauci, "Dr. Fauci Press Briefing on Coronavirus Outbreak," C-Span, Mar. 12, 2020.

63 Michels, "Governor Closes Schools," *Missoulian*, Mar. 16, 2020. Both the *Missoulian* and *Ravalli Republic* are owned by Lee Enterprises, sharing both reporters and a web press;

Bullock, "Montana Directives," • Bullock: "Directive implementing Executive Orders 2-2020 and 3-2020 and providing for measures to combat the spread of COVID-19 Novel Coronavirus," Mar. 15, 2020.

64 Nate Silver, "Coronavirus Case Counts Are Meaningless — Unless You Know Something about Testing. And Even Then, It Gets Complicated.," FiveThirtyEight, Apr. 4, 2020.

65 Pan, "Lung Recruitability in COVID-19–associated Acute Respiratory Distress Syndrome: A Single-Center Observational Study."

66 Deaths and Case Counts from One World Data. https://ourworldindata.org which uses John Hopkins as a source for data.

67 Pan, "Lung Recruitability in COVID-19–associated Acute Respiratory Distress Syndrome: A Single-Center Observational Study."

68 Wang, *In the same breath*; Fang, *Wuhan Diary*.

69 Chapter 1 Bookstore, Main Street, Hamilton: https://www.chapter1bookstore.com

70 Wilhelm, The I Ching, pp.97-100, replacing "to and fro" with "travel."

71 Bullock, "Montana Directives," • Bullock: "State of Montana Office of Governor Executive Order No. 2-202, Executive Order Declaring a State of Emergency to Exist Withing the State of Montana Related to the Communicable Disease COVID-19 Novel Coronvirus, Mar. 12, 2020.

72 Gittleson, "Government response updates: Trump issues stricter guidelines to stop virus spread," abcNews, Mar. 16, 2020; Fauci. *On Call, A Doctors' Journey in Publlic Health*, p. 364.

73 Michels, "Governor Closes Schools," *Missoulian*, Mar. 16, 2020. Bullock, "Montana Directives," • Bullock: "Directive implementing Executive Orders 2-2020 and 3-2020 and providing for measures to combat the spread of COVID-19 Novel Coronavirus," Mar. 15, 2020

74 Michels, "Gov. Two COVID cases in Missoula," *Ravalli Republic*, Mar. 15, 2020.

75 Burrows, "Formation of Ravalli County Coronavirus Incident Management Team," Ravalli County (Ravalli County, March 17, 2020), https://ravalli.us/ArchiveCenter/ViewFile/Item/15542.

76 A pulaski is a tool is named for US Forest Ranger, Edward Pulaski, who is credited with inventing it, through there may be some doubt. https://wildfiretoday.com/2019/08/19/the-true-story-of-the-pulaski-fire-tool/

77 A Fireline is the primary method for controlling wildfires, a dug line to surround the fire. On accessible land it can be scraped with a bull dozer or other machine. It can also, and often is, dug by hand by a fire crew. The primary tool is a Pulaski, an axe on one side and adze on the other, named after the Hero of the Big Burn of 1910, Ranger Ed Pulaski who is said to have invented it. The Big Burn burned 3 million acres and reached the north Bitterroot Valley: PBS, "What's a Pulaski? The Big Burn," Public Broadcasting Service (Undated).

78 Cramer, "First H1N1 Death Reported in Ravalli County," *Missoulian*, Nov. 9, 2009, p. B-3.

79 CDC, "Human Exposures to a Rabid Bat - Montana, 2008," MMWR. Morbidity and mortality weekly report,, May 29, 2009.

80 Erickson, "Whooping Cough Spreads to Ravalli Head Start; Parents Upset by School Ban," *Missoulian*, May 3, 2012; Editor, "Local Doctor Recognized for Handling of Pertussis Outbreak," *Bitterroot Star*, Oct. 30, 2012

81 Backus, "New Ravalli County Commissioner Appointed after Questioning about Survivalist Blog," *Missoulian*, Jul. 12, 2012, p. B2

82 JOHN BAD ELK v. UNITED STATES, No. 350 US Supreme Court, 177 US 529 (1900); Wikipedia, Bad Elk vs United States: https://en.wikipedia.org/wiki/Bad_Elk_v._United_States

83 Deer Lodge is the location the the Montana State Prison. Montana Code Annotated 45-7-301. Resisting arrest; 45-5-210. Assault on peace officer or judicial officer.

84 Backus, "New Ravalli County Commissioner Appointed after Questioning about Survivalist Blog," *Missoulian*, Jul. 12, 2012, p. B2; Backus, "Ravalli County Treasurer Accuses Others of Corruption," *Ravalli Republic*, Sep. 30, 2014; Hoffman, "Biography."; Backus, "Sheriff Hoffman to Challenge Iman for Commission Seat," *Ravalli Republic*, Jan. 15, 2016.

I suppose the quote could have been tongue-in-cheek, but that wasn't how Jeff seemed to react to it.

85 Department of Justice, "Shootings at Ruby Ridge, Idaho During the Arrest of Randy Wearver (1994)." Dept. of Justice, Undated

86 Thorning, "Proceed at Your Peril," *Ravalli Republic*, Apr. 26, 1995 Bethel, "Terror in Montana," *The New York Times*, Jul. 20, 1995, Page A-23;

87 AP, "Gun-Packing Militias Make Stand in Montana Anarchy: What Officials Call Paranoia, Believers Call Patriotism. They Contend That Only an Armed Citizenry Can Defend America from a Corrupt Government," *Los Angeles Times*, Apr. 16, 1995; AP, "Montana Militiaman Greenup Surrenders," *The Lewiston Tribune*, Jun. 8, 1995.

88 Backus, "New Ravalli County Commissioner Appointed after Questioning about Survivalist Blog," *Missoulian*, Jul. 12, 2012;

89 Montana Code Annotated (2021) 2-3-103.

90 Backus, "Ravalli County Commission Settles Lawsuit over Public Meeting Notices," *Missoulian*, Sep. 23, 2014.

91 Backus, "Ravalli County Meeting on Refugees Draws Hundreds," Helena Independent Record, Feb. 7, 2017.

92 Elliott, "New York Judge Throws out State's Quarantine Camp Law Declaring It Unenforceable," *The Epoch Times*, Jul. 28, 2022.

The Japanese American internment in World War II on the West Coast and internment for the pandemic have some parallels, both emergencies; Internment lacked sufficient due process, as did quarantine in New York. Though in the heat of war, the Supreme Court at the time found Japanese American Internment Constitutional, though Korematu's conviction, the event that resulted in the case, was eventually reversed in the 1980s based on new evidence. Korematsu v. United States: https://en.wikipedia.org/wiki/Korematsu_v._United_States

In Ravalli County the process would be a civil action in Justice Court if someone refused an order to quarantine, close a restaurant, etc. As far as I know it was not used for COVID-19.Quantine at home direction from Public Health was sometimes ignored. The same Health Board has and will get a court order for a violation of the septic laws and regulations.

93 Ravalli County COVID-19 Incident Management Team, "Notices Posted By Ravalli County COVID-19 Public Health," Mar. 17, 2020 Press Release;

94 Maki, "Ravalli Co. Issues Emergency Declaration Order," *KECI*, Mar. 21, 2020

95 Nap's Bar and Grill: https://napsgrill.com

96 McConnaha, "Local Bars and Restaurants Struggling," *Ravalli Republic*, March 20, 2020.

97 Big Creek Coffee Roasters: https://bigcreekcoffee.com

98 Bulfone, "Outdoor Transmission of SARS-COV-2 and Other Respiratory Viruses: A Systematic Review.";

  Razani, "Clarification Regarding 'Outdoor Transmission of SARS-COV-2 and Other Respiratory Viruses: A Systematic Review.";

  Editors BBC. "Expert Comments about Outdoor Transmission of SARS-COV-2 and Use of Face Masks Outdoors," BBC Science Media Centre. British Broadcasting Corporation, January 11, 2021

  Leonhardt, "A Misleading C.D.C. Number," The New York Times, May 11, 2021.

  Outdoor work sites where people worked closely together every day in construction sites had a higher transmission rate than the somewhat random other outdoor settings in about 10% of the reported cases — the design of the studies themselves mean that they are only suggestive because they fail to account so many possible confounding conditions — but it still seems a safe bet for me.

  Bhaganagar, "Local Atmospheric Factors That Enhance Air-Borne Dispersion of Coronavirus - High-Fidelity Numerical Simulation of COVID19 Case Study in Real-Time.";

  Editor. "Landmark Study Finds Coronavirus Easily Transmitted Outdoors," University of Texas at San Antonio (UTSA ). *UTSA News*, Jan. 8, 2021

99 Perlis, "Community Mobility and Depressive Symptoms During the COVID-19 Pandemic in the United States."

100 Bulfone, "Outdoor Transmission of SARS-COV-2 and Other Respiratory Viruses: A Systematic Review," ;

  Editor, "Landmark Study Finds Coronavirus Easily Transmitted Outdoors," *UTSA News*, Jan. 8, 2021;

  Bhaganagar, "Local Atmospheric Factors That Enhance Air-Borne Dispersion of Coronavirus - High-Fidelity Numerical Simulation of COVID19 Case Study in Real-Time.";

  Editors BBC, "Expert Comments about Outdoor Transmission of SARS-COV-2 and Use of Face Masks Outdoors," B*BC Science Media Centre*, Jan. 11, 2021.

101 Ravalli County Board of Commissioners, "RES 4093 Emergency Proclamation Covid-19 Recorded - Ravalli.us."

  McConnaha, "MDMH Reports First Case of Virus in Rav. Co. || Commissioners Declare Emergency," *Ravalli Republic*, Mar. 22, 2020.

102 Olsen, Letter to Ravalli County Board of Commissioners. "Responding to COVID-19 (SARS-CoV-2) in Ravalli County."

103 CBS44. "COVID-19 briefing with UAB's Dr. Jeanne Marrazzo," (Embedded Video) CBS42, Apr. 1, 2020, minute 6;

  For more background: McArdle, John, and Jeanne Marrazzo M.D. "Dr. Jeanne Marrazzo on the U.S. Response to the Coronavirus Pandemic" (Video). *C-Span Washington Journal*, Apr. 17, 2020

104 Chaney, "Attorney Datsopoulos Left a Strong Legal Legacy," *Missoulian*, Mar. 20, 2022.

105 Department of Housing and Urban Development. https://www.hud.gov

106 Ravalli County Justice Court, Department 1, Case No. CV-20-385..

107 It is illegal to change the locks in this situation in Montana but we had not gone through the laws yet.

108 Olsen, Texts, Laurel Ferritter, Peter Moore.

109 Bullock, "Montana Directives," • Directive on Evictions, Foreclosures and Utilities 3.31.20.

110 Bullock, "Montana Directives," • Bullock: "Directive Implementing Executive Orders 2-2020 and 3-2020 and providing for measures to combat the spread of COVID-19 via food and beverage services or casinos," Mar. 20, 2020.

  Ibid, "... extending closures and updating social distancing requirements and guidance," Mar. 24, 2020.;

111 Montana News Room, "Governor Bullock Issues Stay at Home Directive to Slow the Spread of Covid-19," State of Montana Newsroom, Mar. 27, 2020;

  Bullock, "Montana Directives," • Bullock: "Directive Implementing Executive Orders 2-2020 and 3-2020 providing measures to stay at home and designating certain essential functions," Mar.26, 2020, Effective Mar. 28.

112 Montana Constitution, Article II, Declaration of Rights, Freedom of Religion, "Section 5, Freedom of religion. The state shall make no law respecting an establishment of religion or prohibiting the free exercise thereof," https://leg.mt.gov/bills/mca/title_0000/article_0020/part_0010/section_0050/0000-0020-0010-0050.html

113 IMDb, "Shane," IMDb, Aug. 14, 1953, Last Scene

114 Olsen, Online comment to Governor Steve Bullock, Mar. 26, 2020, delivered electronically. (Available from author's records).

115 ROMAN CATHOLIC DIOCESE OF BROOKLYN, NEW YORK v. ANDREW M. CUOMO, GOVERNOR OF NEW YORK, USupreme Court,11/5/2020

116 Lee, "How Does The COVID-19 Coronavirus Kill? What Happens When You Get Infected," Forbes Mar. 21, 2020;

  CNN. "Loss of smell or taste could be coronavirus symptoms," *CNN*, Mar. 23, 2020

117 Public Law No. 116-136.

118 Lost Trail Ski Area. https://losttrail.com

119 Beck, MA. "Selenium and vitamin E status: impact on viral pathology.";

  Ren, "Traditional Chinese medicine for COVID-19 treatment.";

  Gombart, "A Review of Micronutrients and the Immune System–Working in Harmony to Reduce the Risk of Infection."

120 Fang, *Wuhan Diary*, p. 143.

121 "West Fork," south of Darby, refers to a road, Highway 473, the area, and the West Fork of the Bitterroot River. The river and the road lead to the Painted Rocks Reservoir, the road forking to go over Nez Perce pass into the wilderness.

122 Shapiro and, Mayor Bill De Blasio, "Mayor Bill De Blasio Describes New York City's Response To Coronavirus," (Audio) NPR, Mar. 10, 2020;

  Cuomo, "New York Gov. Cuomo holds news conference on coronavirus response - 3/16." (Video) *Washington Post*, Mar. 16, 2020.

123 *Outbreak* Trailer, (Video) YouTube. https://www.youtube.com/watch?v=Y5povsMKfT4 (accessed 8/28/2023)

124 Bhatraju, "Covid-19 in Critically Ill Patients in the Seattle Region — Case Series."

# 7. Can these critters be corraled? - Apr 2020 ——————

1 COVID-19 Treatment Guidelines Panel. Coronavirus Disease 2019 (COVID-19) Treatment Guidelines. National Institutes of Health.

2 Madeleine Kunin, Living a Political Life, p. 266.

3 Art City Coop, https://www.facebook.com/profile.php?id=100057527886844

4 McConnaha, "Art City Hosts 22nd Recycled Art Show," *Ravalli Republic*, Apr. 21, 2020.

5 PPE = Personal Protective Equipment such as masks, gloves, face-shields

6 Presentation is a patient's collection of 1) Signs (evidence of disease such as a medical expert examining a rash, a blood test, things like blood pressure and oxygen measurement) and 2) Symptoms (something the patient experiences, like pain and other feelings, gasping for air)

7 Keene and Serino (psudonyms), "Two Doctors Log Their Days inside NYC Emergency Rooms," Slate Magazine, Apr. 1, 2020.

8 Allen, "One Ventilator, Two Patients: New York Hospitals Shift to Crisis Mode," Reuters, Mar. 26, 2020.

9 Keene and Serino (pseudonyms), "Two Doctors Log Their Days inside NYC Emergency Rooms," *Slate Magazine*, Apr. 9, 2020.

10 Gould, "Nurse in New York Claims the City Is Killing Covid-19 Patients by Putting Them on Ventilators," *Daily Mail Online*, Apr. 27, 2020;
   Kertscher, "Politifact - Fact-Checking Viral Video Alleging Some New York Hospitals Are 'Murdering' Covid-19 Patients," @politifact, May 14, 2020

11 Kyle-Sidell, "From NYC ICU: Does Covid-19 Really Cause Ards??!! (Video)," YouTube, Mar. 31, 2020. This appears to have been pulled down by Youtube and reposted. If it is no longer posted I may be able to get a link to an archived version. See References Cited.
   MacKinnon, Michael (Interviewer), Cameron Kyle-Sidell (Interviewee). (Audio) Podcast. Anethesia Deconcruted: Science. Politics. Realities. Episode 9. Apr. 20, 2020.
   Chen, "Airway closure in acute respiratory distress syndrome: an underestimated and misinterpreted phenomenon."

12 Take Back Your Power was still up in 2022 but the post it down as of 2025.

13 Ofgang, "How One Covid-19 Doctor Became a Ventilator Whistleblower," Elemental, Sep. 11, 2020.

14 Richardson, "Presenting Characteristics, Comorbidities, and Outcomes among 5700 Patients Hospitalized with Covid-19 in the New York City Area."

15 Whyte and Cameron Kyle-Sidell, "Video on COVID-19: Do Vent Protocols Need a Second Look?," *WebMD* (Video), Apr. 10, 2020

16 Guérin, "Prone positioning in severe acute respiratory distress syndrome."
   See also: Bellani, "Epidemiology, Patterns of Care, and Mortality for Patients with Acute Respiratory Distress Syndrome in Intensive Care Units in 50 Countries."

17 Kaur, "Early versus late awake prone positioning in non-intubated patients with COVID-19."

18 Whyte and Cameron Kyle-Sidell, "Do Covid-19 Vent Protocols Need a Second Look?," *WebMD* (Video), Apr. 10, 2020.
   Cotton, "PRONING during Covid-19: Challenges and Solutions.";
   Williams, "Nursing Procedure Prone Positioning," VA COVID Training (Veterans Administration, Jul. 2021;
   Papoutsi, "Effect of Timing of Intubation on Clinical Outcomes of Critically Ill Patients with Covid-19: A Systematic Review and Meta-Analysis of Non-Randomized Cohort Studies.";
   Pearce, "Prone Positioning in Acute Respiratory Distress Syndrome."

19 Wollers Klurer: https://www.wolterskluwer.com/en/health;
   COVID-19 Treatment Guidelines Panel. Coronavirus Disease 2019 (COVID-19) Treatment Guidelines. National Institutes of Health.

20 Rodriguez, "Prone Positioning in COVID-19-Related Ards: An Overview and Clinician Opinion," *Pulmonology Advisor*, Oct. 13, 2020.
   Perez-Nieto, "Awake Prone Positioning and Oxygen Therapy in Patients with COVID-19: The Apronox Study."

21 A clear tube with two outlets for the nostrils that a supplemental oxygen to normal breaking. You will see them attached to portable oxygen tanks on in public. Wikipeida, "Nasal cannula" https://en.wikipedia.org/wiki/Nasal_cannula

22 Yeates, "Improved Outcomes over Time for Adult COVID-19 Patients with Acute Respiratory Distress Syndrome or Acute Respiratory Failure."

23 Wikipedia, *Night Nurse*. https://en.wikipedia.org/wiki/Night_Nurse_(1931_film);
   *Night Nurse* Trailer (Video) YouTube: https://www.youtube.com/watch?v=nxo8AV5QGY0

24 Wang, *In The Same Breath*, Minute 55;

25 Wang, *In The Same Breath*, Minute 53;
   Sheiber, "Nurses and Doctors Speaking Out on Safety Now Risk Their Job," New York Times, Apr. 9, 2020;
   Carville, "Hospitals Tell Doctors They'll Be Fired If They Speak Out About Lack of Gear," *Bloomberg*, Mar. 31, 2020;
   Ehling, JNESO .../AFL-CIO, Press Release.. Apr. 10, 2020
   Torres, "There's nothing else to lose but our lives': Nurse describes dire situation at a N.Y. hospital," (Video) . *Washington Post*, Mar. 28, 2020;
   Sengupta, "A N.Y. Nurse Dies. Angry Co-Workers Blame a Lack of Protective Gear," New York Times, Mar. 26, 2020;
   Fonrrouge, "NYC nurse who complained about lack of PPE told to 'shut the f–k up'" New York Post, May 11, 2020;
   Adkins, "Exit only: harms from silencing employee voice."

26 Fang, *Wuhan Diary*, p. 352.

27 Ibid. p. 353.

28 Economist Editors, "The Pandemic Has Exacerbated Existing Political Discontent," *The Economist*, Jul. 29, 2021;
   Killgore, "Mental Health during the First Weeks of the COVID-19 Pandemic in the United States.";
   Singleton, "Doctors Explain Mental Health Impact of Shutdown," Severna Park Voice, Sep. 9, 2020;
   Vestal, "Covid Harmed Kids' Mental Health-and Schools Are Feeling It," The Pew Charitable Trusts, Nov. 9, 2021;
   The impact on Domestic Violence is uncertain. While police calls went down, the real-life issue of a woman having the ability to contact the police safely, outside the view of an abuser, is severely restricted during a lock down.
   Li, "Is Domestic Violence Rising during the Coronavirus Shutdown? Here's What the Data Shows.," The Marshall Project, Apr. 22, 2020;
   Miller, "Effects of Covid-19 Shutdowns on Domestic Violence in US Cities."

29 Backus, "Bitterroot Valley Residents Work Together to Face Pandemic Crisis," *Ravalli Republic*, Apr. 5, 2020.

30 Cornerstone Bible Church: https://cornerstonebiblemt.com

31 Triple-Creek Ranch: https://www.triplecreekranch.com

32 IMDb. "Chopped," IMDb, Jun. 4, 2007

33 Olsen, Interview Alan Ford.

34 Ibid.

35 IndieBound: https://www.indiebound.org/indie-store-finder?q=59840

36 IMDb, "You've Got Mail," IMDb (IMDb.com, Dec. 18, 1998

37 iMDb, "Yellowstone," IMDb, Jun. 20, 2018.

38 Galipeau, "Business of Books: Main Street's Chapter One Bookstore Celebrating 38 Years in Operatio," Apr. 5, 2012

39 Olsen, Inteview Mara Lynn Luther; The $1,200 is the Government stimulus check.

40 Olsen, Letter to Ravalli County Board of Commissioners. "Responding to COVID-19 in Ravalli County," Mar. 23, 2020.
NIH, "Interferons," National Institutes of Health (US HHS, Apr. 20, 2023 (Alpha and Beta);
Reis, "Early Treatment with Pegylated Interferon Lambda for Covid-19.";
Tsuzuki, "Effectiveness of Favipiravir on Nonsevere, Early-Stage COVID-19 in Japan: A Large Observational Study Using the COVID-19 Registry Japan.";
FDA, "Fact Sheet For Healthcare Providers: Emergency Use Authorization For Actemra®.";
RECOVERY Collaborative Group, "Lopinavir-Ritonavir in Hospitalised Patients with COVID-19: Preliminary Report from a Randomised, Controlled, Open-Label, Platform Trial.".

41 Olsen, Letter to Ravalli County Board of Commissioners. "1) Directive Implementing Executive Order 2-2020 and 3-2020 providing measures to stay at home and designating certain essential functions, 2) Medical Rights," Apr. 8, 2020.

42 Ravalli County, "History of Ravalli County," A synopsis of the foundation of the towns in Ravalli County.

43 Granicus. https://granicus.com

44 Membership determined by Ravalli County Board of Commissioners, "Resolution to Amend No. 966 Ravalli County Board of Health Rules and Administrative Procedures."

45 McConnaha, "40 Bitterroot Icons: Hospital Became County Administration Building," *Ravalli Republic*, Aug. 30, 2014.

46 Scholl, "Scholl: Experience Provides Tools to Remain Neutral, Objective and Unbiased," *Ravalli Republic*, Apr. 15, 2018;
McConnaha, "Couple Donates Live Savers," *Missoulian*, August 3, 2014, p. H3.

47 Health Board Meetings 2020, 2021, Apr. 8, 2020, 2:00 PM (Video), Minute 2. {Fillers such as "ah" and pauses omitted}
The shutdown order also included a 14 day quarantine requirement for people entering the state, with some exceptions for essential business.
The law was changed by the legislature in 2021 providing for a following-on meeting with the Commissioners to ratify the decision if they choose to hold a public meeting on the issue.

48 Health Board Meetings 2020, 2021, Apr. 8, 2020, 2:00 PM (Video), 1 Hour.

49 Health Board Meetings 2020, 2021, Apr. 8, 2020, 2:00 PM (Video), 1 Hour.
West House, https://www.wmmhc.org/location/West-House-Crisis-Facility

50 Olsen, "Successes with Integrated Response; Modeling," Letter to Commissioners, Apr. 15, 2020.

51 Sweet tea trio, *All Hat, No Cattle*, Seed Creative Productions (Video), Jan. 3, 2023.

52 Health Board Meetings 2020, 2021, Apr. 8, 2020, 2:00 PM (Video), 1 hour Minute 16, 45 Seconds.
This is not the case in Ravalli County:. See Bourdeaux, "Integrating The US Public Health And Medical Care Systems To Improve Health Crisis Response," *National Affairs*, pp. 310-17. Mar. 2023.

53 Herby, *Did Lockdowns Work?*

54 Bessenyey, Letter to Commissioners, Apr. 15, 2020.

55 Bullock. "Directive implementing Executive Orders 2-2020 and 3-2020 and providing guidance for the phased reopening of Montana and establishing conditions for Phase One," Apr. 22, 2020;
_____. "Reopening the Big Sky Phased Approach," (Slides), Apr. 22, 2020;
Upon declaring and emergency: "Governor of the State of Montana, pursuant to the authority vested in me as Governor under the Constitution and the law of the State of Montana, Title 10, Chapter 3, MCA, and under other applicable statutes, do hereby declare that an emergency exists statewide, as defined in §§ 10-3-103 and 10-3-302, MCA. See Bullock, "Montana Directives,"
• Bullock: EXECUTIVE ORDER No. 2-2020: EXECUTIVE ORDER DECLARING A STATE OF EMERGENCY TO EXIST WITHIN THE STATE OF MONTANA RELATED TO THE COMMUNICABLE DISEASE COVID-19 NOVEL CORONA VIRUS

56 Health Board Meetings 2020, 2021, Apr.22, 2020, 1:00 PM (Video).

57 Bullock, Governor Steve, 2020, Governor Greg Gainforte, 2021, and State Emergency Operations Center, Joint Information Center. "Montana Directives."

58 Some Republicans were unhappy with Farmer's Market and rumors floated around about open-carry through the market which is perfectly legal and often done. But, the political ramifications of disrupting local small businesses.... There was a push from the market members to ban political booths — which at first I was against — until someone said the political arguments in front of their booth was interfering with business — the board's business is to support their business so when the vote came up in the annual membership meeting I went along. Both parties were not happy about it, but they got a permit in the park next to the market and set up — retaining the benefit of the market crowd — now it is between them and the city.

59 McConnaha, "Farmers Market to Open May 2 without Crafters," *Ravalli Republic*, Apr. 29, 2020.

60 Helena is the Capitol of Montana, where the Governor sits and the laws are made.

61 Montana Department of Livestock, "Brucellosis/Montana's Designated Surveillance Area (DSA)," Montana Department of Livestock (Undated).

62 Rhyan, "Transmission of Brucellosis from Elk to Cattle and Bison, Greater Yellowstone Area, USA, 2002–2012.";

63 USDA, "Facts about Brucellosis," US Department of Agriculture (Undated).

64 Potter, "Contagious abortion of cattle. Kansas State Agricultural College," Circular, no 69, August, 1918..

65 USDA, "Facts about Brucellosis - USDA,"U. S. Department of Agriculture (Undated).
*Los Angeles Times*, "Yellowstone Bison Hazing," *Los Angeles Times* (Video), May 21, 2010.
National Park Service, "Yellowstone Bison Management Plan," https://parkplanning.nps.gov/projectHome.cfm?projectId=94496;
Mease, "Introduction to the Yellowstone Buffalo Issue," (Video) YouTube. Buffalo Field Campaign, Nov. 4, 2020
Greater Yellowstone Coalition, Bison: https://greateryellowstone.org/bison

66 Fauci, "Building Trust in Vaccines," National Institutes of Health (US HHS, Dec. 5, 2019;
Fauci, "Why Dr. Fauci Never Misses a Flu Shot," NIAID (Video), Oct.16, 2019.

67 Atske, "US Public Now Divided over Whether to Get COVID-19 Vaccine," Pew Research Center Science & Society (Pew Research Center, Sep. 17, 2020.

68 D'Souza, "Rethinking Herd Immunity and the Covid-19 Response End Game," Johns Hopkins Bloomberg School of Public Health, Sep. 13, 2021;
The math that is commonly used to estimated herd immunity has too many uncertainties to be useful for COVID-19. One set of equations is given in: Kwok, "Will Achieving Herd Immunity Be a Road to Success to End the COVID-19 Pandemic?

$P_{crit} = 1 - 1/R_0$ where $P_{crit}$ is the critical percentage for herd immunity base on the Reproductive number $R_0$, which is the average number of people infected by a person.

$P_{im}$ = this is an equation for portion of the vaccinated community that is vaccinated and the vaccine is effective

The premise of this math is that a community reaches herd immunity when $P_{im} > P_{crit}$

$R_0$ has quite a bit of uncertainty for COVID-19 and depends on the nature of population, including social interactions. Even the citation has a range of $R_0$ from country to country of 1.26 to 4.36. And, I would add that these are not static, since they are computed for a particular time. The point of this note is that numbers thrown out for COVID-19 of 70 to 90% are rough estimates. The requirement like the measles which is 95%. And level of vaccination + immunity at any point in time for COVID-19 herd immunity seems extremely unlikely.

69 Fauci, "NIH, Dr. Fauci Emails, Jan - Apr. 2020 in Response to FOIA Request from *Washington Post* and *Buzz Feed News*," p. NIH-000042.

70 Hyperbaric oxygen therapy, John's Hopkins.

71 Henderson, "Hyperbaric Oxygen Treatment Quickly Resolves Severe Breathing Difficulties in COVID-19 Patients," There does not seem to be any large clinical studies as an outcome. However, there are a number of studies show a lot of promise for treating Long Haul COVID-19 patients:

Ko, "HBOT a Potential Treatment for Those with Long Covid-19," UCLA Health System. Ask the Doctors, Aug. 24, 2022;

Robbins., "Hyperbaric Oxygen Therapy for the Treatment of Long Covid: Early Evaluation of a Highly Promising Intervention."

72 Fauci, "NIH, Dr. Fauci Emails, Jan through Apr. 2020 in Response to FOIA Request from *Washington Post* and *Buzz Feed* News," no page number. Email, Apr. 27, 2020 from Marks, Peter, "Revised Slides."

73 _____, "NIH, Dr. Fauci Emails,....," p. NIH-000030.

74 Frueh, "NAS Annual Meeting: Experts Discuss COVID-19 Pandemic and Science's Response (Text and Video)," National Academy of Science, Apr. 27, 2020.

75 This is evident in the charts: Thompson, "COVID-19 Outbreak - New York City, February 29–June 1, 2020, p. 1726.

76 Hou, "New Insights into Genetic Susceptibility of COVID-19: An ACE2 and TMPRSS2 Polymorphism Analysis," This paper shows differences in AC2 and TMPRSS2 not to their interaction with the SARS-CoV-2 spike protien, but in commobity conditions;

Levine, "RFK Jr. says COVID may have been 'ethnically targeted' to spare Jews" *New York Post*, July.23, 2023. (Article includes video which is more accurate than the senstational headline. RFK Jr seems to be referencing the study above).

77 Desimone, "Why are people of color more at risk of being affected by coronavirus disease 2019 (COVID-19)?" Mayo Clinic, Oct. 12, 2022;

Mude, "Racial disparities in COVID-19 pandemic cases, hospitalisations, and deaths: A systematic review and meta-analysis."

78 Haslett, "Tensions with Trump: Dr. Anthony Fauci on Telling the Truth," *ABC News*, Mar. 23, 2020.;

Fauci. *On Call, A Doctors' Journey in Publlic Health*, p. 382.

79 Backus, "RML: Antiviral Slows Virus in Monkeys," *Ravalli Republic*, Apr. 19, 2020

_____ (of *Ravalli Republic*), "Antiviral Slows Progression of COVID-19 in Monkeys," *The Independent Record* (Hele-

na), Apr. 19, 2020.

80 Munster, "Respiratory Disease and Virus Shedding in Rhesus Macaques Inoculated with SARS-COV-2."

81 *Missoulian*, "6-Day-Old Monkey Going Strong," *Missoulian*, November 17, 1942,

82 *Ravalli Republic*, "Lab's AIDS Work May Expand," *Ravalli Republic*, November 2, 1984.

83 Times-Herad, "Heat Kills Animals at Test Lab," *Times-Herald*, Huron, Michigan, February 14, 2004,

84 Gardner, "Macaque Models of Human Infectious Disease."

Williamson, "Clinical Benefit of Remdesivir in Rhesus Macaques Infected with SARS-COV-2."

85 Wang, "Remdesivir in Adults with Severe COVID-19: A Randomised, Double-Blind, Placebo-Controlled, Multicentre Trial," p. 1574.

86 Beigel, "Remdesivir for the Treatment of Covid-19 — Final Report," p. 1819, Figure E. Sherwat, "Center for Drug Evaluation and Research, Application # 214787Orig1s100, Summary Review, Remdesivir for Hospitalized COVID-19 Patients, Gilread Sciences Applicant," FDA, Oct. 21, 2020

87 Beigel, "Remdesivir for the Treatment of Covid-19 — Final Report," p. 1822, Table 2 first row, first column — Percent computed by taking Recovery/N;

Finelli, "Mortality among US Patients Hospitalized with SARS-COV-2 Infection in 2020."

88 Zimmer, "How the Search for Covid-19 Treatments Faltered While Vaccines Sped Ahead," *New York Times*, Jan. 30, 2021.

89 NIH, "NIH Funds Large Clinical Trial to Repurpose Drugs," Apr. 19, 2021.

90 NIH. "COVID-19 Treatment Guidelines (Updated as Needed)-Ivermectin," National Institutes of Health. US HHS, Feb. 11, 2021;

Schmith, "The Approved Dose of Ivermectin Alone is not the Ideal Dose for the Treatment of COVID-19."

91 Chaccour, "The Effect of Early Treatment with Ivermectin on Viral Load, Symptoms and Humoral Response in Patients with Non-Severe COVID-19: A Pilot, Double-Blind, Placebo-Controlled, Randomized Clinical Trial.";

López-Medina, "Effect of Ivermectin on Time to Resolution of Symptoms among Adults with Mild COVID-19.";

Kory, "Review of the Emerging Evidence Demonstrating the Efficacy of Ivermectin in the Prophylaxis and Treatment of COVID-19," pp. e299-e318. This study is a survey of studies, but may well have selection-bias and misapply fix-effect statistical methods to random effect studies;

In 2024, a small trial from Iran is published which shows that Ivermectin is effective at reducing instances of ICU in hospitalized patients when used in combination with the standard of care which include dexamethesone and Remdesivir: Varnaseri, "Ivermectin as a Potential Addition to the Limited Anti-COVID-19 Arsenal: A Double-Blinded Clinical Trial."

92 Ip, "Hydroxychloroquine in the Treatment of Outpatients with Mildly Symptomatic COVID-19: A Multi-Center Observational Study.";

Prodromos, "Hydroxychloroquine Is Effective, and Consistently so When Provided Early, for COVID-19: A Systematic Review.";

Avezum, "Hydroxychloroquine versus placebo in the treatment of non-hospitalised patients with COVID-19 (COPE – Coalition V): A double-blind, multicentre, randomised, controlled trial.";

Rajasingham, "Hydroxychloroquine as Pre-exposure Prophylaxis for Coronavirus Disease 2019 (COVID-19) in Healthcare Workers: A Randomized Trial, Clinical Infectious Diseases."

93 Alam, "Ivermectin as Pre-exposure Prophylaxis for COVID-19 among Healthcare Providers in a Selected Tertiary Hospital in Dhaka – An Observational Study.";

Behera, "Role of Ivermectin in the Prevention of SARS-COV-2 Infection among Healthcare Workers in India: A Matched Case-Control Study.";

Zaidi, "The mechanisms of action of ivermectin against SARS-CoV-2—an extensive review.";

Temple, "Toxic Effects from Ivermectin Use Associated with Prevention and Treatment of Covid-19." ;

Schmith, "The Approved Dose of Ivermectin Alone is not the Ideal Dose for the Treatment of COVID-19.";

Reardon, "Flawed Ivermectin Preprint Highlights Challenges of Covid Drug Studies.";

Kory, "Review of the Emerging Evidence Demonstrating the Efficacy of Ivermectin in the Prophylaxis and Treatment of COVID-19." The analysis was redone omitting that paper with no change in the conclusions: Marik, "Ivermectin, a Reanalysis of the Data."

Hope, "Ivermectin Saves India," *The Desert Review*, Jun. 14, 2021;

Ray, "Indian State Will Offer Ivermectin to Entire Adult Population - Even as Who Warns against Its Use as Covid-19 Treatment," *Forbes*, May 11, 2021.

Naggie "Ivermectin for Treatment of Mild-to-Moderate COVID-19 in the Outpatient Setting: A Decentralized, Placebo-Controlled, Randomized, Platform Clinical Trial.";;

Chaccour, "The Effect of Early Treatment with Ivermectin on Viral Load, Symptoms and Humoral Response in Patients with Non-Severe COVID-19: A Pilot, Double-Blind, Placebo-Controlled, Randomized Clinical Trial."

Marik, "Ivermectin, a Reanalysis of the Data," American Journal of Therapeutics

Kerr "Ivermectin Prophylaxis Used for COVID-19: A City-wide, Prospective, Observational Study of 223,128 Subjects Using Propensity Score Matchin," NOTE from publisher: Conflicts of interest NOT disclosed.

López-Medina, "Effect of Ivermectin on Time to Resolution of Symptoms among Adults with Mild COVID-19."

Bomze, "Severe Cutaneous Adverse Reactions Associated with Systemic Ivermectin: A Pharmacovigilance Analysis.";

Berg, "Why Ivermectin Should Not Be Used to Prevent or Treat COVID-19." American Medical Association, September 2, 2021.

94 Wang, (for President Obama). "FACT SHEET: Progress in Our Ebola Response at Home and Abroad," Press Release. February 11, 2015.

95 Fauci, "Fauci Financial Disclosure 2019."

96 Going from the government to the industry they regulate and do business with and back again.

97 Smith, *An Inquiry into the Nature and Causes of the Wealth of Nations*.

98 Wikipedia, Alex Azar. https://en.wikipedia.org/wiki/Alex_Azar

99 Baker, *New York Times*, "Health Secretary Tom Price Resigns After Drawing Ire for Chartered Flights ." Sep. 29, 2017.

100 Bright, "Retaliation for Whistleblowing" Office of Special Counsel: OSC Form-14 - Complaint of Prohibited Personnel Practice or Other Prohibited Activity. May 5, 2020, Addendum, p. 2.

101 Carson, *Silent Spring*.

102 Nobel Prize Outreach AB 2022, "https://www.nobelprize.org/Prizes/Medicine/1948/Summary/,"

103 Carson, *Silent Spring*, pp. 28-30, 97, 101-104,160-161.

104 DDT, Wikipedia: https://en.wikipedia.org/wiki/DDT#cite_note-no-bel-6

Sonnenberg, "Shoot to Kill: Control and Controversy in the History of DDT".

105 EPA, "Updated Human Health Risk Analyses for Chlorpyrifos," Nov. 10, 2016;

106 Earth Justice, "The EPA Has Finally Banned the Toxic Pesticide Chlorpyrifos From Food," Feb. 8, 2022

107 FDA, CFR Title 21 - Food and Drugs: Parts 1 to 1499: https://www.accessdata.fda.gov/scripts/cdrh/cfdocs/cfcfr/cfrsearch.cfm

21 CFR 312.84: https://www.accessdata.fda.gov/scripts/cdrh/cfdocs/cfcfr/CFRSearch.cfm?fr=312.84

108 10 US Code § 4171 - Operational test and evaluation of defense acquisition programs.

Acquisition Notes Operation Test & Evaluation: https://acqnotes.com/acqnote/careerfields/operational-test-and-eval-uation-ot;

DOD Instruction 5000.89, Test and Evaluation.

109 Scott, "How the UK Found the First Effective Covid-19 Treatment - and Saved a Million Lives," Vox, Apr. 26, 2021.

110 Ibid.

111 Mullard, "Recovery 1 Year on: A Rare Success in the COVID-19 Clinical Trial Landscape," N*ature News*, Apr. 16, 2021..

112 Middle East Respiratory Syndrome, viral respiratory illness first reported in Saudi Arabia in 2012

113 Mullard, "Recovery 1 Year on: A Rare Success in the COVID-19 Clinical Trial Landscape," N*ature News*, Apr. 16, 2021..

114 Welcome Trust, https://wellcome.org

115 Scott, "How the UK Found the First Effective Covid-19 Treatment - and Saved a Million Lives," Vox, Apr. 26, 2021.;

Pessoa-Amorim, "Making trials part of good clinical care: lessons from the RECOVERY trial."

116 RECOVERY Trial. https://www.recoverytrial.net. "Randomise" is the British spelling of Randomize.

117 Introduction to the Recovery Trial (Video). RECOVERY.

118 De Francesco, "COVID-19 antibodies on trial," *Nature Biotechnology*, Jan. 8, 2021.

119 PRINCIPLE Home Page, https://www.principletrial.org

120 Hernandez, "Activ-6 Study," Activ6 Study (The Duke Clinical Research Institute and Vanderbilt University Medical Center), Start date, April 1, 2021.

121 Timmermans, "A General Introduction to Glucocorticoid Biology. " Scott, "How the UK Found the First Effective Covid-19 Treatment - and Saved a Million Lives," *Vox*, Apr. 26, 2021;

FDA, "Orange Book: Approved Drug Products with Therapeutic Equivalence Evaluations: DEXAMETHA-SONE SODIUM PHOSPHATE," Aug. 4, 2021;

Beigel, "Remdesivir for the Treatment of Covid-19 — Final Report.";

RECOVERY Collaborative Group, "Dexamethasone in Hospitalized Patients with Covid-19."

122 Lammers, "Early Hydroxychloroquine but Not Chloroquine Use Reduces ICU Admission in COVID-19 Patients.";

Reis, "Effect of Early Treatment with Hydroxychloroquine or Lopinavir and Ritonavir on Risk of Hospitalization among Patients with Covid-19." ;

Naggie, "Effect of Ivermectin vs Placebo on Time to Sustained Recovery in Outpatients with Mild to Moderate COVID-19".

# 7. Keep your distance - May 2020 ——————————

1 Carson, *Silent Spring*, pp. 107-108, Some words omitted in first sentence of quote.

2 Supplemental Nutrition Assistance Program (SNAP) and the Special Supplemental Nutrition Program for Women, Infants, and Children (WIC)
  Bullock, "Directive implementing Executive Orders 2-2020 and 3-2020 and providing for measures to combat the impacts of COVID-19 by responding to immediate food and energy needs," May 7, 2020.

3 Petrosky-Nadeau, "Did the $600 Unemployment Supplement Discourage Work?," Federal Reserve Bank of San Francisco, Sep. 21, 2020.

4 Health Board Meetings 2020, 2021, May 13, 2020, 2:00 PM (Video).

5 Neuman, "Virus Cluster Linked to Stock Farm," *Ravalli Republic*, May 24, 2020.

6 Stock Farm Club. https://stockfarm.com/Home

7 Bitterroot Performing Arts Council. https://bitterrootperformingarts.org

8 Smith. "Out of Shadows - Cast and Crew," IMDb

9 Savage, "Ex-C.I.A. Agent Goes Public with Story of Mistreatment on the Job," *New York Times*, Feb. 11, 2011.

10 Wikipedia, "Why We Fight," https://en.wikipedia.org/wiki/Why_We_Fight

11 Committee on Assessment of the Possible Health Effects of Ground Wave Emergency Network, Assessment of the Possible Health Effects of Ground Wave Emergency Network, Chapter 3, pp.25-44,

12 Gherardini, "Searching for the Perfect Wave: The Effect of Radiofrequency Electromagnetic Fields on Cells."

13 Johns Hopkins Center for Health Security, "Event 201, a Pandemic Exercise to Illustrate Preparedness Efforts," Oct. 18, 2019.

14 ID2020, "ID2020 | Digital Identity Alliance," Digital Identity Alliance (Identity2020 Systems, Inc), accessed January 14, 2022, https://id2020.org/.
  From Digital Identity Alliance FAQ page. "So your digital identifiers are, for the most part, not in your control. In fact, they are more often than not stored in siloes. And the more siloed and numerous your digital identifiers become the less control you have over them. So while you do have a digital identity, you probably don't have control over it. And that's just you. Over 1 billion people worldwide do not have access to any form of identification. This lack can make it difficult, if not impossible, to access basic critical services like education or healthcare. So, a large group of individuals have no digital identity and those of us that do don't enjoy our rights to privacy, security, and choice. Your digital identity should be yours, but it isn't."

15 Ambrason, "Cryptocurrency System Using Body Activity" - Application WO 2020/060606 A1 (International Publication) based on US Patent US2020/0097951A1, Assigned to Microsoft Technology Licensing, LLC issued May 26, 2020.

16 Zhong, "China's Covid Apps: A Primer," DigiChina, Stanford University, Jul.14, 2022.
  Wikipedia, "QR Code," https://en.wikipedia.org/wiki/QR_code

17 Xiangwei, "China's Health QR Code System Is Ripe for Abuse and Must Not Outlast Covid," *South China Morning Post*, Jun.18, 2022.

18 Olsen, "Bridge game scoring and display computer," US Patent 4130871.

19 Ambrason, Cryptocurrency System Using Body Activity - Application WO 2020/060606 A1 (International Publication) .
  Abramson, "US20200097951A1 - Cryptocurrency System Using Body Activity Data (Abandoned)."

20 Carlson, "Civilization Is Unraveling," *Fox News* (Video), Apr. 18, 2023.

21 Editor, "Editorial: Face Masks Are about Compassion, Not Fear," The Daily Courier, Prescott Valley, AZ, May 11, 2020.

22 Kim, FDA, Letter to Ashely Rhoades, Gilead Sciences, "Emergency Use Authorization for VEKLURY (Remdesivir)."

23 Lovelace, "Trump Wants FDA to Move 'as Quickly as They Can' on Remdesivir Coronavirus Approval," CNBC, Apr. 29, 2020.

24 Sanger, "Profits and Pride at Stake, the Race for a Vaccine Intensifies," *New York Times*, May 2, 2020.

25 NIAID, "Niaid Strategic Plan for Covid-19 Research," April 22, 2020;
  Moderna, "Moderna Announces Expansion of BARDA Agreement to Support Larger Phase 3 Program for Vaccine (mRNA-1273) Against COVID-19," Moderna, Investor relations overview, Jul. 26, 2020.

26 Fauci, "NIH, Dr. Fauci Emails, Jan through April 2020 in Response to FOIA Request from *Washington Post* and *Buzz Feed* News," p. NIH-002434.

27 mRNA = **m**essenger **Ribo**Nucleic **A**cid.

28 NIAID, "Niaid Strategic Plan for Covid-19 Research," April 22, 2020. Note the 2020 plan is no longer posted on the NIH website and was founding in another location. It has been replaced by NIAID, "Niaid Strategic Plan for Covid-19 Research - 2021 Update," Apr. 23, 2021.

29 Deoxyribonucleic acid.

30 Kiszewski, "NIH Funding for Vaccine Readiness before the COVID-19 Pandemic.";
  Aldosari, "Lipid Nanoparticles as Delivery Systems for RNA-Based Vaccines."

31 Dolgin, "The Tangled History of mRna Vaccines," *Nature News*, Sept. 14, 2021

32 Moderna, "Patents," Moderna, Inc., accessed February 1, 2022, https://www.modernatx.com/patents;
  de Fougerolles, Patent US 10,898,574, "Delivery and Formulation of Engineered Nucleic Acids," issued January 26, 2021, filed 2018 — Assigned to Moderna.

33 Simões, "Classical and next-Generation Vaccine Platforms to SARS-COV-2: Biotechnological Strategies and Genomic Variants."

34 McArdl and Bloom. "Dr. Marshall Bloom on Investigating Infectious Diseases (Video)," minutes 19 to 24.

35 Ibid.

36 Norgaard, *Development Betrayed: The End of Progress and a Co-Evolutionary Revisioning of the Future.*

37 Prodrug, Wikipedia, https://en.wikipedia.org/wiki/Prodrug
  Rautio, "The expanding role of prodrugs in contemporary drug design and development."

38 Cuffari, "What Is a Pseudovirus?," Medical News, Mar. 5, 2021.

39 COVID-19 Pseudovirus Service, https://www.raybiotech.com/covid-19-pseudovirus-service.

40 Bilska, "Short Communication: Potential Risk of Replication-Competent Virus in HIV-1 Env-Pseudotyped Virus Preparations."

41 Interferes with Enzyme ability to break down a protein. What come to Protease Inhibitor. https://www.cancer.gov/publications/dictionaries/cancer-terms/def/protease-inhibitor

42 Mitchell, "Engineering Precision Nanoparticles for Drug Delivery."

43 Batty, "Historical Perspective of Clinical Nano and Microparticle Formulations for Delivery of Therapeutics.";
  Bartlett, "The Uptake of Human Papillomavirus (HPV) Vaccine among Adolescent Females in the US (Abstract).";
  Cancer Research UK, "Does HPV Cause Cancer?," Cancer Research UK Cancer Research UK, Apr. 12, 2023

44 O'Shaughnessy, Letter to Merck Sharp & Dohme Corp. "Emergency Use Authorization 108," Feb. 11, 2022.

45 Olsen, Texts, Laurel Ferritter, Peter Moore.

46 Sullivan et al., Datsopoulos, MacDonald, & Lind, ____ (Plaintiff) vs Laurel Ferriter (Defendant), In Justice Court State of Montana, Case No. CV.20.385, Dept. No. 1. "Complaint for Possession," Apr. 15, 2020, pp. 4-5, Para.17.

47 Since it is not a court of record, the narrative in Justice Ray's court, was based on the author's memory. The text was shared with the defendants and their attorney and a draft was given to the clerk of the court to be given to the judge. No corrections have been suggested.

Justice Jennifer Ray's words seem politically partisan to me and appears to ignore the law as it stands, orders implementing a declared emergency. Arbitrary behavior by judges is what bothers most people I talk to who mistrust of the judiciary in this town.

48 Local colloquialism for Stevensville.

49 Stand Together for Freedom website. https://standtogether-forfreedom.com;

Stand Together for Freedom was registered in Montana as a Limited Liability Company in Sep. 2020. Records indicate that Stand Together For Freedom was an ad-hoc group until the end of September when an LLC was registered;

Olsen, Interview, Alan Lackey.

50 The local colloquialism for the Turnoff from Highway 93, going east to Stevensville, which is about a mile away.

51 Olsen, Interview, Alan Lackey.

52 Ibid.

53 Montana Farmer's Union, "CSKT Water Compact Explained," CSKT Water Compact explained (Montana Farmer's Union, Feb. 15, 2019.

54 Wenzel and Chiefs Victor, Alexander, and Michelle et al., "Hell Gate Treaty (1855).";

Devlin, "Flathead Reservation Marks Century of White Settlement," *Missoulian*, Sep. 28, 2010.

55 Confederated Salish and Kootenai Tribes, "CSKT Constitution and Bylaws.pdf," Oct. 28, 1935.

Smith, "Tribal Jurisdiction over Nonmembers: A Legal Overview," Congressional Research Service, Nov. 26, 2013.

CSKT, "The Rez We Live on," CSKT (Video)

56 Montana Department of Natural Resources, "Stream Permitting," Montana Department of Natural Resources, 2023.

Branch, "The Legend of the Horned Rabbit of the West," High Country News, Feb. 24, 2022.

57 Olsen, Interview, Alan Lackey.

58 Nullification, Wikipedia: https://en.wikipedia.org/wiki/Nullification_(US_Constitution);

The prosecutor cut a deal with the defendant charges with selling marijuana after the people called for the people in jury pool said in open court they did not believe the law was fair: Mckinley, "Montana Jurors Raise Hopes of Marijuana Advocates," *New York Times*, Dec. 23, 2010.

59 Jury Nullification, Wikipedia: https://en.wikipedia.org/wiki/Jury_nullification

60 Olsen, Interview, Alan Lackey.

61 Wall Builders, https://wallbuilders.com

62 John Birch Society, https://jbs.org.

The John Birch Society was a political force in the 1960s at the peak of its membership, very anti-communist, seeing collectivism and a welfare state as stages in the march to communism. It is named after a Christian minister in China and an intelligence officer for General Claire Chennault, commander of the "Flying Tigers" during World War II.

63 Olsen, Interview, Alan Lackey.

Montana Code Annotated, Title 49, Chapter 2, Part 3.

64 Trefousse, *Carl Schurz: A Biography*, p. 180.

65 Olson, "'I can't breathe': George Floyd dies after being detained by Minneapolis police," (Video) FOX 9 Minneapolis-St. Paul. May 26, 2020;

*10 Tampa Bay*. "RAW: George Floyd Minneapolis police body camera footage," (Video). 10 Tampa Bay, CBS. Aug.10, 2020.

# 8. A Crack in the Wall - Jun. 2020

1 King, "The Other America," Mar. 14, 1968

2 Oppel, "What to Know about Breonna Taylor's Death," *New York Times*, May 30, 2020

3 Associated Press, "'That's Just Not Taught': Police, Experts Condemn Knee Restraint on George Floyd," WBNS, May 29, 2020..

4 PBS Voices and Organian, "An Unseen Look inside the Portland Protests," PBS Voices (Video), Oct. 29, 2020;

Now This News, "Peaceful Black Lives Matter Protesters Get Creative," *Now This News* (Video), Jun. 2, 2020.

New York Times, "How US Police Took a Hands-off Approach to Armed Groups in 2020," *New York Times* (Video), Nov. 2, 2020;

Eyewitness News ABC7NY, "Protesters in Smithtown Met by Large Group of Counter-Protesters," *Eyewitness News ABC7NY*, Jun. 8, 2020;

Now This News, "Police across the US Unleash Violence on Peaceful Protesters," *NoFw This News* (Video), Jun. 8, 2020;

Police Executive Research Forum, "Response Mass Demonstrations - Police Executive Research Forum," Police Executive Research Forum, Feb. 2022

5 Behrmann, "White House Was Locked down as Protests over Floyd's Death Reach Nation's Capital," *USA Today*, May 30, 2020;

US Secret Service, "Secret Service Statement on Pennsylvania Avenue Demonstrations," May 31, 2020.

Westcott, "May 30 George Floyd Protests News," *CNN*, May 31, 2020.

Office of Inspector General, "Review of US Park Police Actions at Lafayette Park."

6 Cotton, "Tom Cotton: Send in the Troops," Opinion. *New York Times*, Jun. 5, 2020;

News Guild, "Our Statement on The New York Times' op-ed 'Send in the Tropps.'.". Jun. 3, 2020;

Bennet, "When the New York Times lost its way," *Economist*. Dec. 14, 2023;

King. "The Other America," Mar,14, 1968.

7 Wikipedia, Antifa, https://en.wikipedia.org/wiki/Antifa_(United_States);

Jones, "Examining Extremism: Antifa."

Larson, "Police: 'No Credible' Proof Outside Agitators En Route to Missoula," *Missoulian*, Jun. 3, 2020

8 Olsen, Interview, Alan Lackey.

9 McConnaha, "Protesters Rally for Black Lives Matter in Hamilton," *Ravalli Republic*, Jun. 10, 2020

Maki, "Supporters of Black Lives Matter Movement Protest in Hamilton," (Video) *KECI*, Jun.7, 2020

10 Maki, "Supporters of Black Lives Matter Movement Protest in Hamilton," (Video) *KECI*, Jun.7, 2020.

11 Conversation with a reliable source.

12 Larson, "Update: Teen: Armed Group Wanted 'Reason to Hunt Me down'," *Missoulian*, Jun. 13, 2020.

13 Harkins, "Salt Lake City Is under Curfew as a Rally to Protest Death of George Floyd Turns Violent. Gov. Herbert Activates National Guard.," The Salt Lake Tribune, May 30, 2020.

14 Speltz, "Darby coach ... suspended without pay for social media post," *Ravalli Republic*, Jul. 1, 2020.

15 Ibid.

16 Mobley, "James Crow" (Video).

17 Appleton, "Martin Luther King In Life...and Memory, Changing Public Assessment."

18 CBS News. "1968: Robert F. Kennedy Assassinated." June 5, 1968.

19 Swearingen, FBI Secrets: An Agent's Exposé, pp.88-89.
    Roos, "The Raid That Killed Black Panther Leader Fred Hampton," History Stories, Jan. 29, 2021.

20 Buffalo Springfield. "For What It's Worth," (Video), 1967.

21 Heat = Police. This is a quote from Buffalo Springfield, "For What It's Worth," (Video), 1967

22 Tart, "In Cold Blood -- James Rector," Berkeley Barb, May 29, 1969;
    Wikipedia, "People's Park (Berkeley)," https://en.wikipedia.org/wiki/People%27s_Park_(Berkeley);
    Ronald Reagan was Governor of California and had aspirations for President. Student demonstrators were a favorite target of his rhetoric. I wasn't at the street protest, though I was going to classes on campus. I had heard a day or so earlier from someone who worked in the emergency room that they had been treating people for buckshot.

23 Rolling Stone, "Flashback: Ronald Reagan and the Berkeley People's Park Riots," Rolling Stone, May 15, 2017.

24 IMDb, "Portlandia," IMDb, Jan. 11, 2011.

25 Kuoferschmidt, "Three Big Studies Dim Hopes That Hydroxy-chloroquine Can Treat or Prevent COVID-19," Science, Jun. 9, 2020

26 Quotes use the British spelling for Randomize and Hospitalize.

27 Chief Investigators of the Randomised Evaluation of COVid-19 therapy (RECOVERY) Trial, Press Release: "No Clinical Benefit from Use of Hydroxychloroquine in Hospitalised Patients with COVID-19," Jun. 5, 2020.

RECOVERY Collaborative Group, "Effect of Hydroxychloroquine in Hospitalized Patients with Covid-19."

28 Mitjà., "A Cluster-Randomized Trial of Hydroxychloroquine for Prevention of Covid-19," pp. 422 - 423.

29 Hinton, Letter to Gary Disbrow, Dept. Asst. Secretary, US Dept. of Health and Human Services dated 15 June 2020 and attached "Memorandum Explaining Basis for Revocation of Emergency Use Authorization for Emergency Use of Chloroquine Phosphate and Hydroxychloroquine Sulfate.";
    Tang, "Hydroxychloroquine in patients with mainly mild to moderate coronavirus disease 2019: open label, randomised controlled trial."; The FDA cites the Tang study were HCQ is compared to the standard of care in China, which at the time included other antivials: China National Health Commission. "Chinese Clinical Guidance for COVID-19 Pneumonia Diagnosis and Treatment ." 7th Edition. China National Health Commission. March 4, 2020.
    Rosenberg, "Association of Treatment with Hydroxychloroquine or Azithromycin with in-Hospital Mortality in Patients with COVID-19 in New York State.";
    Lacout, "Correspondence: Hydroxychloroquine in Hospitalized Patients with Covid-19" and response by Horby et al. Here Peter Horby states, "There was no evidence of cardiovascular toxicity in either trial.";
    Fauci. On Call, A Doctors' Journey in Public Health, p. 393..

30 Olsen, James R. "Food and Drug Administration (FDA) Decision on Chloroquine," Commissioners Correspondence. Ravalli County, Aug.20, 2020

31 Hooks, "Effects of Hydroxychloroquine Treatment on QT Interval."

32 Mitjà., "A Cluster-Randomized Trial of Hydroxychloroquine for Prevention of Covid-19," Figure 2, pp. 424.

33 Health Board Meetings 2020, 2021, June 9, 2020, 2:00 PM (Video), Minute 6.

34 Ibid. Hour 2, minute 35.

35 Backus, "Ravalli Co. Sees 1st Case of Community Spread," *Ravalli Republic*, Jun. 26, 2020.

# 9. Pull Up Your Bandanna - Jul. 2020 ——————————

1 Gandhi, Non-Violent Resistance (Satyagraha), p. 269.

2 McConnaha, "Ravalli County Tallies 8 New Cases," *Ravalli Republic*, Jul. 1, 2020.

3 _____, "Marcus Daly Viral Clinic set in permanent location," *Ravalli Republic*, Jul. 1, 2020.

4 4-H, https://4-h.org; Ravalli County 4-H, https://www.facebook.com/Ravalli4H.

5 McConnaha, "Ravalli Co. Fair set for Sept. 2-5," *Ravalli Republic*, Jul. 1, 2020

6 Ibid.

7 Hamilton Players and Hamilton Playhouse, (Local Community Theater) https://hamiltonplayers.com

8 IMDb, "Steel Magnolias," IMDb, Nov.22, 1989.

9 McConnahah, "26th Annual Brewfest Postponed," *Ravalli Republic*, Jul. 5, 2020.

10 Larson, "Sheriff Says Posse Unlikely," *Ravalli Republic*, Jul. 5, 2020.

11 Whittecar Rifle and Pistol Range, https://www.whittecarrange.com

12 DD-214, Department of Defense Form 214, military discharge form which is evidence of military service.

13 Larson, "Sheriff Says Posse Unlikely," *Ravalli Republic*, Jul. 5, 2020.

14 Manzella, "Theresa Manzella Campaign Letter," circa 2014;

15 Cohan, "Group Forms to Help Neglected and Abused Horses," Billings Gazette, Section C, pp.1C & 5C, Nov. 24, 2008;

Non Profit Facts, Willing Servants Inc, In Care of Theresa Manzella, 2015. http://www.nonprofitfacts.com/MT/Willing-Servants-Inc.html. It was, according to the Montana Secretary of State website, dissolved in 2016.
    Manzella,

16 "The Red Pill Festival: Theresa Manzella Montana Representative (Video)," Jul. 29, 2021, minute. 17;
    Animal Rights groups are quite different than "environmental" groups in mission and outlook. PETA (People for the Ethical Treatment of Animals) the most well known of the animal rights groups sees animals in nature in terms of suffering — which in the case of abused pets, a lab monkey, or forced fed duck, may well be in sync with an environmentalist in the mountain west. But, PETA is anti-hunting presumably because of the suffering.
    Many environmentalists are hunters, seeing it, when done with respect for the animal, as participating in the natural cycle, which inherently and inevitably has predator/prey relationships and the violence that goes with it— a fact recognized in the Deep Ecology ethic. So, some well known, life-long wildlands advocates can be found working hard to protect threatened species, preserve the natural landscape, while hunting deer — sporting a bumper sticker for PETA: People who Eat Tasty Animals.

17 John Birch Society shop, https://shopjbs.org/products/jbs/constitution/ (Accessed 2/25/2022)

18 Theresa Mazella for Senate District 44, https://manzellafor-montana.com. (Accessed 2/25/2022)

19 Backus, "Crowd Urges Ravalli County Not to Attempt Takeover of Federal Lands," Missourian, December 10, 2013;
Ouellet, "Cliven Bundy, Senator Fielder Speak to Packed House in Sanders County," *Montana Public Radio*, Nov. 2, 2021.

20 Theresa Mazellla for Senate District 44, https://manzellafor-montana.com. (Accessed 2/25/2022).

21 Ravalli Republic, "Theresa Manzella Seeking Re-Election to State Legislature,"*Ravalli Republic*, Apr. 4, 2017.
Kamarck, "Did Trump Damage American Democracy?," Brookings Institute, Jul. 9, 2021.

22 Bedey, "Manzella Letter Troubling," *Ravalli Republic* Letter to Editor, April 26, 2020.

23 Backus, "Hamilton School District Responds to Social Media Dust-up over Planned School Walkout," *Ravalli Republic*, Mar. 8, 2018.

24 Ibid.

25 I would not vote for Trump or Clinton in 2016. I was a Montana Sanders delegate and delegate at the platform convention. I thought Bernie was stuck in the '60s and do not think a full-up socialist economy works. But, many of us needed a protest candidate that year. I helped with the local ground campaign. Bernie Sanders won the Montana Democratic primary in 2016.

26 The law does not use the word "posse," MCA § 7-32-2121(6):
"command the aid of as many inhabitants of the county as are necessary in the execution of the sheriff's duties."

27 Howell, "Sheriff Responds to Social Media Comments," *Bitterroot Star*, Jul. 7, 2020;

28 Reilly, "Hillary Clinton Transcript: 'Basket of Deplorables' Comment," *Time*, Sep. 16, 2016.

29 Bedey, David, https://www.davidbedey.com/political-philosophy.

30 _____, https://www.davidbedey.com

31 "The *Ravalli Republica*n (Stevensville, Mont.) 1894-1899," Library of Congress. https://www.loc.gov/item/sn85053167/..

32 Gilluly, "Nichols Edges Romny by 52-Vote Margin for Sixth Term in House," *Ravalli Republica*n, Nov. 4, 1964.

33 National Grange of the Order of Patrons of Husbandry, Wikipedia: https://en.wikipedia.org/wiki/National_Grange_of_the_Order_of_Patrons_of_Husbandry

34 There was an surge of an unusually large election day turnout that favored Trump and Republicans in 2016.

35 GOTV = Get out the Vote, often with a phone bank or canvassing, now social media. Rarer now, Poll Watching is an end game where someone sits with the Election Officials and marks down who votes. Then calls into a phone bank who then calls people who haven't voted yet. Early voting has made this tactic much less effective.

36 Bitterroot Valley Board of Realtors® https://www.bvbor.org

37 Health Board Meetings 2020, 2021, Jul. 8, 2020, 1:00 PM (Video), Minute 7.

38 Health Board Meetings 2020, 2021, Jul. 8, 2020, 2:00 PM (Hearing) (Video).
Backus, "Flap over Social Distancing Postpones Health Board Meeting," *Ravalli Republic*, Jul. 10, 2020.

39 Ibid.

40 Ravalli County Public Health quoted by KPAX, "Ravalli County Reports 1st Covid-19 Related Death," *KPAX Mountain News*, Jul. 10, 2020;
Ravalli County Public Health Facebook: https://m.facebook.com/story.php?story_fbid=2663779340544432&id=1539730922949285&m_entstream_source=timeline (Accessed 4/12/22).

41 Backus, "RML's Bloom Shares COVID Experience," *Ravalli Republic*, Jan.13, 2021

42 Morawska, "Size Distribution and Sites of Origin of Droplets Expelled from the Human Respiratory Tract during Expiratory Activities.";
Pan, "Characterization and size distribution of initial droplet concentration discharged from human breathing and speaking.";
Lee, "Minimum Sizes of Respiratory Particles Carrying SARS-COV-2 and the Possibility of Aerosol Generation.";
Asadi, "Aerosol Emission and Superemission during Human Speech Increase with Voice Loudness.";
Fennelly, "Particle Sizes of Infectious Aerosols: Implications for Infection Control.";
Han, "Characterizations of Particle Size Distribution of the Droplets Exhaled by Sneeze.";
Dhand, "Coughs and Sneezes: Their Role in Transmission of Respiratory Viral Infections, Including SARS-COV-2.";
Leung, N.H.L. "Transmissibility and transmission of respiratory viruses."

43 Buonanno, "Estimation of Airborne Viral Emission: Quanta Emission Rate of SARS-COV-2 for Infection Risk Assessment."
A test using human volunteers measured over 30 minutes showed dramatic reductions in exhaled aeroslols. Longer duratiouns may have a number of confounding effects: Leung, "Respiratory Virus Shedding in Exhaled Breath and Efficacy of Face Masks.";
Li, "Probable Airborne Transmission of SARS-COV-2 in a Poorly Ventilated Restaurant."

44 CDC, "Masks and Respirators," Jan. 28, 2022

45 Department of Labor, "Standard Number 1910.134, Particulate Respirators Certified under 42 CFR Part 84".

46 Leonard, Acting Director, Officer of Research Facilities Development and Operations, NIAID. "Final Environmental Impact Statement, Rocky Mountain Laboratories, Integrated Research Facility"

47 Kevin Albers, "What Is a HEPA Filter; How Does It Work?: Iso-Aire," ISO-Aire Air Purifiers, Sep. 2022.

48 Sender, "The Total Number and Mass of SARS-COV-2 Virions.";
Ueki, "Effectiveness of Face Masks in Preventing Airborne Transmission of SARS-COV-2."

49 1 Inch = 2.54 cm (centimeters). 1 μm (micrometer) = 10,000 cm .
Sender, "The Total Number and Mass of SARS-COV-2 Virions.";

50 Ueki, "Effectiveness of Face Masks in Preventing Airborne Transmission of SARS-COV-2," p. 3. Fig. 2B. % ranges, the first is a difference in plaque assay in titers the second the viral RNA count from RT-PCR test.

51 Sender, "The Total Number and Mass of SARS-COV-2 Virions.";
Ueki, "Effectiveness of Face Masks in Preventing Airborne Transmission of SARS-COV-2."

52 Simplyscubadotcom, "How to Fit a Mask," (Video) YouTube, Jan. 22, 2018.

53 Fischer, "Low-Cost Measurement of Face Mask Efficacy for Filtering Expelled Droplets during Speech."

54 The Science Brief for cloth masks was taken down. An archived copy: https://www.jamesrolsen.com/maskbrief

55 Jefferson, "Physical Interventions to Interrupt or Reduce the Spread of Respiratory Viruses," p. 2;"

56 Zhou, "Unmasking the Surgeons: The Evidence Base behind the Use of Facemasks in Surgery.";
Loeb, "Medical Masks versus N95 Respirators for Preventing COVID-19 among Health Care Workers.";
van Beusekom, "Study on Masks vs N95 Respirators for Health Workers Spurs Concerns," *CIDRAP News*, Nov. 29, 2022;

CDC "Healthcare Personnel Use of N95 Respirators or Medical/ Surgical Masks for Protection Against Respiratory Infections: A Systematic Review and Meta-Analysis," DRAFT Undated (File dated Nov.1, 2023).

57 Vaughan, *Influenza: An Epidemiologic Study*, pp. 36-39 including Charts III to VII.

58 Ibid, p. 241.

59 Saw, "The Myth of Air Purifier in Mitigating the Transmission Risk of SARS-COV-2 Virus";

EPA, "Air Cleaners, HVAC Filters, and Coronavirus (COVID-19)," Environmental Protection Agency, Jul, 7, 2020.

Ueki, "Effectiveness of HEPA Filters at Removing Infectious SARS-COV-2 from the Air."

Cawcutt, "3 Common Mask Myths Debunked by an Infectious Diseases Expert," Nebraska Medicine, May 14, 2021.

60 Rosner, "Adverse Effects of Prolonged Mask Use among Healthcare Professionals during COVID-19."

61 Huang, "The Effectiveness of Government Masking Mandates on COVID-19 County-Level Case Incidence across the United States, 2020." The "tipping point" analysis for unknown confounders showed results that assured the reader there was a difference, but it did not seem to be a tipping point analysis as is commonly defined. The math was not shown, so it is hard to say how it was done, but the explanation was not convincing.

62 Ibid;

63 Shein., "The Effects of Wearing Facemasks on Oxygenation and Ventilation at Rest and during Physical Activity.";

Zhu, "Evaluation of Nasal Functions While Wearing N95 Respirator and Surgical Facemask."

64 Giorgi, "The Impact of Face-Mask Mandates on All-Cause Mortality in Switzerland;

65 Jefferson, "Physical interventions to interrupt or reduce the spread of respiratory viruses: systematic review."

66 Choi, "Dairy Consumption and Risk of Type 2 Diabetes Mellitus in Men," p. 1000;

Johns, "The Ice-Cream Conspiracy," The Atlantic, May 2023, pp. 17-22;

Korat, "Dairy Products and Cardiometabolic Health Outcomes" dissertation, Harvard University, 2018, Table 2.2, p. 53, p. 54. Table 2.6. p. 60

67 See "CDC Mask Brief Contents and Commentary" at https://www.jamesrolsen.com/maskbrief. The CDC replaced the Mask Science Brief (Science Brief: Community Use of Cloth Masks to Control the Spread of SAR-CoV-2) webpage with different content about a month after posting. I had archived the original and it is include tin the link.

68 Republic of Korea, "All about Korea's Response to COVID-19," Ministry of Foreign Affairs, Republic of Korea," Oct. 2020.

69 Bullock for US Senate: https://stevebullock.com;

Steve Bullock, Wikipedia. https://en.wikipedia.org/wiki/Steve_Bullock_(American_politician)

Siegler, "Steve Bullock's COVID-19 Response May Boost His Senate Run in Montana," *NPR*, Jun. 25, 2020.

70 Oran, "Prevalence of Asymptomatic SARS-CoV-2 Infection.";

Pollock, "Asymptomatic Transmission of Covid-19.";

Subramanian, "Quantifying Asymptomatic Infection and Transmission of COVID-19 in New York City Using Observed Cases, Serology and Testing Capacity."

71 Helena Montana, Wikipedia, https://en.wikipedia.org/wiki/Helena,_Montana

72 Claremont McKenna College. College Consensus, https://collegeconsensus.com/rankings/best-small-colleges/;

73 2020 Raw Data Law School Rankings. https://www.ilrg.com/rankings/law/.

74 Supporters for an Abuse Free Environment: https://www.safeinthebitterroot.org

75 Soroptimist International of Hamilton: https://www.sihamilton.org/community-organizations

76 SAFE, https://www.safeinthebitterroot.org.

77 Backus, "Emma's House: Center a Model in Helping Victims of Child Abuse," *Missoulian*, Sep. 4, 2009;

Emma's House. https://www.emmashousecac.org

78 2016 United States presidential election in Montana, Wikipedia. https://en.wikipedia.org/wiki/2016_United_States_presidential_election_in_Montana.

79 Plettenberg, "Federal General Election Official Results," Federal General Election Official Results - Nov. 8, 2016.

I decided at the last minute before the filing deadline to run for state senate that year, to make a point in the Democratic primary, which I won with 64% of the vote in the primary. While I ran a classic campaign in 2000 getting in the mid 30's, I didn't do any work for the general in 2016, getting 30%. The two Dems who worked hard in a state senate house and county commissioner races got 33% and 34%. Jo Young, running against Manzella got 27%. The lesson is that the effectiveness of local campaigns swings the needle between 4% and, on a good day, 10%: Plettenberg, "Federal General Election Official Results," Federal General Election Official Results - Nov. 8, 2016.

80 Steve Daines. https://www.daines.senate.gov

81 Raile, "Western States Coronavirus Survey - Montana State University," Apr. 2020;

Rinfret, "Summer 2020 - Aggregate - University of Montana," Jun. 2020.

82 Montana Nurses Association, https://www.mtnurses.org

83 Samuels, "Records: Montana Governor Feared Backlash over Mask Mandate," Associated Press, Jul. 17, 2020

Backus, "Ravalli County Toes Line on Governor's Face Mask Directive," *Ravalli Republic*, Jul. 17, 2020.

84 Ibid.

85 Bullock, "Montana Directives," • Bullock: "Directive implementing Executive Orders 2-2020 and 3-2020 and providing for the mandatory use of face coverings in certain settings," Jul. 15, 2020, p. 4.

86 Ibid. p. 5.

87 Ibid. p. 5.

88 Ibid. p. 5.

89 Editor, "Ravalli County Sheriff's Response to Governor's Mask Directive," *Bitterroot Star*, Jul. 17, 2020

90 Christian, "Ravalli County Sheriff Explains Covid 19 Enforcement Policy," *Newstalk KGVO*, Jul. 21, 2020.

91 Johnson, "Press Release CANCELING FAIR," Corona Virus COVID 19, Ravalli County, Jul. 13, 20.

Backus, "Ravalli County Fair's Main Portion Canceled," *Ravalli Republic*, Jul. 13, 2020.

92 Brew Fest, Bitterroot Valley Chamber of Commerce, https://bitterrootchamber.com/?s=brew+fest

93 Bitterroot Celtic Society: http://www.bcgg.org.

94 MT Shakespeare in the Parks: https://shakespeareintheparks.org

95 McConnaha, "Numerous Jul. Events in the Bitterroot Canceled Due to Pandemic," *Ravalli Republic*, Jul. 14, 2020.

96 Howell, "County Fair Victim of Covid-19," *Bitterroot Star*, Jul. 22, 2020.

Maki, "Ravalli Co.. Fair Board Hears Support, Opposition to Canceling Major Portion of Fair," *KECI*, Jul. 15, 2020;

97 Backus, "FFA, 4-H Leaders Looking for Alternatives after Fair Cancellation," *Ravalli Republic*, Jul. 16, 2020.

Howell, "County Fair Victim of Covid-19," *Bitterroot Star*, Jul. 22, 2020.

98 Maki, "Ravalli Co.. Fair Board Hears Support, Opposition to Canceling Major Portion of Fair," *KECI*, Jul. 15, 2020.

99 Ibid.;Backus, "FFA, 4-H Leaders Looking for Alternatives after Fair Cancellation," *Ravalli Republic*, Jul. 16, 2020.
   Howell, "County Fair Victim of Covid-19," *Bitterroot Star*, Jul. 22, 2020.
   Note that the *Bitterroot Star* article said the vote was for cancellation, contrary to other sources.
100 Hu, "When Public Health Becomes the Public Enemy," High Country News, Sep. 24, 2021.
101 Calderwood, Dr. Carol to Ravalli County Elected Officials and To Whom It May Concern, "Letter: Resignation Letter Ravalli County Health Officer," 18 Jul. 2020
102 McConnaha, "County Public Health Officer Resigns," *Ravalli Republic*, Jul. 22, 2020;
_____. "Medical Professional Respond to Health Officer Resignation," R*avalli Republic*, Jul. 22, 2020..
103 _____, "Medical Professional Respond to Health Officer Resignation," R*avalli Republic*, Jul. 22, 2020.
104 _____, "County Public Health Officer Resigns," *Ravalli Republic*, Jul. 22, 2020;
105 _____, "Interim Public Health Director Has Heart for the Community," R*avalli Republic*, Aug. 9, 2020;

106 Health Board Meetings 2020, 2021, Jul. 24, 2020, 1:00 PM (Video), Minute 16 30 seconds.
107 Health Board Meetings 2020, 2021, Jul. 28, 2020, 12:30 PM (Video).
108 A standard Letter/number code for the diagnosis and treatment, used, among other things, for billing.
109 Health Board Meetings 2020, 2021, Jul. 28, 2020, 12:30 PM (Video). 2 Hours 50 minutes.
110 Olsen, Interview Alan Ford.
111 Ibid.
112 According Luther: Later one of her staff got COVID after they were vaccinated.
113 Olsen, Interview Mara Lynn Luther.
114 River Rising Bakery, https://www.facebook.com/RiverRisingBakery/
115 Haimerl, A "Mask Mandate? in a Montana Town, It 'Puts Us at Odds with Customers'." *New York Times*, Oct. 19, 2020.
116 Kodiak Jax, https://kodiakjax.com/pizzas
117 Larson "Community of Stevensville speaks on both sides of mask debate," *Ravalli Republic*, Jun. 24, 2020.

# 10. What A Rodeo - Aug. 2020 ——————————————

1 Schurman, "Freedom Show to Replace 4H/FFA Livestock Show," *Bitterroot Star*, Jul. 28, 2020;
   McConnaha, "4 H'Ers Showcase Project at Fairgrounds," *Ravalli Republic*, Aug. 23, 2020.
2 _____, "Hamilton Players Present Steel Magnolias Live-Streamed," *Ravalli Republic*, Aug. 5, 2020
3 Darby Rodeo Association, https://www.darbyrodeo.org/arena
4 Speltz, "Roping the Moment," R*avalli Republic*, Aug. 16, 2020;
   The rope is tied to the saddle horn linked by a string so that it breaks away when the calf pulls on it, unlike calf roping or tie-down roping where the cowhand does a three legged tie of the calf after roping it.
5 Orin Larson, https://prorodeo.com/bio/contestant/orin-larsen/59635
6 Mannering, "Darby Rodeo Packs Bleachers during Yellowstone Riggin' Rally," KPAX (Video), Sep. 8, 2020.
7 Backus, "Yellowstone Riggin' Rally," R*avalli Republic*, Aug. 28, 2020;
   Speltz, "'Electric' Atmosphere at Riggin Rally," R*avalli Republic*, Sep. 9, 2020.
8 Even when we don't have our own fires, our North South Valley often has inversions, trapping smoke that blows on fronts from around the region.
9 Backus, "Firefighters Make Progress on Six Bitterroot Fires," *Ravalli Republic*, Aug. 6, 2020;
   _____, "Cinnabar fire on Lolo Forest jumps to 2000 acres," R*avalli Republic*, Aug. 26, 2020;
10 _____, "Rain Keeps Cinnibar in check," R*avalli Republic*, Aug. 28, 2020.
11 _____, "Fast Work by Victor Couple Knocks down Potential Wildfire," R*avalli Republic*, Aug. 13, 2020.
12 _____, "Victor woman drowned in the Bitterroot River," R*avalli Republic*, Aug. 9, 2020.
13 _____, "Ravalli County 'Community Snapshot' COVID Tests All Negative," *Ravalli Republic*, Aug. 7, 2020.
14 _____, "Stevensville Mayor Asks Schools to Reconsider Mask Decision," R*avalli Republic*, Aug. 7, 2020;
   McConnaha, "Masks Required for Darby and Stevi Schools," R*avalli Republic*, Aug. 21, 2020.
15 Backus, "Stevi Protesters Say Virtual Meeting Locking Public Out," R*avalli Republic*, Aug. 30, 2020.
16 Health Board Meetings 2020, 2021, Aug. 12, 2020, 12:30 PM (Video). Agenda Item V-d. Discussion with possible decision on joint public statement - Katie, 1 Hour 55 minutes to 4 Hours 5 minutes.

17 Katie cited the Hair Salon study I comment on in Appendix 3. Alan's citations include a letter from a group of scientists: Morawska, "It Is Time to Address Airborne Transmission of Coronavirus Disease 2019 (Covid-19)."
18 CITIZENS UNITED v. FEDERAL ELECTION COMMISSION (United States Supreme Court January 21, 2010).
19 Zack Colman, "Green Group Denies Funding Montana Senate AD," The Hill, October 29, 2012;
   Open Secrets. "PAC Profile: Montana Hunters & Anglers Leadership Fund."
20 Sharon, "In Montana, Dark Money Helped Democrats Hold a Key Senate Seat," The Smokey Wire, Jan. 8, 2013.
   Open Secrets. "PAC Profile: Montana Hunters & Anglers Leadership Fund.";
21 Samuels, "US High Court Denies Bid to Restore Greens on Montana Ballot," Associated Press, Aug. 25, 2020;
   Silvers, "How Montana's Green Party Found Itself at the Center of a Political Circus. Again," *Montana Free Press*, Jul. 2, 2020.
22 Root, "Voter Suppression during the 2018 Midterm Elections.";
   Heritage Foundation, "Voter Fraud."
23 I have found nothing definitive showing that If mail voting was unavailable, what percentage would vote in person anyway — the percentage is higher if the political motivation was high. (Griffin, "America's Electoral Future: The Coming Generational Transformation.")
24 Levine, "Trump Campaign Fails to Show Evidence of Vote-by-Mail Fraud, Filing Reveals," T*he Guardian*, Aug. 21, 2020.
25 Bullock, "Directive implementing Executive Orders 2-2020 and 3-2020 and providing for measures to implement the 2020 November general election safely," Aug. 6, 2020.
26 Backus, "County to start process for mail in ballots," R*avalli Republic*, Aug. 9, 2020.
27 Bragg, Dennis. "Ravalli County Voters Express Concerns over Mail-in Voting (Text & Video)." KPAX, Aug. 17, 2020.
28 Ravalli County Board of Commissioner Meetings 2020. 2021, Regular Meeting, Aug. 25, 2020, Agenda Item 4, "Designate Polling Place for November 3, General Election. (Video)"
29 Scheck, "How Private Money from Facebook's CEO Saved the 2020 Election," All Things Considered, *NPR*, December 8, 2020
   Hemingway, "Team Zuckerberg Masks the Heavily Pro-Democrat Tilt of 2020 Election 'Zuck Bucks,' Study Finds," Real Clear Investigations, June 7, 2022.
   Daily Wire, "Daily Wire - Zuckerberg Admits to FB Liberal

Bias," Daily Wire (Video), April 11, 2018. Ironically also posted in facebook.

30 KLYQ. https://klyq.com/shows/friday/.

31 McConnaha, "Poetry Contest Broadcast Live on KLYQ," *Ravalli Republic*, Sep. 13, 2020.

32 _____, "Emma's House Fundraising 'Show' Features Valley Locals," *Ravalli Republic*, Sep. 20, 2020;
_____, "Chamber retools brewfest," R*avalli Republic*, Sep. 20, 2020.

33 Meet Montana's Bitterroot Valley, the Backdrop of the TV Series Yellowstone https://www.visitbitterrootvalley.com/meet-montanas-bitterroot-valley-the-backdrop-of-the-tv-series-yellow-stone/; (Accessed 2/12/2022).
'Yellowstone' cast on Montana's 'breathtaking' Bitter-root Valley. (Video) YouTube. https://www.youtube.com/watch?v=OpRfmEbcGqE (Accessed 2/12/2022).

34 Chief Joseph Ranch: https://www.chiefjosephranch.net/history

35 Barnes, "A Farming Community in Transition: The Impact of Sprawl on the Agricultural Legacy in the Bitterroot Valley.";
Swanson, "Growth and Change in the Bitterroot Valley and Implications for Area Agriculture and Ag Lands."

36 Barkey, "The Economic Impact of Filming Yellowstone in Ravalli County," Montana Business Quarterly, Spring 2022 edition, pp. 2-5;
A message asking for the source to the author through the University of Montana in July 2022 was not answered.

37 Olsen, Inteview Mara Lynn Luther

38 Backus, "Downtown Hamilton Becomes Set for Popular 'Yellow-stone' Television Series," *Ravalli Republic*, Sep. 25, 2020.

39 Daniel, "Family Affair: Local Department Store Win Chamber Award after Clothing the Bitterfoot for Half a Century," *Ravalli Republic*, Apr. 14, 2004;
Downtown Hamilton Business Improvement District, "The Hamilton Record, a Walking Tour 2006 (Flyer)," p. 16;
The original JCPenney would be the next shop back from the Roxy in the shots in Yellowstone, Season 4, Episode 1,

40 *Missoulian*, "Tom Ford," *Missoulian*, Feb. 14, 2021.
*Ravalli Republica*n, "C And C Committees Named by Shults," *Ravalli Republica*n, Jan. 10, 1957;
*Ravalli Republica*n, "Earl Malone given Live Membership at Wildlife Assn. at Annual Banquet," *Ravalli Republican*,

Apr. 27, 1959;
Ravalli County Fish and Wildlife Association: https://rcfwa.org;(Accessed 2/15/2022).

41 Saint Thomas Military Academy: https://www.cadets.com.

42 Heller, *Catch 22;*
IMDb, "Catch-22," Movie, 1970, IMDB, https://www.imdb.com/title/tt0065528/;
IMDb, "Catch-22," Miniseries 2019, IMDB, https://www.imdb.com/title/tt50561

43 IMDb, "Thirty Seconds over Tokyo," IMDb, https://www.imdb.com/title/tt00373

44 *Missoulian*. "Tom Ford," M*issoulian*, Feb, 14, 2021;
Wikipedia, "15th Air Force," https://en.wikipedia.org/wiki/Fifteenth_Air_Force; Foggia: https://en.wikipedia.org/wiki/Foggia_Airfield_Complex

45 Wikipedia, "Norden Bombsight," https://en.wikipedia.org/wiki/Norden_bombsight

46 Add flak and a few German fighters and this was Tom Ford's ride (Video) YouTube: https://www.youtube.com/watch?v=rBRla1IaBD0

47 Graff, "Unbreakable," Air and Space Magazine, Nov. 1, 2001

48 Valley Furniture, "Display Ad, Best Wishes to John and Tom Ford," *Ravalli Republica*n, p. 6., October 22, 1947.;

49 Valley Furniture, https://www.homefurnishings.com/retailer/valley-furniture-montana

50 Olsen, Interview Alan Ford.

51 Backus, "Bowhunter Encounters Grizzly in Bitterroot Mountains," *Ravalli Republic*, Sep. 27, 2020.

52 Dax, *Grizzly West: A Failed Attempt to Reintroduce Grizzly Bears in the Mountain West,* Kindle location, 3603.

53 Knibb, *Grizzly Wars: The Public Fight over the Great Bear.*

54 Malone, *The Battle for Butte: Mining and Politics on the Northern Frontier, 1864-1906,* pp. 44-45.

55 Backus, "Pioneering Journalist, Publisher Leaves Legacy," *Ravalli Republic*, Sep. 23, 2020.
Burk, *The Clearcut Crisis: Controversy in the Bitterroot.*

56 Blondie ( Debbie Harry), "Heart of Glass," (Video), 1978.

57 A drip torch is a hand held canister with flammable liquid used to start controlled burns and backfires: Porter, "Drip Torches Facilitate Effective Prescribed Burns," Noble Research Institute, Mar. 2014.

## 11. Time Out For Public Health - Sep. 2020

1 BBC. "US Election: Trump Won't Commit to Peaceful Transfer of Power," B*BC News*, Sep 24, 2020.

2 Ravalli County Board of Commissioner Meetings, Sep. 16 2020, "(Video). Agenda Item 4, Tiffany Webber interview for Public Health Director."

3 Editor, "Bitterroot Forest Lowers Fire Danger," *Bitterroot Star*, Aug. 25, 2021.

4 Health Board Meetings 2020, 2021, Sep. 9, 2020, 1:00 PM (Video)..

5 Christina, "Harris Wants to Woo Irish Nurses Home, but Only 36 Turned up on First Day of Recruitment Drive," The Journal Online News Ireland, Dec. 29, 2016

6 Ravalli County Board of Commissioner Meetings 2020. 2021. (Video). Agenda Item 4: Tiffany Webber Interview," (Video), Sep. 16, 2020, "

7 Backus, "Hamilton Schools Report 1st COVID Case," *Ravalli Republic*, Sep. 30, 2020.

8 HUD, *PIT Count Methodology Guide*, 2014, Standard No. 11, p. 8..

9 Bitterroot Affordable Housing Coalition, https://www.facebook.com/BitterrootAHC/

10 Olsen and Bitterroot Affordable Housing Coalition, "Homelessness in the Bitterroots."

11 Backus, "Real Estate Prices Soar," *Ravalli Republic*, October 18, 2020.

12 Ravalli County Board of Commissioner Meetings 2020. 2021, Regular Meeting. Aug. 31, 2020, Agenda "5. Presentation on Ravalli County homelessness and possible solutions from Homeless Coalition. (Video)";
McConnaha, "Housing Coalition Predicts Increase in Homelessness," *Ravalli Republic*, September 4, 2020.

13 Nomadland (movie), Wikipedia. https://en.wikipedia.org/wiki/Nomadland_(film);
IMDb, "Nomadland," IMDb, Feb. 19, 2021.

## 12. Threadbare - Oct. 2020

1 Health Board Meetings 2020, 2021, Oct. 14, 2020, 1:00 PM (Video), Minute 50.

2 C-Span, "President Trump to Give Acceptance Speech from White House next Week," C-Span, Aug. 17, 2020, minute 57, 40 seconds;

FOX 13 News, "Trump Claims the Only Way He'll Lose Is If Election Is 'Rigged,' Says He Should Be Able to Run for 3rd Term," FOX 13 Tampa Bay, Aug. 18, 2020.

3 C-Span, "President Trump to Give Acceptance Speech from White House next Week," C-Span, Aug. 17, 2020, minute 23, 15 seconds.

4 Bullock, "Bullock Ad Front Page Fold Over," *Ravalli Republic*, Oct. 2, 2020.

5 McConnaha, "Health Dept. not approving, denying events," R*avalli Republic*, Oct. 2, 2020.

6 Backus, "Early Voting Under Way, Ballots to Be Mailed Oct. 9," *Ravalli Republic*, Oct. 4, 2020.

7 ____, "Official Warns of Giving Ballots to Strangers," *Ravalli Republic*, Oct. 23, 2020;
   _____, "Election Official Urges People to Vote Soon," *Ravalli Republic*, Oct. 28, 2020.

8 Morris, "Smoking-Gun Email Reveals How Hunter Biden Introduced Ukrainian Businessman to VP Dad," New York Post, Oct. 14, 2020.
   Bump, "Now Warning about Hunter Biden-Laptop Disinfo: The Guy Who Leaked It," *Washington Post*, Apr. 12, 2020.

9 Everett, "'Jon Poked the Bear': Tester Braces for Trump's Revenge," POLITICO, Apr. 28, 2018.

10 Sympathy with the Devil, Rolling Stones (Video) You-Tube, https://www.youtube.com/watch?v=Jwtyn-L-2gQ

11 Backus, "Former UN Ambassador stumps for Daines, Gianforte in Hamilton," *Ravalli Republic*, Oct. 21, 2020;
   Rogers, "Ted Cruz Stumps in Hamilton," *Ravalli Republic*, Oct. 30, 2020.

12 Backus, "Treasurer Turns Away out of County Requests," *Ravalli Republic*, p. A-3, Oct. 11, 2020.

13 Behrmann, "Here's a Timeline of President Donald Trump's and Dr. Anthony Fauci's Relationship," *USA Today*, Oct. 28, 2020.

14 Backus, "Local Health Care Resources Strained," *Ravalli Republic*, Oct. 9, 2020.

15 Ibid.

16 Ibid.

17 Backus, "RML to Build New Animal Facility," *Ravalli Republic*, Oct. 11, 2020.

18 Ibid.

19 IMDb, "M*A*S*H," IMDb, Sep.17, 1972;
   Wikipedia, Mobile Army Surgical Hospital, https://en.wikipedia.org/wiki/Mobile_Army_Surgical_Hospital

20 Backus, "Ravalli Co. Records Two More Deaths," *Ravalli Republic*, Oct. 16, 2020.

21 _____, "Busy Big Game Season Expected This Year," *Ravalli Republic*, Oct. 21, 2020;
   Orange Army — Hunter orange is required for a hunter and recommended for anyone in the woods during rifle season.

22 Bob Ward's, https://www.bobwards.com.

23 Health Board Meetings 2020, 2021, Oct. 14, 2020, 1:00 PM (Video), Minute 4.

24 Marcus Daly is shown as having 25 beds, 10 full with non-COVID patients, 5 with COVID patients, and 10 empty in an Oct. 19 report. No reports of staff or staffing limits seems to have ever been reported to the state health department. Montana Dept. of Public Health and Human Services, "Covid-19 Hospital Occupancy and Capacity in Montana," Oct. 19, 2020, p. 3.

25 *New York Times* staff, "Ravalli County, Montana Covid Case and Exposure Risk Tracker," *New York Times*, Jan. 27, 2021;
   Ravalli County Web Site Press Releases and Covid Information. INFORMATION USED IS NO LONGER ACTIVE. The county archives records for some a defined legal period if requested. https://ravalli.us/207/Public-Health
   The case counts for Ravalli County at are reported in places like *New York Times* and other sites is an average of new cases — a new case can go on for sometime until it is closed. The new plus ongoing cases are what

public health is reporting, the current active case count, a better metric for characterizing the workload of Public Health. So, that is what is shown in the graphic.

26 Commissioners Correspondence 2020. https://ravalli.us/Archive.aspx?AMID=106, Oct 20-23.

27 Hoffman, "Released to the Public - COVID-19 Statement," Ravalli County Commissioners Correspondence, Oct. 28, 2020.

28 "Ravalli County Board of Commissioner Meetings 2020." Commissioners Meetings (Video), Oct. 22, 2020

29 Ravalli County Board of Commissioner Meetings 2020. 2021." Commissioners Meetings (Video), Oct. 22, 2020;
   Backus, "Commissioners Get Pushback on COVID Message," *Ravalli Republic*, Oct. 25, 2020.

30 Ravalli County Board of Commissioner Meetings 2020. 2021." Commissioners Meetings (Video), Oct. 22, 2020.

31 Editors New England Journal of Medicine. "Dying in a Leadership Vacuum." *NEJM* (Editorial), Oct. 7, 2020.
   There was an effort to make up for lack of pretrained contract tracers in New York City: New York City Mayor's Office. "What is Test and Trace Corps?" (Video) New York City. Jul. 27, 2020; New York City Mayor's Office. "Learn How the Test and Treat Corps Works," (Video) New York City. Jul. 27, 2020.)

32 McLean, "American Pie," Don McLean (Video), May 9, 2017.

33 Jasmin, "George Chip Albert Jasmin III," *Ravalli Republic*, Nov. 4, 2020, p. A2.

34 Cove Jasmin, https://www.reverbnation.com/musician/covejasmin

35 Lawrence, "Bitterroot Original – Big Sky Mudflaps Celebrate 40 Years," *Bitterroot Star*, Jul. 21, 2015

36 Jenn Adams, http://www.jennadams.com

37 Hard Time Bluegrass Festival. http://hardtimesbluegrass.com;
   Daly Mansion, Events: https://www.dalymansion.org/events/;
   Annual Celtic Games and Gathering, Events: http://www.bcgg.org/bitterroot-celtic-games-and-gathering.html;
   MTN News, "Annual Memorial Day Celebration Returning to Corvallis," *KPAX*, May 27, 2022;
   *Ravalli Republic*, "The 102nd Annual Corvallis Memorial Day Parade Set for Monday, May 30," *Ravalli Republic*, May 22, 2022.

38 Spice of Life, https://www.thespicemt.com

39 Wikipedia, "Eden Atwood," https://en.wikipedia.org/wiki/Eden_Atwood

40 Holien, "So Long, Hoyt Axton," *Missoulian*, Oct. 27, 1999l.

41 Schott, "Huey Lewis Interview," Whitefish Review, Jul. 6, 2022;
   Cohen, "Slough of Controversy," *Missoulian*, Aug.4, 2003..

42 Rider, "A Lawman's Life," *Missoulian*, Nov. 14, 1998;
   PRINTZ, SHERIFF/CORONER, RAVALLI COUNTY, MONTANA v. UNITED STATES, 521 US 898 (1997).

43 Backus, "Snow and Cold Make for Challenging Big Game Opener in Montana," *Ravalli Republic*, Oct. 26, 2020;
   Chaney, "FWP Extends Elk Hunting," *Missoulian*, Nov. 28, 2020.
   Orange Army: Hunters in the woods as the law requires the wearing of orange clothing or vest as a safety measure.

44 The National Association for Amateur Radio (ARRL), http://www.arrl.org/upgrading-to-an-extra-license

45 McConnaha, "Thrills Chills and Treats (Front Page Photo)," *Ravalli Republic*, Oct. 30, 2019;
   Brigade = A military ground unit, Army or Marine, 3000 to 5000 strong. https://www.defense.gov/Multimedia/Experience/Military-Units/Army/#army

46 AWOL =Absent Without Leave: Uniform Code of Military Justice, Article 86. (10 US Code § 886).

47 IMDb. "The Mist," IMDb., 2007;

"The Mist Trailer," (Video) YouTube. https://www.youtube.    com/watch?v=LhCKXJNGzN8

# 13. Trying Times in Ravalli County- Nov. 2020————————————.

1 Hamby (Billings Gazette), "Montana Added 887 New Cases,; 11 More Deaths," *Ravalli Republic*, p. A2, Nov. 1, 2020..

2 Health Board Meetings 2020, 2021, Nov. 10, 2020, 1:00 PM (Video), Minute 14:50.

3 Backus, "Firth Ravalli County Resident Dies," *Ravalli Republic*, Nov. 1, 2020.

4 Hamby (Billings Gazette), "Montana Added 887 New Cases,; 11 More Deaths," *Ravalli Republic*, p. A2, Nov. 1, 2020.

5 Lawless, "Connery, 'Original' Bond, Dies," *The Billings Gazette*, Associated Press, Nov. 1, 2020

6 Dylan, "The Times They Are A-Changin' (Official Audio)," Bob Dylan (Video), Mar, 11, 2019.

7 Dietrict, "National Groups Fuel Senate Race," *The Billings Gazette*, Montana Free Press, Nov. 1, 2020.

8 McConnaha, "Election Runs Smoothly," *Ravalli Republic*, Nov. 4, 2020.

9 Backus, "American-as-Apple-Pie Party Hopes to Cross Party Lines," *Ravalli Republic*, Nov. 4, 2020.

10 Cultivating Connections, https://www.cultivatingconnectionsmt.org

11 Homestead Organic Farm, https://www.facebook.com/homesteadorganicsfarmmt/

12 CSA = Community Based Agriculture. Generally a CSA arrangement with a farm involves paying some amount at the beginning of the growing season — which serves to help with the cash flow needed to buy seeds, soil, labor, etc. to grow the plants. Then, every week, usually starting in spring and ending with the final harvest, a box of food is made up with your name on it, Around here, you usually can pick it up at the farm,

13 Trump, "Tweets of Nov. 3, 2020," Tweets of Nov. 3, 2020 | The American Presidency Project, Nov. 3, 2020 07:01:28.

14 270toWin, "Real Time Live Presidential Election Results," 270towin.com, Nov. 14, 2020.

15 Backus, "Republicans Sweep Record-Setting Election in Ravalli County," *Ravalli Republic*, Nov. 3, 2020;

Ravalli County Clerk and Recorder, "2020 General Election Results (Working Copy) - Ravalli.";

Jo Jorgensen, the Libertarian who was qualified in 50 states, got a little over 3% of the vote in Montana — In Ravalli County, 349 ballots with no Presidential vote, 252 ballots with no US Senate vote, 408 fewer votes for Daines than for Trump. Bullock got 1,108 more votes than Biden.

16 Trump, "Tweets of Nov. 3, 2020," Tweets of Nov. 3, 2020 |... Nov. 3, 2020 23:15:55

17 Atlantic Council's DFRLab, "#Stopthesteal: Timeline of Social Media and Extremist Activities Leading to 1/6 Insurrection," Atlantic Council, Feb. 10, 2021.

18 Cummings, "By the Numbers: President Donald Trump's Failed Efforts to Overturn the Election," *USA Today*, Jan. 6, 2021..

19 Backus, "'We Are on the Brink of Disaster," *Ravalli Republic*, Nov. 11, 2020

Backus, "Local food banks face challenges because of pandemic," *Ravalli Republic*, Nov. 11, 2020;

_____, "Hunting Harvest Up," *Ravalli Republic*, Nov. 20, 2020.

20 Dough Boy Memorial, Hamilton, MT, https://centennial.legion.org/montana/post47/1921/08/01/doughboy-memorial-ravalli-county

21 McConnaha, "Nearly 60 flags fly at Ravalli Co, Museum for Veterans Day,." *Ravalli Republic*, Nov. 11, 2020.

22 Facebook, Marias Restaurant: https://www.facebook.com/pages/category/Product-service/Marias-restaurant-104185388042265/

23 McConnaha, "Hamilton Restaurant Giving Away Turkey Basket Dinners," *Ravalli Republic*, Nov. 22, 2020;

24 Health Board Meetings 2020, 2021, Nov. 10, 1:00 PM (Video), Minutes 0-23.

25 Health Board Meetings 2020, 2021, Nov. 10, 1:00 PM (Video), Minute 55.

26 Ibid.

27 See Federal Laws including 28 US Code § 2679 - Exclusiveness of remedy;

For a dissertation on Montana law and common law limiting public official liability see: Johnstone, "Confusion Over Sovereign Immunity : What Is Article II, Section 18 About?," pp. 19-22.

28 Health Board Meeting, Nov. 10, 1:00 PM (Video), 1 hr - 1 hr, 9 min.

29 Backus, "Mixed Messages: Ravalli County Struggles to Find Consensus on Addressing Pandemic," *Ravalli Republic*, Nov. 14, 2020. (Paragraphing changed in quote).

30 Florio, "Montana's Face-Off Over Face Masks," The Nation, Jan.11, 2021.

31 Bullock, "Montana Directives," • Bullock: "Directive implementing Executive Orders 2-2020 and 3-2020 and limiting size for public gatherings and events and limiting bar and restaurant capacity and hours," Nov. 17, 2020.

32 Ravalli County Public Health, "Corona Virus Covid 19: Ravalli County, MT - Official Website," Press Release, 11/30/20.

33 Edelberg, "Recession Remedies," Brookings Institute, Apr. 2022.

Garner, "Receipt and Use of Stimulus Payments in the Time of the Covid-19 Pandemic," *Beyond the Numbers*, Jul. 2020.

Edelberg, "Inflation-Related Updates to 'Recession Remedies'." Brookings. Brookings, May 18, 2022;

Mitchell, "Income Rise Tees Up Fast Growth," *Wall Street Journal*, Jan. 28, 2021.

# 14. The Coming Fury - Dec. 2020 ————————————————

1 Trump, "President Trump Statement on 2020 Election Results," C-Span (Transcribed by Rev AI), Dec. 2, 2020.

2 Ibid.

3 Jones, "Alex Jones Show," Prison Planet TV (Video), Dec. 20, 2020;

Smith, "If Biden Is Elected, Then Expect 'Obey Authority' Covid Collectivism," *Epoch Times*, Nov. 17, 2020

4 Wheeler, "Trump's Judicial Campaign to Upend the 2020 Election: A Failure, but Not a Wipe-Out.";

The win: A court tossed out a few votes. Bowden, "Trump Campaign Enjoys Minor Victory in Pennsylvania Lawsuit over Voter ID," *New York Post*, Nov. 13, 2020;

Boebert, "Video: 'Trump Should Fight with Everything He Has'-Rep.-Elect Lauren Boebert Talks Election Fraud, Gun Rights," *Epoch TV* (Video), Dec. 6, 2020.

5 Catton, The Coming Fury, pp. 78-109.

6 Ibid, p. 111. It is sometimes argued that the Democrat Candidate conceded to Lincoln. But, there were two Democrat Candidates. The Democrat Party had already split, a separate convention picking another candidate. Most southern states did not even print Lincoln's name on the ballot. The canons were primed. The announcement of Lincoln's election was the spark that lit the canon fuses at Charleston.

7 Lane, "Annie's Mailbox, Party Is way too risky," *Ravalli Republic*, Dec. 30, 2020, p. A2.

8 Montana Bioscience Alliance, Marshall Bloom: https://montanabio.org/hall-of-fame/2020-marshall-bloom-m-d/

9 Backus, "RML's Bloom Shares COVID Experience," *Ravalli Republic*, Jan.13, 2021

10 Ibid.

11 Leonhardt, "Why Masks Work, but Mandates Haven't," *New York Times*, May 31, 2022.

12 Briand, "COVID-19 Deaths A Look at U.S. Data FEB 2021 WORKING PAPER Genevieve Briand" Research Gate. Economics Dept, Johns Hopkins, Februrary 2021
____ "Age Distribution per Cause U.S. Monthly Deaths 1999-2021 March 2022 ..." March 2022.
The author my have still been the process of getting the PdD when the webinar was recorded.

13 Briand, "Covid-19 Deaths: A Look at US Data," Genevieve Briand (Video), Nov. 13, 2020, Minute 40:15.

14 Ibid.

15 Gu, "A Closer Look at US Deaths Due to Covid-19," Johns Hopkins News Letter, Nov. 27, 2020.

16 CDC, "Monthly Provisional Counts of Deaths by Select Causes, 2020-2022," CDC, Jan. 4, 2023.
Macrotrends, "US Population 1950-2023";
Karlinsky, "Tracking excess mortality across countries during the COVID-19 pandemic with the World Mortality Dataset," eLife, Jun. 30, 2021.
Mathieu, "Excess mortality during the Coronavirus pandemic (COVID-19)";
Katz, "574,000 More US Deaths than Normal since Covid-19 Struck," *New York Times*, Jan. 14, 2021.
On the Cumulative Deaths Per Capita 2020 chart, the dashed line, Cumulative Excess Mortality United States, is higher than the charts shown in OurWorldInData. Mathieu describes how several sources including WHO and *the Economist* compute excess mortality noting that they are all consistent that it is significantly higher than confirmed-COVID-19 mortality. OurWorldInData uses the method described by Karlinsky in the reference above for most nations in the world. It is not clear from Karlinsky's discussion whether they adjusted their baseline averages for population changes from year to year. Since I am doing a Per Capita Chart, I use the population for each year as the divisor for the raw mortality numbers reported by the CDC.

Rahhal, "Johns Hopkins Apologizes after Retracting Article Claiming Covid Has Not Caused Excess Deaths," *Daily Mail Online*, Dec. 4, 2020

17 Papayil, "Public Health Experts and Biostatisticians Weigh in on 'Covid-19 Deaths: A Look at US Data' Webinar,"*Johns Hopkins News Letter*, Dec. 14, 2020.

18 Briand, "No Increase in Death Rate by Covid-19 in the United States," *The Conejo Guardian*, Nov. 7, 2021.

19 Wikipidia, Vietnam War Body Count, https://en.wikipidia.org/wiki/Vietnam_War_body_count_controversy. — especially when it is backed up by a programmer I know who worked on the body count software in country,

20 FDA. "Emergeny Use Authorization";
Puthumana "Speed, Evidence, and Safety Characteristics of Vaccine Approvals by the US Food and Drug Administration."

21 McConnaha, 'Vaccines Begin in Ravalli County," *Ravalli Republic*, Dec. 30, 2020.

22 Ibid.

23 Gianforte, "Gov. Gianforte Announces Opening of Monoclonal Antibody Clinic in Missoula," State of Montana Newsroom, Nov.8, 2021.

24 Super 8 Hotel, Hamilton, https://www.wyndhamhotels.com/super-8/hamilton-montana/super-8-hamilton/overview

25 Ludvigsson, "How Sweden Approached the COVID-19 Pandemic: Summary and Commentary on the National Commission Inquiry."

26 Czeisler "Delay or Avoidance of Medical Care Because of COVID-19–Related Concerns — United States."

27 French, "Impact of Hospital Strain on Excess Deaths during the COVID-19 Pandemic — United States, July 2020–July 2021." Morbidity and Mortality Weekly Report 70, no. 46, Nov. 19, 2021, pp. 1613-1616;
WHO, "Covid-19 Pandemic Triggers 25% Increase in Prevalence of Anxiety and Depression Worldwide," World Health Organization, Mar. 2, 2022;
John, "Editorial: Trends in Suicide during the COVID-19 Pandemic," Nov. 12, 2020.

28 Caraballo, "Excess Mortality and Years of Potential Life Lost Among the Black Population in the US, 1999-2020.";
Arrazola "COVID-19 Mortality Among American Indian and Alaska Native Persons — 14 States, January–June 2020."

29 *Memories* (Official Video), Maroon 5, Oct. 8, 2019..

## 15. Trying Times In Washington - Jan. 6, 2021 ——————————

1 BBC, "Capitol Riots Timeline: What Happened on 6 Jan One Year Ago?," BBC, Jan. 6, 2022;
Congressional Record, "Congressional Record, Proceedings and Debates of the 117th Congress, 1st Session, Vol. 169, No. 4," Jan. 6, 2021;
*New York Times*, "Day of Rage: How Trump Supporters Took the US Capitol (Video)," Jul. 1, 2021;
Biden, "At This Hour Our Democracy Is under Unprecedented Assault (Video)," CSPAN-2, Jan. 6, 2021;
CBS News Online, "Protesters Swarm Capitol, Abruptly Halting Electoral Vote Count (Video)," Jan. 6, 2021;
CSPAN, "Joint Session of Congress for Counting of Electoral College Ballots (Video)," Jan. 7, 2021;
Nishimura, "Never-before-Seen Jan. 6 Footage from Our Photographer (Video)," LA Times, Jan. 4, 2022;
Raskin, "House Impeachment Managers' Video Compilation of Jan. 6 Attack on the US Capitol (Video)," US House of Representatives Impeachment Managers, Feb. 9, 2021;
Peters, "Examining the US Capitol Attack: A Review of the Security ..," US Senate Committee on Homeland Security...";
Groeger, "What Parler Saw during the Attack on the Capitol," *ProPublica* (Videos), Jan. 17, 2021;
Thompson, "Select Committee to Investigate the January 6th Attack on the United States Capitol," United States Congress, 117th Congress Second Session House Report 117-663, Dec. 22, 2022;

2 Arkin, "Exclusive: Classified Documents Reveal the Number of Jan. 6 Protesters," Newsweek, Dec. 23, 2021;
*Washington Post*, "D.C. Police Requested Backup at Least 17 Times in 78 Minutes during Capitol Riot | Visual Forensics (Video)," *Washington Post*, Apr. 15, 2021;

3 Catton, The Coming Fury, p. 313. (Paraphrase).

4 Connelly, "Proud Boys Leader Visits White House Ahead of DC Rally," New York Post, Dec. 12, 2020.
German, "Proud Boys Ex-Leader Charged With Jan. 6 Conspiracy," *Wall Street Journal*, Jun. 7, 2022, p. A4.

5 Davis, "DC Police Post Signs Banning Guns during Jan. 6 Maga Rally," The Western Journal, Jan. 4, 2021/.

6 Carallee, "Katelyn 'Let's Have Trial by Combat!' Rudy Giuliani Riles up Crowd before Riot," *Daily Mail*. Post associated with story "New York State (Video) Bar Association considers removing Giuliani for telling MAGA rally crowd to use 'trial by combat' before they stormed Capitol", Jan.11, 2021.

7 Women for America First (Producer), "Live: Trump Delivers Remarks at the 'Save America Rally' in Washington," D.C. (Video)," Bloomberg Quicktake, Jan. 6, 2021: Amy Kramer {"Hi Deplorables"} minute 28:49;
  Giuliani {"trial by combat"} 1 hr: 27 min; President Trump's Speech {We gather...} 2 hrs, 34 min.

8 Even those that didn't go the capitol planned gathering. Stand Together for Freedom, which had been non-partisan announced an event to support Trump and stop corruption at 1:00 PM at the Courthouse.Event, Jan. 6, 2021, Stand Together for Freedom: https://standtogetherfor-freedom.com/event/rally-in-ravalli/

9 Cadre is used here in its original meaning, a small trained group able to assume control of others: https://www.merriam-webster.com/dictionary/cadre. The communist movement adopted "cadre" the word as a designation for organizing within the movement, "cadre" is sometimes associated with the organization of communist movements and was adopted by the leaders of the movement. Here it is used in its original meaning.

10 PBS News Hour, "The Jan. 6 Insurrection, 1 Year Later (Video)," PBS, Jan. 8, 2022;
  Washington Post, "D.C. Police Requested Backup at Least 17 Times... (Video)," Washington Post, Apr. 15, 2021.

11 Washington Post, "How Trump Supporters View What Happened after Mob Attack on Capitol (Video)," Jan. 7, 2021

12 Arkin, "Exclusive: Classified Documents Reveal the Number of Jan. 6 Protesters," Newsweek, Dec. 23, 2021.
  PBS News Hour, "Watch: Del. Plaskett Reconstructs Insurrection w..," PBS, Feb. 10, 2021, Minute 24:55.

13 Cameron, "These Are the People Who Died in Connection with the Capitol Riot," New York Times, Jan. 5, 2022;
  New York Times states, "Rosanne Boyland appeared to have been crushed in a stampede of fellow rioters as they surged against the police," The Medical Examiner determined some time later that she died of a drug overdose: McEvoy, "D.C. Medical Examiner Reveals Causes of Capitol Insurrection Deaths...," Forbes Magazine, Apr. 7, 2021.

14 Smock, "Ray Smock on House Speaker's Lobby (Video)," CSPAN3/American History TV, Jan. 15, 2015.

15 Daugherty, "House Chamber Evacuated as Rioters Storm the Capitol (Video)," Miami Herald, Jan. 6, 2021. In this video is is clear that members are still in the chamber when the officer shot. You can hear the gunshot.

16 NBC New York, "Video Shows Fatal Shooting of Ashli Babbitt at US Capitol (Video)," NBC New York, Jan. 8, 2021;

WUSA9, "Graphic: Video Shows Moment Woman Was Shot, Killed in Capitol Riots (Video)," WUSA9, Jan. 8, 2021;
  Attkisson, "Ashli Babbitt | Full Measure (Video)," Full Measure, May 3, 2021.

17 Peters, "Examining the US Capitol Attack: A Review of the Security, - Senate," US Senate Committee Homeland Security & Gov. Affairs.

18 FBI_SIR_000001–2, quoted in Peters, "Examining the US Capitol Attack: A Review of the Security, - Senate," US Senate Committee on Homeland Security...", p. 31.

19 Nirappil, "'Black Lives Matter': In Giant Yellow Letters, D.C. Mayor Sends Message to Trump," Washington Post, Jun. 5, 2020.

20 Garamone, "DOD Details National Guard Response to Capitol Attack," DoD News, Jan. 8, 2021;
  DeMarche, "Trump Says He Requested 10K National Guard Troops at Capitol on Day of Riot (Video)," Fox News Mar. 1, 2021;
  Himmelman, "Fact Check: Did Pelosi Reject Trump's Request for National Guard Troops..?," The Dispatch, Oct.31, 2022.

21 Thompson, "Select Committee to Investigate the January 6th Attack on the United States Capitol," Dec. 23, 2022, p. 709

22 Pence, Dear Colleague Letter, Jan. 6, 2020.,

23 CSPAN, "Joint Session of Congress for Counting of Electoral College Ballots," CSPAN, Jan. 6, 2021;

24 Pence, So Help Me God, pp. 441-462.

25 Colvin, "What We Know about Trump's Actions as Insurrection Unfolded," Ravalli Republic (AP Article), Jun. 9, 2022.
  Pence, So Help Me God, p. 462.

26 Larson, "Claims That Led to D.C. Insurrection Echoed in Montana," Ravalli Republic, Jan. 10, 2021, p. B1.

27 Larson, "Claims That Led to D.C. Insurrection Echoed in Montana," Ravalli Republic, Jan. 10, 2021, p. B4.

28 Ibid.

29 Ho, "The Arrests of Capitol Rioters per Million People in Each State, Mapped," Digg, May 7, 2021.

30 Viswanatha, "Within the FBI, Jan. 6 Spurs Protests, Conspiracy Theories," Wall Street Journal, p. A4, Mar. 20, 2023.

31 Reeve, "'What Are We Supposed to Do?': Rioter Speaks to CNN's Elle Reeve," (Video), CNN, Jan. 8, 2021, minute 3:42.

32 Arkin, "Exclusive: Classified Documents Reveal the Number of Jan. 6 Protesters," Newsweek, Dec.- 23, 2021;

## 16. Epilogue

1 Frueh, "NAS Annual Meeting: Experts Discuss COVID-19 Pandemic and Science's Response (Text and Video)," National Academy of Science, Apr. 27, 2020.

## Acknowledgments

1 Health Meetings 2020 and 2021, Special Meeting July 24, 2020, (Video), 1 hr 1 minute.

# Bibliography

Compete *References Cited* is at
**https://www.jamesrolsen.com/covid-wars-references** (with hyperlinks).

Over 1,000 references are cited in the end-notes and as sources list in Artwork. They include peer reviewed studies, books, news articles, and other sources. In a few cases a reference may no longer be available for some reason. A few have been redacted from a journal or pulled from social media. I archived most citations and can make one available upon request, if available. Contract Jim@JamesROlsen.com.

The sources used for this book include:

- Official records of public comment and public discussions during meetings of the Ravalli County Health Board, Ravalli County Board of County Commissioners and the City of Hamilton City Council. Comments from Florance and Darby school board meetings and the Ravalli County Fair Board are taken from news sources.

- Video records from C-Span, YouTube, and a few cases of website hosted videos of public comment at FDA Review Committee meetings, White House, Health and Human Services, National Institutes of Health press conferences, Congressional Hearings, and video interviews.

- Interviews with Alan Lackey, Mara Lynn Luther, Alan Ford, and released texts, emails, and conversations with Laurel Ferriter, Peter Moore, and Hollie Rose Conger. The proceedings in Ravalli County Justice Court, Dept. 1, are taken from my observations, verified with other people present.

- Federal public records for Congressional Hearings. Congressional Committee Reports, General Administration Office (GAO).

- State and County public records, including Emergency Declarations and Directive from the Montana Governors Office, reports from Health and Human Services for Montana and other states. Ravalli County Correspondence and Department Reports.

- Scientific Medical Journals.
  » When presenting something as fact rather than a claim, I have used peer reviewed sources from reputable journals whenever possible — including *New England Journal of Medicine* (NEJM), *Journal of the American Medical Association* (JAMA) and *The BMJ*. The citations include both studies and articles from *Science* and *Nature* magazines, Lancet and various Springer and Elsevier journals as well as specialist association journals.
  » I cite a large number of open access journals in which the author pays the editing and publication cost $2,000 to $4,500, though some of these cost $5,000 or more. Universities and grants often help the author pay some or all of the cost — BMJ journal, Lancet Journals, Elsevier Science Direct Journals, Springer, Wiley Online Library Journals, Oxford University Press Journals, PLOS ONE. Also *Nature* and *Science* (AAAS), and JAMA publish open source journals.
  » Journals from societies in other professional fields.
  » The peer review process is far from perfect, with the quality varying quite a bit. Thus, I check even highly reputable papers using the methods in the Research Chapter of Book 3 in this series; I may have made mistakes or missed something.
  » When using preprint studies (not yet peer reviewed), I generally note it, but not always. When unreputable or non-peer reviewed is usually the case that it has become newsworthy or picked up and an counter-narrative site for the purpose of editorial comment and analysis in this book. A few are used as valid evidence after a review for quality and transparency.

- FDA records from the drug approval process including drug-maker submission and staff analysis.

- Extensive use of the website OurWorldInData for COVID-19 statistics. Also used the CDC, Johns Hopkins, the US Census Bureau, Stastica, and other sources.

- News. For news reports that make claims about trials and studies, I have, when feasible, found the original source material as well.

- Textbooks on Medical Microbiology, Meta-analysis, and Public Health.

- Sources for insight into the events in Wuhan in January 2020 are *Wuhan Diary* book by Fang Fang, *In the Same Breath* movie by Nanfu Wang, and video footage smuggled out of China to *Al Jazeera*.

- Sources for insight into the events within the Federal Government that were not known until later are Freedom of Information Act Request (FOIA) responses, the movie *Totally Under Control* directed by Alex Gibney and retrospective articles on the CDC test kit failure from the *Wall Street Journal*, ProPublica, and *Washington Post*.

- Social media, particularly YouTube, websites, and blogs, with few instances of Facebook and Twitter (now X).

# INDEX

Artwork is shown in SMALLER TYPE SMALL CAPS. References to movies and books are in *Italic*

# ABOUT THE AUTHOR

Photo by Mary Byers

James R. Olsen, Jim, is an engineer by trade, having successfully led large-scale, multi-company defense and air traffic control projects. He developed a double-bottom line business inspired by Patagonia. The company engaged in a variety of markets, delivering embedded teams into large projects, property management, and a native plant greenhouse operation. The company engaged with the community, for example, producing the Hamilton Performing Arts Series for a year to transition from the school district to a non-profit.

For the last 30 years, Jim has also been a volunteer grassroots activist engaged in forest and wilderness advocacy, effective domestic abuse intervention, local food systems, community-based mental health crisis management, subdivisions and water quality, and biosafety.

For the last ten years, Jim has focused on well-researched nonfiction books for the popular market. His writing hero is Erik Larson.

www.ingramcontent.com/pod-product-compliance
Lightning Source LLC
Chambersburg PA
CBHW051311020426
42333CB00027B/3294